Lecture Notes in Mathematics 1871

Editors:
J.-M. Morel, Cachan
F. Takens, Groningen
B. Teissier, Paris

C.I.M.E. means Centro Internazionale Matematico Estivo, that is, International Mathematical Summer Center. Conceived in the early fifties, it was born in 1954 and made welcome by the world mathematical community where it remains, in good health and spirit. Many mathematicians from all over the world have been involved in a way or another in C.I.M.E.'s activities during the past years.

So they already know what the C.I.M.E. is all about. For the benefit of future potential users and co-operators the main purposes and the functioning of the Centre may be summarized as follows: Every year, during the summer, Sessions (three or four as a rule) on different themes from pure and applied mathematics are offered by application to mathematicians from all countries. Each Session is generally based on three or four main courses (24–30 hours over a period of 6-8 working days) held from specialists of international renown, plus a certain number of Seminars.

A C.I.M.E. Session, therefore, is neither a Symposium, nor a School, but maybe a blend of both. The aim is that of bringing to the attention of younger researchers the origins, later developments, and perspectives of some branch of live mathematics.

The topics of the courses are generally of international resonance and the participant of the courses is covering the expertise of different countries and continents. Such combination, gave an excellent opportunity to young participants to be acquainted with the most advanced researches in the topics of the courses and the possibility of an interchange with the world famous specialists. The full immersion atmosphere of the courses and the daily exchange among participants are a first brick in building of international collaboration in Mathematics research.

For more information see CIME's homepage: http://www.cime.unifi.it

CIME's activity is supported by:

– Istituto Nationale di Alta Matematica "F.Severi"
– Ministero dell'Istruzione, dell'Università e della Ricerca
– Ministero degli Affari Esteri - Direzione Generale per la Promozione e la
    Cooperazione - Ufficio V
– Università Franco-Italiana.

P. Constantin · G. Gallavotti ·
A.V. Kazhikhov · Y. Meyer · S. Ukai

Mathematical Foundation of Turbulent Viscous Flows

Lectures given at the
C.I.M.E. Summer School
held in Martina Franca, Italy,
September 1–5, 2003

Editors: M. Cannone, T. Miyakawa

 Springer

Fondazione
C.I.M.E.

Authors and Editors

Peter Constantin

Department of Mathematics
University of Chicago
5734 University Avenue
Chicago, IL 60637, USA
e-mail: const@math.uchicago.edu

Giovanni Gallavotti

Dipartimento di Fisica
Università di Roma "La Sapienza"
P.le Moro 2, 00185 Roma, Italia
e-mail: giovanni.gallavotti@roma1.infn.it

Alexandre V. Kazhikhov

Lavrentyev Institute of Hydrodynamics
Siberian Branch of RAS
pr. ac. Lavrentyeva 15
630090, Novosibirsk, Russia
e-mail: kazhikhov@hydro.nsc.ru

Yves Meyer

Centre de mathématiques
et de leurs applications
ENS-Cachan
61, avenue du Président-Wilson
F-94235 Cachan Cedex, France
e-mail: ymeyer@cmla.ens-cachan.fr

Seiji Ukai

Department of Applied Mathematics
Yokohama National University
79-5 Tokiwadai, Hodogaya
Yokohama 240-8501
Japan
e-mail: ukai@mathlab.sci.ynu.ac.jp

Marco Cannone

Laboratoire d'Analyse
et de Mathématiques Appliquées,
Université de Marne-la-Vallée
5, boulevard Descartes
Champs-sur-Marne
77454 Marne-la-Vallée cedex 2
France
e-mail: marco.cannone@univ-mlv.fr

Tetsuro Miyakawa

Department of Mathematics
Kanazawa University
Kakuma-machi
Kanazawa, Ishikawa 920-1192 Japan
e-mail: miyakawa@kenroku.kanazawa-u.ac.jp

Library of Congress Control Number: 2005593631

Mathematics Subject Classification (2000): 35-XX, 35-02; 35QXX, 42-X, 42-02, 76-XX, 76BXX, 76DXX, 76P05

ISSN print edition: 0075-8434
ISSN electronic edition: 1617-9692
ISBN-10 3-540-28586-5 Springer Berlin Heidelberg New York
ISBN-13 978-3-540-28586-1 Springer Berlin Heidelberg New York
DOI 10.1007/b11545989

Springer is a part of Springer Science+Business Media
springer.com
© Springer-Verlag Berlin Heidelberg 2006

Typesetting: by the authors and TechBooks using a Springer LATEX package

Cover design: *design & production* GmbH, Heidelberg

Printed on acid-free paper   SPIN: 11545989   41/TechBooks   5 4 3 2 1 0

Preface

Over two centuries ago, L. Euler (1750) derived an ideal model equation describing the evolution of fluids. Later on, this model was revised under a more realistic basis by H. Navier (1822) and G. Stokes (1845). Finally, with his eponymous equation L. Boltzmann (1872) introduced the foundation of Gas Dynamics. Since then, much progress has been made in the understanding of these physical models. But many fundamental mathematical questions still remain unresolved, such as the existence, uniqueness and stability of solutions to the corresponding equations in three dimensions.

Due to the large number of applications to different fields (such as meteorology, astrophysics, aeronautics, thermodynamics, lasers and plasma physics), the study of these model equations from a purely mathematical point of view plays a crucial role in Applied Mathematics.

The series of lectures contained in this volume reflects five different and complementary approaches to several fundamental questions arising in the study of the Fluid Mechanics and Gas Dynamics equations. These lectures were presented by five well-known mathematicians at the International CIME Summer School held in Martina Franca, Italy, from 1 to 5 September 2003.

P. Constantin presents the Euler equations of ideal incompressible fluids and discusses the blow-up problem for the Navier-Stokes equations of viscous fluids, also describing some of the major mathematical questions of turbulence theory.

These questions are intimately connected to the Caffarelli-Kohn-Nirenberg theory of singularities for the incompressible Navier-Stokes equations, that is explained in detail in *G. Gallavotti*'s lectures.

A. Kazikhov introduces the reader to the theory of strong approximation of weak limits via the method of averaging, applied to the Navier-Stokes equations.

On the other hand, *Y. Meyer*'s lectures focus on several nonlinear evolution equations – in particular the Navier-Stokes ones – and some related unexpected cancellation properties, that are either imposed on the initial condition, or satisfied by the solution itself, whenever it is localized in space or in time variable.

Finally, *S. Ukai* presents the asymptotic analysis theory of fluid equations. More precisely, he discusses the Cauchy-Kovalevskaya technique for the Boltzmann-Grad limit of the Newtonian equation, the multi-scale analysis, giving the compressible and incompressible limits of the Boltzmann equation, and the analysis of their initial layers.

Many Ph. D. students and researchers from all over the world attended the summer school, thereby contributing to its success.

The Apulian landscape with its Romanesque and Baroque cathedrals, castles, rocky settlements, trullis and caves, and the city of Martina Franca, with its Ducal Palace – where the lectures were held – contributed to creating an attractive and pleasant working atmosphere. The summer school would not have taken place without the contagious optimism of Vincenzo Vespri, the efficient coordination of Elvira Mascolo and Pietro Zecca and the precious help of Marco Romito and Veronika Sustik. We would like also to thank here Carla Dionisi, who took care of the final typesetting of the lectures notes.

Finally, let us remember that the CIME summer school benefited from the financial support of Ministero degli Affari Esteri – Direzione Generale per la Promozione e la Cooperazione – Ufficio V, M.U.R.S.T., INdAM and Università Franco-Italiana.

July 2004   *Marco Cannone*
Paris, Kanazawa   *Tetsuro Miyakawa*

During the process of proofreading, we learned with sorrow that our colleague and friend Alexandre V. Kazhikhov, one of the authors of this volume, passed away on November 3, while at work in his office in Novosibirsk.

Contents

Oscillating Patterns in Some Nonlinear Evolution Equations

Asymptotic Analysis of Fluid Equations

Euler Equations, Navier-Stokes Equations and Turbulence

Peter Constantin

Department of Mathematics, University of Chicago
5734 University Avenue Chicago, IL 60637, USA
const math.uchicago.edu

1 Introduction

In 2004 the mathematical world will mark 120 years since the advent of turbulence theory ([80]). In his 1884 paper Reynolds introduced the decomposition of turbulent flow into mean and fluctuation and derived the equations that describe the interaction between them. The Reynolds equations are still a riddle. They are based on the Navier-Stokes equations, which are a still a mystery. The Navier-Stokes equations are a viscous regularization of the Euler equations, which are still an enigma. Turbulence is a riddle wrapped in a mystery inside an enigma ([11]).

Crucial for the determination of the mean in the Reynolds equation are Reynolds stresses, which are second order moments of fluctuation. The fluctuation requires information about small scales. In order to be able to compute at high Reynolds numbers, in state-of-the-art engineering practice, these small scales are replaced by sub-grid models. "Que de choses il faut ignorer pour 'agir' ! " sighed Paul Valéry ([88]). ("How many things must one ignore in order to 'act' ! ") The effect of small scales on large scales is the riddle in the Reynolds equations. In 1941 Kolmogorov ([65]) ushered in the idea of universality of the statistical properties of small scales. This is a statement about the asymptotics: long time averages, followed by the infinite Reynolds number limit. This brings us to the mystery in the Navier-Stokes equations: the infinite time behavior at finite but larger and larger Reynolds numbers. The small Reynolds number behavior is trivial (or "direct", to use the words of Reynolds himself). Ruelle and Takens suggested in 1971 that deterministic chaos emerges at larger Reynolds numbers ([83]). The route to chaos itself was suggested to be universal by Feigenbaum ([49]). Foias and Prodi discussed finite dimensional determinism in the Navier-Stokes equations already in 1967 ([55]), four years after the seminal work of Lorenz ([68]). The dynamics have indeed finite dimensional character if one confines oneself to flows in bounded regions in two dimensions ([2], [31], [32], [34], [56], [69]). In three dimensions, however, the long time statistics question is muddied by the blow up problem.

Leray ([67]) showed that there exist global solutions, but such solutions may develop singularities. Do such singularities exist? And if they do, are they relevant to turbulence? The velocities observed in turbulent flows on Earth are bounded. If one accepts this as a physical assumption, then, invoking classical results of Serrin ([84]), one concludes that Navier-Stokes singularities, if they exist at all, are not relevant to turbulence. The experimental evidence, so far, is of a strictly positive energy dissipation rate $0 < \epsilon = \langle \nu |\nabla u|^2 \rangle$, at high Reynolds numbers. This is consistent with large gradients of velocity. The gradients of velocity intensify in vortical activity. This activity consists of three mechanisms: stretching, folding and reconnection of vortices. The stretching and folding are inviscid mechanisms, associated with the underlying incompressible Euler equations. The reconnection is the change of topology of the vortex field, and it is not allowed in smooth solutions of the Euler equations. This brings us to the enigma of the Euler equations, and it is here where it is fit we start.

2 Euler Equations

The Euler equations of incompressible fluid mechanics present some of the most serious challenges for the analyst. The equations are

$$D_t u + \nabla p = 0 \tag{2.1}$$

with $\nabla \cdot u = 0$. The function $u = u(x, t)$ is the velocity of an ideal fluid at the point x in space at the moment t in time. The fluid is assumed to have unit density. The velocity is a three-component vector, and x lies in three dimensional Euclidean space. The requirement that $\nabla \cdot u = 0$ reflects the incompressibility of the fluid. The material derivative (or time derivative along particle trajectories) associated to the velocity u is

$$D_t = D_t(u, \nabla) = \partial_t + u \cdot \nabla. \tag{2.2}$$

The acceleration of a particle passing through x at time t is $D_t u$. The Euler equations are an expression of Newton's second law, $F = ma$, in the form $-\nabla p = D_t u$. Thus, the only forces present in the ideal incompressible Euler equations are the internal forces at work keeping the fluid incompressible. These forces are not local: the pressure obeys

$$-\Delta p = \nabla \cdot (u \cdot \nabla u) = \mathrm{Tr}\left\{ (\nabla u)^2 \right\} = \partial_i \partial_j (u_i u_j).$$

If one knows the behavior of the pressure at boundaries then the pressure satisfies a nonlocal functional relation of the type $p = F([u \otimes u])$. For instance, in the whole space, and with decaying boundary conditions

$$p = R_i R_j (u_i u_j)$$

where $R_i = \partial_i(-\Delta)^{-\frac{1}{2}}$ are Riesz transforms. (We always sum repeated indices, unless we specify otherwise. The pressure is defined up to a time dependent constant; in the expression above we have made a choice of zero average pressure.)

Differentiating the Euler equations one obtains:

$$D_t U + U^2 + \mathrm{Tr}\,(R \otimes R)\,U^2 = 0$$

where $U = (\nabla u)$ is the matrix of derivatives. We used the specific expression for p written above for the whole space with decaying boundary conditions. This equation is quadratic and it suggests the possibly of singularities in finite time, by analogy with the ODE $\frac{d}{dt}U + U^2 = 0$. In fact, the distorted Euler equation

$$\partial_t U + U^2 + \mathrm{Tr}\,(R \otimes R)\,U^2 = 0$$

does indeed blow up ([15]). The incompressibility constraint $\mathrm{Tr}\,U = 0$ is respected by the distorted Euler equation. However, the difference between the Eulerian time derivative ∂_t and the Lagrangian time derivative D_t is significant. One may ask whether true solutions of the Euler equations do blow up. The answer is yes, if one allows the solutions to have infinite kinetic energy. We will give an example in Section Three. The blow up is likely due to the infinite supply of energy, coming from infinity. The physical question of finite time local blow up is different, and perhaps even has a different answer.

In order to analyze nonlinear PDEs with physical significance one must take advantage of the basic invariances and conservation laws associated to the equation. When properly understood, the reasons behind the conservation laws show the way to useful cancellations.

Smooth solutions of the Euler equations conserve total kinetic energy, helicity and circulation. The total kinetic energy is proportional to the L^2 norm of velocity. This is conserved for smooth flows. The Onsager conjecture ([72], [48]) states that this conservation occurs if and only if the solutions are smoother than the velocities supporting the Kolmogorov theory, (Hölder continuous of exponent $1/3$). The "if" part was proved ([28]). The "only if" part is difficult: there is no known notion of weak solutions dissipating energy but with Hölder continuous velocities. The work of Robert ([81]) and weak formulations of Brenier and of Shnirelman are relevant to this question ([85], [6]).

In order to describe the helicity and circulation we need to talk about vorticity and about particle paths. The Euler equations are formally equivalent to the requirement that two first order differential operators commute:

$$[D_t, \Omega] = 0.$$

The first operator $D_t = \partial_t + u \cdot \nabla$ is the material derivative associated to the trajectories of u. The second operator

$$\Omega = \omega(x,t) \cdot \nabla$$

is differentiation along vortex lines, the lines tangent to the vorticity field ω. The commutation means that vortex lines are carried by the flow of u, and is equivalent to the equation

$$D_t\omega = \omega \cdot \nabla u. \tag{2.3}$$

This is a quadratic equation because ω and u are related, $\omega = \nabla \times u$. If boundary conditions for the divergence-free ω are known (periodic or decay at infinity cases) then one can use the Biot-Savart law

$$u = \mathcal{K}_{3DE} * \omega = \nabla \times (-\Delta)^{-1}\omega \tag{2.4}$$

coupled with (2.3) as an equivalent formulation of the Euler equations, the vorticity formulation used in the numerical vortex methods of Chorin ([13], [14]). The helicity is

$$h = u \cdot \omega.$$

The Lagrangian particle maps are

$$a \mapsto X(a,t), \quad X(a,0) = a.$$

For fixed a, the trajectories of u obey

$$\frac{dX}{dt} = u(X,t).$$

The incompressibility condition implies

$$det\,(\nabla_a X) = 1.$$

The Euler equations can be described ([63], [1]) formally as Euler-Lagrange equations resulting from the stationarity of the action

$$\int_a^b \int |u(x,t)|^2\,dxdt$$

with

$$u(x,t) = \frac{\partial X}{\partial t}(A(x,t),t)$$

and with fixed end values at $t = a,b$ and

$$A(x,t) = X^{-1}(x,t).$$

Helicity integrals ([71])

$$\int_T h(x,t)dx = c$$

are constants of motion, for any vortex tube T (a time evolving region whose boundary is at each point parallel to the vorticity, $\omega \cdot N = 0$ where N is the normal to ∂T at $x \in \partial T$.) The constants c have to do with the topological complexity of the flow.

Davydov, and Zakharov and Kuznetsov ([42], [93]) have formulated the incompressible Euler equations as a Hamiltonian system in infinite dimensions in Clebsch variables. These are a pair of active scalars θ, φ which are constant on particle paths,

$$D_t\varphi = D_t\theta = 0$$

and also determine the velocity via

$$u^i(x,t) = \theta(x,t)\frac{\partial\varphi(x,t)}{\partial x_i} - \frac{\partial n(x,t)}{\partial x_i}.$$

The helicity constants vanish identically for flows which admit a Clebsch variables representation. Indeed, for such flows the helicity is the divergence of a field that is parallel to the vorticity $h = -\nabla \cdot (n\omega)$. This implies that not all flows admit a Clebsch variables representation. But if one uses more variables, then one can represent all flows. This is done using the Weber formula ([90]) which we derive briefly below.

In Lagrangian variables the Euler equations are

$$\frac{\partial^2 X^j(a,t)}{\partial^2 t} = -\frac{\partial p(X(a,t),t)}{\partial x_j}. \tag{2.5}$$

Multiplying this by $\frac{\partial X^j(a,t)}{\partial a_i}$ we obtain

$$\frac{\partial^2 X^j(a,t)}{\partial t^2}\frac{\partial X^j(a,t)}{\partial a_i} = -\frac{\partial \tilde{p}(a,t)}{\partial a_i}$$

where $\tilde{p}(a,t) = p(X(a,t),t)$. Forcing out a time derivative in the left-hand side we obtain

$$\frac{\partial}{\partial t}\left[\frac{\partial X^j(a,t)}{\partial t}\frac{\partial X^j(a,t)}{\partial a_i}\right] = -\frac{\partial \tilde{q}(a,t)}{\partial a_i}$$

with

$$\tilde{q}(a,t) = \tilde{p}(a,t) - \frac{1}{2}\left|\frac{\partial X(a,t)}{\partial t}\right|^2$$

Integrating in time, fixing the label a we obtain:

$$\frac{\partial X^j(a,t)}{\partial t}\frac{\partial X^j(a,t)}{\partial a_i} = u^i_{(0)}(a) - \frac{\partial \tilde{n}(a,t)}{\partial a_i}$$

with

$$\tilde{n}(a,t) = \int_0^t \tilde{q}(a,s)ds$$

where

$$u_{(0)}(a) = \frac{\partial X(a,0)}{\partial t}$$

is the initial velocity. We have thus:

$$(\nabla_a X)^* \partial_t X = u_{(0)}(a) - \nabla_a \tilde{n}.$$

where we denote M^* the transpose of the matrix M.

Multiplying by $\left[(\nabla_a X(a,t))^* \right]^{-1}$ and reading at $a = A(x,t)$ with

$$A(x,t) = X^{-1}(x,t)$$

we obtain the Weber formula

$$u^i(x,t) = \left(u_{(0)}^j(A(x,t)) \right) \frac{\partial A^j(x,t)}{\partial x_i} - \frac{\partial n(x,t)}{\partial x_i}.$$

This relationship, together with boundary conditions and the divergence-free requirement can be written as

$$u = W[A,v] = \mathbf{P}\left\{ (\nabla A)^* v \right\} \tag{2.6}$$

where $\mathbf{P}$ is the corresponding projector on divergence-free functions and v is the virtual velocity

$$v = u_{(0)} \circ A.$$

We will consider the cases of periodic boundary conditions or whole space. Then

$$\mathbf{P} = I + R \otimes R$$

holds, with R the Riesz transforms. This procedure turns A into an *active scalar system*

$$\begin{cases} D_t A = 0, \\ D_t v = 0, \\ u = W[A,v]. \end{cases} \tag{2.7}$$

Active scalars ([17]) are solutions of the passive scalar equation $D_t \theta = 0$ which determine the velocity through a time independent, possibly non-local equation of state $u = U[\theta]$.

Conversely, and quite generally: Start with two families of labels and virtual velocities

$$A = A(x,t,\lambda), \quad v(x,t,\lambda)$$

depending on a parameter λ such that

$$D_t A = D_t v = 0$$

with $D_t = \partial_t + u \cdot \nabla_x$. Assume that u can be reconstructed from A, v via a generalized Weber formula

$$u(x,t) = \int \nabla_x A(x,t,\lambda) v(x,t,\lambda) d\mu(\lambda) - \nabla_x n$$

with some function n, and some measure $d\mu$. Then u solves the Euler equations

$$\frac{\partial u}{\partial t} + u \cdot \nabla u + \nabla \pi = 0$$

where

$$\pi = D_t n + \frac{1}{2}|u|^2.$$

Indeed, using the kinematic commutation relation

$$D_t \nabla_x f = \nabla_x D_t f - (\nabla_x u)^* \nabla_x f$$

and differentiating the generalized Weber formula we obtain:

$$D_t u = D_t(\int \nabla_x A v d\mu - \nabla_x n) =$$

$$- \int ((\nabla_x u)^* \nabla_x A) v d\mu - \nabla_x(D_t n) + (\nabla_x u)^* \nabla n =$$

$$-\nabla_x(D_t n) - (\nabla_x u)^* \left[\int (\nabla_x A) v d\mu - \nabla_x n \right] =$$

$$-\nabla_x(D_t n) - (\nabla_x u)^* u = -\nabla_x(\pi).$$

The circulation is the loop integral

$$C_\gamma = \sqrt{\int_\gamma u \cdot dx}$$

and the conservation of circulation is the statement that

$$\frac{d}{dt} C_{\gamma(t)} = 0$$

for all loops carried by the flow. This follows from the Weber formula because

$$u^j(X(a,t)) \frac{\partial X^j}{\partial a_i} = u^i_{(0)}(a) - \frac{\partial \tilde{n}(a,t)}{\partial a_i}.$$

The important thing here is that the right hand side is the sum of a time independent function of labels and a label gradient. Viceversa, the above formula follows from the conservation of circulation. The Weber formula is equivalent thus to the conservation of circulation.

Differentiating the Weber formula, one obtains

$$\frac{\partial u^i}{\partial x_j} = \mathbf{P}_{ik} \left(Det \left[\frac{\partial A}{\partial x_j}; \frac{\partial A}{\partial x_k}; \omega_{(0)}(A) \right] \right).$$

Here we used the notation

$$\omega_{(0)} = \nabla \times u_{(0)}.$$

Taking the antisymmetric part one obtains the Cauchy formula:

$$\omega_i = \frac{1}{2}\epsilon_{ijk}\left(Det\left[\frac{\partial A}{\partial x_j}; \frac{\partial A}{\partial x_k}; \omega_{(0)}(A)\right]\right)$$

which we write as

$$\omega = \mathcal{C}[\nabla A, \zeta] \tag{2.8}$$

with ζ the Cauchy invariant

$$\zeta(x, t) = \omega_{(0)} \circ A.$$

Therefore the active scalar system

$$\begin{cases} D_t A = 0, \\ D_t \zeta = 0, \\ u = \nabla \times (-\Delta)^{-1}\left(\mathcal{C}[\nabla A, \zeta]\right) \end{cases} \tag{2.9}$$

is an equivalent formulation of the Euler equations, in terms of the Cauchy invariant ζ. The purely Lagrangian formulation (2.5) of the Euler equations is in phrased in terms of independent label variables (or ideal markers) a, t, except that the pressure is obtained by solving a Poisson equation in Eulerian independent variables x, t. The rest of PDE formulations of the Euler equations described above were: the Eulerian velocity formulation (2.1), the Eulerian vorticity formulation (2.3), the Eulerian-Lagrangian virtual velocity formulation (2.7) and the Eulerian-Lagrangian Cauchy invariant formulation (2.9). The Eulerian-Lagrangian equations are written in Eulerian coordinates x, t, in what physicists call "laboratory frame". The physical meaning of the dependent variables is Lagrangian.

The classical local existence results for Euler equations can be proved in either purely Lagrangian formulation ([45]), in Eulerian formulation ([70]) or in Eulerian-Lagrangian formulation ([19]). For instance one has

Theorem 2.1. *([19]) Let $\alpha > 0$, and let u_0 be a divergence free $C^{1,\alpha}$ periodic function of three variables. There exists a time interval $[0, T]$ and a unique $C([0, T]; C^{1,\alpha})$ spatially periodic function $\ell(x, t)$ such that*

$$A(x, t) = x + \ell(x, t)$$

solves the active scalar system formulation of the Euler equations,

$$\frac{\partial A}{\partial t} + u \cdot \nabla A = 0,$$

$$u = \mathbf{P}\left\{(\nabla A(x, t))^* u_0(A(x, t))\right\}$$

with initial datum $A(x, 0) = x$.

A similar result holds in the whole space, with decay requirements for the vorticity. As an application, let us consider rotating three dimensional incompressible Euler equations

$$\partial_t u + u \cdot \nabla u + \nabla \pi + 2\Omega e_3 \times u = 0.$$

The Weber formula for relative velocity is:

$$u(x,t) = \mathbf{P}(\partial_i A^m(x,t) u_0^m(A(x,t),t))$$

$$+ \Omega \mathbf{P} \{(\widehat{z}; A(x,t), \partial_i A(x,t)) - (\widehat{z}; x; e_i)\}.$$

Here Ω is the constant angular velocity (not $\omega \cdot \nabla$), and e_i form the canonical basis of $\mathbf{R}^3$. We consider the Lagrangian paths $X(a,t)$ associated to the relative velocity u, and their inverses $A(x,t) = X^{-1}(x,t)$, obeying

$$(\partial_t + u \cdot \nabla) A = 0.$$

As a consequence of the Cauchy formula for the total vorticity $\omega + 2\Omega e_3$ one can prove that the direct Lagrangian displacement

$$\lambda(a,t) = X(a,t) - a$$

obeys a time independent differential equation. The Cauchy formula for the total vorticity (the vorticity in a non-rotating frame) follows from differentiation of the Weber formula above and is the same as in the non-rotating case

$$\omega + 2\Omega e_3 = \mathcal{C}[\nabla A, \zeta + 2\Omega e_3]$$

Composing with X the right hand side is

$$\mathcal{C}[\nabla A, \zeta + 2\Omega e_3] \circ X = (\omega_{(0)} + 2\Omega e_3) \cdot \nabla_a X.$$

Rearranging the Cauchy formula we obtain

$$\partial_{a_3}\lambda(a,t) + \frac{1}{2}\rho_0(a)\xi_0(a) \cdot \nabla_a \lambda(a,t) =$$

$$= \frac{1}{2}(\rho_t(a)\xi(a,t) - \rho_0(a)\xi(a,0))$$

where

$$\rho_t(a) = \frac{|\omega(X(a,t),t)|}{\Omega}$$

is the local Rossby number and $\xi = \frac{\omega}{|\omega|}$ is the unit vector of relative vorticity direction. This fact explains directly ($\partial_{a_3}\lambda = O(\rho)$) the fact that strong rotation inhibits vertical transport ([24]). In particular, one can prove rather easily

Theorem 2.2. *Let $u_0 \in H^s(\mathbf{T}^3)$, $s > \frac{5}{2}$ and let $T > 0$ be small enough . For each Ω, consider the inverse and direct Lagrangian displacements*

$$\ell(x,t) = A(x,t) - x$$

and

$$\lambda(a_1, a_2, a_3, t) = X(a_1, a_2, a_3, t) - (a_1, a_2, a_3).$$

They obey

$$\|\partial_{x_3}\ell(\cdot, t)\|_{L^\infty(dx)} \leq C\rho$$

and

$$\|\partial_{a_3}\lambda(\cdot, t)\|_{L^\infty(da)} \leq C\rho$$

with $\rho = \Omega^{-1} \sup_{0 \leq t \leq T} \|\omega(\cdot, t)\|_{L^\infty(dx)}$.

Let $\Omega_j \to \infty$ be an arbitrary sequence and let $X^j(a_1, a_2, a_3, t)$ denote the Lagrangian paths associated to Ω_j. Then, there exists a subsequence (denoted for convenience by the same letter j) an invertible map $X(a_1, a_2, a_3, t)$, and a periodic function of two variables $\lambda(a_1, a_2, t)$ such that

$$\lim_{j \to \infty} X^j(a_1, a_2, a_3, t) = X(a_1, a_2, a_3, t)$$

holds uniformly in a, t and

$$X(a_1, a_2, a_3) = (a_1, a_2, a_3) + \lambda(a_1, a_2, t)$$

This represents a nonlinear Taylor-Proudman theorem in the presence of inertia. It also implies, at positive Rossby number, that the vertical transport, in a relative vorticity turnover time is of the order of the local Rossby number. The nonlinear Taylor Proudman statement and derivation of effective equations are usually addressed via analysis of resonant interactions ([46], [3]).

3 An Infinite Energy Blow Up Example

A classical stagnation point ansatz for the solution of the Euler equations was shown to lead to blow up in three dimensions by J.T. Stuart ([87]), and two dimensions by Childress, Iearly, Spiegel and Young ([12]). A new ansatz, and numerics for the singularity are due to Gibbon and Ohkitani ([76]). The ansatz requires infinite energy, separation of variables, and demands the three dimensional velocity in the form

$$\mathbf{u}(x, y, z, t) = (u(x, y, t), z\gamma(x, y, t))$$

with $\gamma(x, y, t)$ and $u(x, y, t) = (u_1(x, y, t), u_2(x, y, t))$ periodic functions of two variables with fundamental domain Q. The divergence-free condition for $\mathbf{u}$ becomes the two dimensional

$$\nabla \cdot u = -\gamma.$$

The Euler equations respect this ansatz and the dynamics reduce to a pair of equations, one for the vorticity in the vertical direction

$$\omega_3(x,y,t) = \frac{\partial u_2(x,y,t)}{\partial x} - \frac{\partial u_1(x,y,t)}{\partial y},$$

and one for the variable γ which represents the vertical derivative of the vertical component of velocity. The velocity is recovered using constitutive equations for a stream function ψ and a potential h. This entire system is

$$\begin{cases} \partial_t \omega_3 + u \cdot \nabla \omega_3 = \gamma \omega_3. \\ \frac{\partial \gamma}{\partial t} + u \cdot \nabla \gamma = -\gamma^2 + \frac{2}{|Q|} \int_Q \gamma^2(x,t)dx \\ u = \nabla^\perp \psi + \nabla h \\ -\Delta h = \gamma, \\ -\Delta \psi = \omega_3. \end{cases}$$

The two dynamical equations for (ω_3, γ) coupled with the constitutive equations for u form the nonlocal Riccati system. This blows up from all nontrivial initial data:

Theorem 3.1. *([18]) Consider the nonlocal conservative Riccati system. For any smooth, mean zero initial data $\gamma_0 \neq 0$, ω_0, the solution becomes infinite in finite time. Both the maximum and the minimum values of the component γ of the solution diverge to plus infinity and respectively to negative infinity at the blow up time.*

One can determine explicitly the blow up time and the form of γ on characteristics, without having to actually integrate the characteristic equations, which may be rather difficult. The solution is given on characteristics $X(a,t)$ in terms of the initial data $\gamma(x,0) = \gamma_0(x)$ by

$$\gamma(X(a,t),t) = \alpha(\tau(t)) \left(\frac{\gamma_0(a)}{1 + \tau(t)\gamma_0(a)} - \overline{\phi}(\tau(t)) \right)$$

where

$$\overline{\phi}(\tau) = \left\{ \int_Q \frac{\gamma_0(a)}{(1 + \tau\gamma_0(a))^2} da \right\} \left\{ \int_Q \frac{1}{1 + \tau\gamma_0(a)} da \right\}^{-1},$$

$$\alpha(\tau) = \left\{ \frac{1}{|Q|} \int_Q \frac{1}{1 + \tau\gamma_0(a)} da \right\}^{-2}$$

and

$$\frac{d\tau}{dt} = \alpha(\tau), \quad \tau(0) = 0.$$

The function $\tau(t)$ can also be obtained implicitly from

$$t = \left(\frac{1}{|Q|}\right)^2 \int_Q \int_Q \frac{1}{\gamma_0(a) - \gamma_0(b)} \log\left(\frac{1 + \tau\gamma_0(a)}{1 + \tau\gamma_0(b)}\right) da\,db.$$

The blow up time $t = T_*$ is given by

$$T_* = \frac{1}{|Q|^2} \int \int \frac{1}{\gamma_0(a) - \gamma_0(b)} \log\left(\frac{\gamma_0(a) - m_0}{\gamma_0(b) - m_0}\right) da\,db$$

where

$$m_0 = \min_Q \gamma_0(a) < 0.$$

The Jacobian $J(a,t) = \text{Det}\left\{\frac{\partial X(a,t)}{\partial a}\right\}$ is given by

$$J(a,t) = \frac{1}{1 + \tau(t)\gamma_0(a)} \left\{\frac{1}{|Q|} \int_Q \frac{da}{1 + \tau(t)\gamma_0(a)}\right\}^{-1}$$

Because the Jacobian can be found explicitly, one can get some of the Eulerian information as well: integrals of powers of γ can be computed. The moments of γ are given by

$$\int_Q (\gamma(x,t))^p dx =$$

$$(\alpha(\tau))^p \int_Q \left\{\frac{\gamma_0(a)}{1 + \tau(t)\gamma_0(a)} - \overline{\phi}(\tau(t))\right\}^p J(a,t)da.$$

The proof of these facts is based on several auxilliary constructions. Let ϕ solve

$$\partial_\tau\phi + v \cdot \nabla\phi = -\phi^2$$

together with

$$\nabla \cdot v(x,\tau) = -\phi(x,\tau) + \frac{1}{|Q|} \int_Q \phi(x,\tau)dx.$$

We take the curl $\zeta = \frac{\partial v_2}{\partial x} - \frac{\partial v_1}{\partial y}$ and demand that

$$\partial_\tau\zeta + v \cdot \nabla\zeta = \left(\phi - \frac{3}{|Q|} \int_Q \phi(x,\tau)dx\right)\zeta.$$

Note that this is a linear equation when ϕ is known, which allows a bootstrap argument to guarantee that the construction does not breakdown before the blow up time derived below. Passing to characteristics

$$\frac{dY}{d\tau} = v(Y,\tau)$$

we integrate and obtain

$$\phi(Y(a,\tau),\tau) = \frac{\phi_0(a)}{1 + \tau\phi_0(a)}$$

valid as long

$$\inf_{a \in Q}(1 + \tau\phi_0(a)) > 0.$$

We need to compute

$$\overline{\phi}(\tau) = \frac{1}{|Q|}\int_Q \phi(x,\tau)dx.$$

The Jacobian

$$J(a,\tau) = Det\left\{\frac{\partial Y}{\partial a}\right\}$$

obeys

$$\frac{dJ}{d\tau} = -h(a,\tau)J(a,\tau)$$

where

$$h(a,\tau) = \phi(Y(a,\tau),\tau) - \overline{\phi}(\tau).$$

Initially the Jacobian equals to one, so

$$J(a,\tau) = e^{-\int_0^\tau h(a,s)ds}.$$

Consequently

$$J(a,\tau) = e^{\int_0^\tau \overline{\phi}(s)ds}\exp\left(-\int_0^\tau \frac{d}{ds}\log(1 + s\phi_0(a))ds\right)$$

and thus

$$J(a,\tau) = e^{\int_0^\tau \overline{\phi}(s)ds}\frac{1}{1 + \tau\phi_0(a)}.$$

The map $a \mapsto Y(a,\tau)$ is one and onto. Changing variables one has

$$\int_Q \phi(x,\tau)dx = \int_Q \phi(Y(a,\tau),t)J(a,\tau)da$$

and therefore

$$\overline{\phi}(\tau) = e^{\int_0^\tau \overline{\phi}(s)ds}\frac{1}{|Q|}\int_Q \frac{\phi_0(a)}{(1 + \tau\phi_0(a))^2}da.$$

Consequently

$$\frac{d}{d\tau}e^{-\int_0^\tau \overline{\phi}(s)ds} = \frac{d}{d\tau}\frac{1}{|Q|}\int_Q \frac{1}{1 + \tau\phi_0(a)}da.$$

Because both sides at $\tau = 0$ equal one, we have

$$e^{-\int_0^\tau \overline{\phi}(s)ds} = \frac{1}{|Q|}\int_Q \frac{1}{1 + \tau\phi_0(a)}da$$

and

$$\overline{\phi}(\tau) = \left\{ \int_Q \frac{\phi_0(a)}{(1+\tau\phi_0(a))^2} da \right\} \left\{ \int_Q \frac{1}{1+\tau\phi_0(a)} da \right\}^{-1}.$$

Note that the function $\delta(x,\tau) = \phi(x,\tau) - \overline{\phi}(\tau)$ obeys

$$\frac{\partial \delta}{\partial \tau} + v \cdot \nabla \delta = -\delta^2 + 2\frac{1}{|Q|} \int_Q \delta^2 dx - 2\overline{\phi}\delta.$$

We consider now the function

$$\sigma(x,\tau) = e^{2\int_0^\tau \overline{\phi}(s)ds} \delta(x,\tau)$$

and the velocity

$$U(x,\tau) = e^{2\int_0^\tau \overline{\phi}(s)ds} v(x,\tau).$$

Multiplying the equation of δ by $e^{4\int_0^\tau \overline{\phi}(s)ds}$ we obtain

$$e^{2\int_0^\tau \overline{\phi}(s)ds}\frac{\partial \sigma}{\partial \tau} + U \cdot \nabla \sigma = -\sigma^2 + \frac{2}{|Q|} \int \sigma^2 dx.$$

Note that

$$\nabla \cdot U = -\sigma.$$

Now we change the time scale. We define a new time t by the equation

$$\frac{dt}{d\tau} = e^{-2\int_0^\tau \overline{\phi}(s)ds},$$

$t(0) = 0$, and new variables

$$\gamma(x,t) = \sigma(x,\tau)$$

and

$$u(x,t) = U(x,\tau).$$

Now γ solves the nonlocal conservative Riccati equation

$$\frac{\partial \gamma}{\partial t} + u \cdot \nabla \gamma = -\gamma^2 + \frac{2}{|Q|} \int \gamma^2 dx$$

with periodic boundary conditions,

$$u = (-\Delta)^{-1} \left[\nabla^\perp \omega + \nabla \gamma \right]$$

and

$$\frac{\partial \omega}{\partial t} + u \cdot \nabla \omega = \gamma \omega$$

The initial data are

$$\gamma_0(x) = \delta_0(x) = \phi_0(x)$$

and

$$t = \left(\frac{1}{|Q|}\right)^2 \int_Q \int_Q \frac{1}{\phi_0(a) - \phi_0(b)} \log \frac{1 + \tau\phi_0(a)}{1 + \tau\phi_0(b)} da \, db.$$

Note that the characteristic system

$$\frac{dX}{dt} = u(X, t)$$

is solved by

$$X(a, t) = Y(a, \tau)$$

This implies the relationship

$$\gamma(X(a, t), t) = \alpha(\tau) \left(\frac{\phi_0(a)}{1 + \tau\phi_0(a)} - \overline{\phi}(\tau)\right)$$

$$\alpha(\tau) = e^{2 \int_0^\tau \overline{\phi}(s) ds}.$$

If the initial smooth function $\gamma_0(a) = \phi_0(a)$, of mean zero has minimum $m_0 < 0$, then the blow up time is

$$\tau_* = -\frac{1}{m_0}$$

Lets consider now a simple example, in order to determine the blow up asymptotics. Let ϕ_0 attain the minimum m_0 at a_0, and assume locally that

$$\phi_0(a) \geq m_0 + C|a - a_0|^2$$

for $0 \leq |a - a_0| \leq r$. Then it follows that the integral

$$\frac{1}{|Q|} \int \frac{da}{\epsilon^2 + \phi_0(a) - m_0}$$

behaves like

$$\frac{1}{|Q|} \int \frac{da}{\epsilon^2 + \phi_0(a) - m_0} \sim \log \left\{\sqrt{1 + \left(\frac{Cr}{\epsilon}\right)^2}\right\}$$

for small ϵ. Taking

$$\epsilon^2 = \frac{1}{\tau} - \frac{1}{\tau_*}$$

it follows that

$$e^{-\int_0^\tau \overline{\phi}(s) ds} \sim \log \left\{\sqrt{1 + \frac{C}{\tau_* - \tau}}\right\}$$

and for small $(\tau_* - \tau)$

$$\frac{1}{|Q|} \int_Q \frac{\phi_0(a)}{(1+\tau\phi_0(a))^2} \, da \sim -\frac{C}{\tau_* - \tau}$$

and thus $t(\tau)$ has a finite limit $t \to T_*$ as $\tau \to \tau_*$. The average $\overline{\phi}(\tau)$ diverges to negative infinity,

$$\overline{\phi}(\tau) \sim -\frac{C}{\tau_* - \tau} \left[\log \left\{ \sqrt{1 + \frac{C}{\tau_* - \tau}} \right\} \right]^{-1}.$$

The prefactor α becomes vanishingly small

$$\alpha(\tau) \sim (\log(\tau_* - \tau))^{-2}$$

and so

$$\gamma(X(a,t),t) \sim (\log(\tau_* - \tau))^{-2} \left(\frac{\phi_0(a)}{1+\tau\phi_0(a)} - \overline{\phi}(\tau) \right).$$

If $\phi_0(a) > 0$ then the first term in the brackets does not blow up and γ diverges to plus infinity. If the label is chosen at the minimum, or nearby, then the first term in the brackets dominates and the blow up is to negative infinity, as expected from the ODE. From

$$(\alpha(\tau))^{-1} d\tau = dt$$

it follows that the asymptotic behavior of the blow up in t follows from the one in τ:

$$T_* - t \sim (\tau_* - \tau) \left(1 + \log \left(\frac{1}{\tau_* - \tau} \right) \right)^2$$

4 Navier-Stokes Equations

The Navier-Stokes equations are

$$D_\nu u + \nabla p = 0, \qquad (4.1)$$

together with the incompressibility condition $\nabla \cdot u = 0$. The operator D_ν

$$D_\nu = D_\nu(u, \nabla) = \partial_t + u \cdot \nabla - \nu \Delta \qquad (4.2)$$

describes advection with velocity u and diffusion with kinematic viscosity $\nu > 0$. When $\nu = 0$ we recover formally the Euler equations (2.1), and $D_{\nu|\nu=0} = D_t$. The vorticity $\omega = \nabla \times u$ obeys an equation similar to (2.3):

$$D_\nu \omega = \omega \cdot \nabla u. \qquad (4.3)$$

The Eulerian-Lagrangian equations (2.7) and (2.9) have also viscous counterparts ([20]). The equation corresponding to (2.7) is

$$\begin{cases} D_\nu A = 0, \\ D_\nu v = 2\nu C \nabla v, \\ u = W[A, v] \end{cases} \tag{4.4}$$

The $u = W[A, v]$ is the Weber formula (2.6), same as in the case of $\nu = 0$. The right hand side of (4.4) is given terms of the connection coefficients

$$C_{k;i}^m = \left((\nabla A)^{-1} \right)_{ji} \left(\partial_j \partial_k A^m \right).$$

The detailed form of virtual velocity equation in (4.4) is

$$D_\nu v_i = 2\nu C_{k;i}^m \partial_k v_m.$$

The connection coefficients are related to the Christoffel coefficients of the flat Riemannian connection in $\mathbf{R}^3$ computed using the change of variables $a = A(x, t)$:

$$C_{k;i}^m(x, t) = -\Gamma_{ji}^m(A(x, t)) \frac{\partial A^j(x, t)}{\partial x_k}$$

The equation $D_\nu(u, \nabla) A = 0$ describes advection *and diffusion* of labels. Use of traditional $(D_t A = 0)$ Lagrangian variables when $\nu > 0$ would introduce third order derivatives of A in the viscous evolution of the Cauchy invariant, making the equations ill posed: the passive characteristics of u are not enough to reconstruct the dynamics.

The diffusion of labels is a consequence of the physically natural idea of adding Brownian motion to the Lagrangian flow. Indeed, if $u(X(a, t), t)$ is known, and if

$$dX(a, t) = u(X(a, t), t)dt + \sqrt{2\nu}dW(t), \ X(a, 0) = a,$$

with $W(t)$ standard independent Brownian motions in each component, and if

$$Prob\{X(a, t) \in dx\} = \rho(x, t; a)dx$$

then the expected value of the back to labels map

$$A(x, t) = \int \rho(x, t; a) a \, da$$

solves

$$D_\nu(u, \nabla) A = 0.$$

In addition to being well posed, the Eulerian-Lagrangian viscous equations are capable of describing vortex reconnection. We associate to the virtual velocity v the Eulerian-Lagrangian curl of v

$$\zeta = \nabla^A \times v \tag{4.5}$$

where

$$\nabla_i^A = \left((\nabla A)^{-1} \right)_{ji} \partial_j$$

is the pull back of the Eulerian gradient. The viscous analogue of the Eulerian-Lagrangian Cauchy invariant active scalar system (2.9) is

$$\begin{cases} D_\nu A = 0, \\ D_\nu \zeta^q = 2\nu G_p^{qk} \partial_k \zeta^p + \nu T_p^q \zeta^p, \\ u = \nabla \times (-\Delta)^{-1} \left(\mathcal{C}[\nabla A, \zeta] \right) \end{cases} \tag{4.6}$$

The Cauchy transformation

$$\mathcal{C}[\nabla A, \zeta] = (det(\nabla A))(\nabla A)^{-1} \zeta.$$

is the same as the one used in the Euler equations, (3.1). The specific form of the two terms on the right hand side of the Cauchy invariant's evolution are

$$G_p^{qk} = \delta_p^q C_{k;m}^m - C_{k;p}^q, \tag{4.7}$$

and

$$T_p^q = \epsilon_{qji} \epsilon_{rmp} C_{k;i}^m C_{k;j}^r. \tag{4.8}$$

The system (4.4) is equivalent to the Navier-Stokes system. When $\nu = 0$ the system reduces to (2.7). The system (4.6) is equivalent to the Navier-Stokes system, and reduces to (2.9) when $\nu = 0$.

The pair (A, v) formed by the diffusive inverse Lagrangian map and the virtual velocity are akin to charts in a manifold. They are a convenient representation of the dynamics of u for some time. When the representation becomes inconvenient, then one has to change the chart. This may (and will) happen if ∇A becomes non-invertible. Likewise, the pair (A, ζ) formed with the "back-to-labels" map A and the diffusive Cauchy invariant ζ are convenient charts. In order to clarify this statement let us introduce the terminology of "group expansion" for the procedure of resetting. More precisely, the group expansion for (4.4) is defined as follows. Given a time interval $[0, T]$ we consider resetting times

$$0 = t_0 < t_1 < \cdots < t_n \cdots \leq T.$$

On each interval $[t_i, t_{i+1}]$, $i = 0, \ldots$ we solve the system (4.4):

$$\begin{cases} D_\nu(u, \nabla) A = 0, \\ D_\nu(u, \nabla) v = 2\nu C \nabla v, \\ u = \mathbf{P} \left((\nabla A)^* v \right). \end{cases}$$

with resetting values

$$\begin{cases} A(x, t_i) = x, \\ v(x, t_i + 0) = ((\nabla A)^* v)(x, t_i - 0). \end{cases}$$

We require the strong resetting criterion that $\nabla \ell = (\nabla A) - \mathbf{I}$ must be smaller than a preassigned value ϵ in an analytic norm: $\exists \lambda$ such that for all $i \geq 1$ and all $t \in [t_i, t_{i+1}]$ one has

$$\int e^{\lambda |k|} \left| \widehat{\ell}(k) \right| dk \leq \epsilon < 1.$$

If there exists N such that $T = \sum_{i=0}^{N}(t_{i+1} - t_i)$ then we say that the group expansion *converges* on $[0, T]$. A group expansion of (4.6) is defined similarly. The resetting conditions are

$$\begin{cases} A(x, t_i) = x, \\ \zeta(x, t_i + 0) = \mathcal{C}[(\nabla A))(x, t_i - 0), \zeta(x, t_i - 0)]. \end{cases}$$

The strong analytic resetting criterion is the same. The first interval of time $[0, t_1)$ is special. The initial value for v is u_0 (the initial datum for the Navier-Stokes solution), and the initial value for ζ is ω_0, the corresponding vorticity. The local time existence is used to guarantee invertibility of the matrix ∇A on $[0, t_1)$ and Gevrey regularity ([57]) to pass from moderately smooth initial data to Gevrey class regular solutions. Note that the resetting data are such that both u and ω are time continuous.

Theorem 4.1. *([21]) Let $u_0 \in H^1(\mathbf{R}^3)$ be divergence-free. Let $T > 0$. Assume that the solution of the Navier-Stokes equations with initial datum u_0 obeys $\sup_{0 \leq t \leq T} \|\omega(\cdot, t)\|_{L^2(dx)} < \infty$. Then there exists $\lambda > 0$ so that, for any $\epsilon > 0$, there exists $\tau > 0$ such that both group expansions converge on $[0, T]$ and the resetting intervals can be chosen to have any length up to τ, $t_{i+1} - t_i \in [0, \tau]$. The velocity u, solution of the Navier-Stokes equation with initial datum u_0, obeys the Weber formula (2.6). The vorticity $\omega = \nabla \times u$ obeys the Cauchy formula (3.1).*

Conversely, if one group expansion converges, then so does the other, using the same resetting times. The Weber and Cauchy formulas apply and reconstruct the solution of the Navier-Stokes equation. The enstrophy is bounded $\sup_{0 \leq t \leq T} \|\omega(\cdot, t)\|_{L^2(dx)} < \infty$, and the Navier-Stokes solution is smooth.

The quantity λ can be estimated explicitly in terms of the bound of enstrophy, time T, and kinematic viscosity ν. Then τ is proportional to ϵ, with a coefficient of proportionality that depends on the bound on enstrophy, time T and ν. The converse statement, that if the group expansion converges, then the enstrophy is bounded, follows from the fact that there are finitely many resettings. Indeed, the Cauchy formula and the near identity bound on ∇A imply a doubling condition on the enstrophy on each interval. It is well-known that the condition regarding the boundedness of the enstrophy implies regularity of the Navier-Stokes solution. Our definition of convergent group expansion is very demanding, and it is justified by the fact that once the enstrophy is bounded, one could mathematically demand analytic norms. But the physical resetting criterion is the invertibility of the matrix ∇A. The Euler equations require no resetting as long as the solution is smooth. The Navier-Stokes equations, at least numerically, require numerous and frequent resettings. There is a deep connection between these resetting times and vortex reconnection ([74], [75]). In the Euler equation, as long as the solution is smooth, the Cauchy invariant obeys $\zeta(x, t) = \omega_{(0)}(A(x, t))$ with $\omega_{(0)} = \omega_0$, the initial vorticity. The topology of vortex lines is frozen in time. In the Navier-Stokes system the topology changes. This is the phenomenon of vortex reconnection. There is ample numerical and physical evidence for this phenomenon. In the more complex, but similar case of magneto-hydrodynamics, magnetic reconnection occurs, and has powerful physical implications. Vortex reconnection is a dynamical dissipative process. The solutions of the Navier-Stokes equations obey a space time average bound ([16], [21])

$$
\int\limits_0^T \int\limits_{\mathbf{R}^3} |\omega(x,t)| \left| \nabla_x \left(\frac{\omega(x,t)}{|\omega(x,t)|} \right) \right|^2 dx dt \leq \frac{1}{2} \nu^{-2} \int\limits_{\mathbf{R}^3} |u_0(x,t)|^2 dx.
$$

This bound is consistent with the numerically observed fact that the region of high vorticity is made up of relatively straight vortex filaments (low curvature of vortex lines) separated by distances that vanish with viscosity. The process by which this separation is achieved is vortex reconnection. When vortex lines are locally aligned, a geometric depletion of nonlinearity occurs, and the local production of enstrophy drops. Actually, the Navier-Stokes equations have global smooth solutions if the vorticity direction field $\frac{\omega}{|\omega|}$ is Lipschitz continuous ([29]) in regions of high vorticity. So, vortex reconnection is a regularizing mechanism.

In the case of the Navier-Stokes system the virtual velocity and the Cauchy invariant in an expansion can be computed using the back to labels map A, but unlike the case of the Euler equations, they are no longer frozen in time: they diffuse. Let us recall that, using the smooth change of variables $a = A(x, t)$ (at each fixed time t) we compute the Euclidean Riemannian metric by

$$
g^{ij}(a, t) = (\partial_k A^i)(\partial_k A^j)(x, t) \tag{4.9}
$$

The equations for the virtual velocity and for the Cauchy invariant can be solved by following the path A, i.e., by seeking

$$
\begin{aligned}
v(x, t) &= \upsilon(A(x, t), t), \\
\zeta(x, t) &= \xi(A(x, t), t)
\end{aligned} \tag{4.10}
$$

The equations for υ and ξ become purely diffusive. Using $D_\nu A = 0$, the operator D_ν becomes

$$
D_\nu(f \circ A) = \left((\partial_t - \nu g^{ij} \partial_i \partial_j) f\right) \circ A \tag{4.11}
$$

The equation for υ follows from (4.4):

$$\partial_t v_i = \nu g^{mn} \partial_m \partial_n v_i - 2\nu V_i^{mj} \partial_m v_j \tag{4.12}$$

with

$$V_i^{mj} = g^{mk} \Gamma_{ik}^j$$

The derivatives are with respect to the Cartesian coordinates a. The equation reduces to $\partial_t v = 0$ when $\nu = 0$, and in that case we recover $v = u_{(0)}$, the time independent initial velocity. For $\nu > 0$ the system is parabolic and well posed. The equation for ξ follows from (4.6):

$$\partial_t \xi^q = \nu g^{ij} \partial_i \partial_j \xi^q + 2\nu W_n^{qk} \partial_k \xi^n + \nu T_p^q \xi^p \tag{4.13}$$

with

$$\begin{cases} W_n^{qk} = -\delta_n^q g^{kr} \Gamma_{rp}^p + g^{kp} \Gamma_{pn}^q, \\ T_p^q = \epsilon_{qji} \epsilon_{rmp} \Gamma_{\alpha j}^r \Gamma_{\beta i}^m g^{\alpha \beta} \end{cases}$$

Again, when $\nu = 0$ this reduces to the invariance $\partial_t \xi = 0$. But in the presence of ν this is a parabolic system. Both the Cauchy invariant and the virtual velocity equations start out looking like the heat equation, because $g^{mn}(a, 0) = \delta^{mn}$ and $\Gamma_{jk}^i(a, 0) = 0$. The equation for the determinant of ∇A is

$$D_\nu \left(\log(det(\nabla A)) \right) = \nu \; C_{k;s}^i C_{k;i}^s \tag{4.14}$$

The initial datum vanishes. When $\nu = 0$ we recover conservation of incompressibility. In the case $\nu > 0$ the inverse time scale in the right hand side of this equation is significant for reconnection. Because the equation has a maximum priciple it follows that

$$det(\nabla A)(x, t) \geq exp \left\{ -\nu \int_{t_i}^t \sup_x \; C_{k;s}^i C_{k;i}^s \; d\sigma \right\}$$

Considering

$$g = det(g_{ij}) \tag{4.15}$$

where g_{ij} is the inverse of g^{ij} and observing that

$$g(A(x, t)) = (det(\nabla A))^{-2}$$

we deduce that the equation (4.14) becomes

$$\partial_t (\log(\sqrt{g})) = \nu g^{ij} \partial_i \partial_j \log(\sqrt{g}) - \nu g^{\alpha\beta} \Gamma_{\alpha p}^m \Gamma_{\beta m}^p \tag{4.16}$$

The initial datum is zero, the equation is parabolic, has a maximum priciple and is driven by the last term. The form (2.64) of the equation (4.14) has the same interpretation: the connection coefficients define an inverse length scale associated to A. The corresponding inverse time scale

$$\nu \; C_{k;s}^i C_{k;i}^s = \nu \; g^{mn} \Gamma_{ms}^i \Gamma_{ni}^s \; A$$

decides the time interval of validity of the chart A, and the time to reconnection. Let us denote

$$L = \log(\sqrt{g})$$
$$V^i = g^{jk}\Gamma^i_{jk},$$
$$F = g^{jk}\Gamma^i_{js}\Gamma^s_{ki}.$$

The equation (2.64) for L can be written as

$$\partial_t L = \nu g^{ij}\partial_i\partial_j L - \nu F.$$

We will make use now of a few well-known facts from Riemannian geometry. The first fact is

$$\partial_j \log(\sqrt{g}) = \Gamma^m_{jm}. \tag{4.17}$$

The second fact is the vanishing of the curvature of the Euclidean Riemannian connection in $\mathbf{R}^3$

$$\partial_k \Gamma^i_{ql} + \Gamma^i_{pk}\Gamma^p_{ql} = \partial_l\Gamma^i_{qk} + \Gamma^i_{pl}\Gamma^p_{qk}. \tag{4.18}$$

The third fact is that the connection is compatible with the metric

$$\partial_i g^{jk} + \Gamma^j_{ip}g^{pk} + \Gamma^k_{ip}g^{jp} = 0. \tag{4.19}$$

Taking the sum $i = l$ in (4.18) we deduce that

$$\Gamma^i_{pk}\Gamma^p_{qi} = \partial_i\Gamma^i_{qk} - \partial_k\Gamma^i_{qi} + \Gamma^i_{pi}\Gamma^p_{qk}.$$

Using (4.17) and multiplying by g^{kq} we deduce

$$F = -g^{ij}\partial_i\partial_j L + V^j\partial_j L + g^{\alpha\beta}\partial_i\Gamma^i_{\alpha\beta}.$$

But, using (4.19) we deduce

$$g^{\alpha\beta}\partial_i\Gamma^i_{\alpha\beta} = 2F - V^i\partial_i L + \mathrm{div}V$$

with

$$\mathrm{div}V = \frac{1}{\sqrt{g}}\partial_i\left(\sqrt{g}V^i\right).$$

Therefore

$$g^{ij}\partial_i\partial_j L - F = \mathrm{div}V, \tag{4.20}$$

and we obtained an alternative form of (2.64):

$$\partial_t\sqrt{g} = \nu\partial_i\left(\sqrt{g}g^{\alpha\beta}\Gamma^i_{\alpha\beta}\right). \tag{4.21}$$

Let us introduce now

$$f^q = \frac{1}{\sqrt{g}}\xi^q. \tag{4.22}$$

Writing $\xi = \sqrt{g}f$ in (4.13) and using (4.21) it follows that

$$\partial_t f^q = \nu g^{\alpha\beta} \partial_\alpha \partial_\beta f^q + \nu D_n^{qk} \partial_k f^n + \nu F_p^q f^p \tag{4.23}$$

with

$$D_n^{qk} = 2g^{kp} \Gamma_{pn}^q$$

and

$$F_p^q = T_p^q + 2g^{kj} \Gamma_{jp}^q \partial_k L + \delta_p^q g^{\alpha\beta} \left(\Gamma_{\alpha j}^i \Gamma_{\beta i}^j - \Gamma_{\alpha i}^i \Gamma_{\beta j}^j \right).$$

In the derivation of the last expression we used the fact that $\sqrt{g} = e^L$ and the equation (4.20). Let us note here what happens in two spatial dimensions. In that case $A^3 = x^3$, $\Gamma_{ij}^3 = 0$, and $f^1 = f^2 = 0$. Then the free term F_3^3 in (4.23) vanishes because T_3^3 cancels the δ_3^3 term. The rest of the terms involve Γ_{ij}^3 and vanish. The equation becomes thus the scalar equation

$$\partial_t f^3 = \nu g^{\alpha\beta} \partial_\alpha \partial_\beta f^3$$

which means that

$$\omega^3 = f^3 \circ A.$$

Because the rest of components are zero, this means that in two dimensions $\omega = f \circ A$. This represents the solution of the two dimensional vorticity equation

$$D_\nu \omega = 0$$

as it is readily seen from (4.11). The three dimensional situation is more complicated, f obeys the nontrivial well posed parabolic system (4.23), and the Cauchy formula in terms of f reads

$$\omega = (\nabla A)^{-1} (f \circ A). \tag{4.24}$$

The function $f \circ A = (det(\nabla A)) \zeta$ deserves just as much the name "Cauchy invariant" as does $\zeta = \nabla^A \times v$. In the inviscid case these functions coincide, of course. The Eulerian evolution equation of $\tilde{\zeta} = (det(\nabla A))\zeta$ is similar to (4.6). There are two more important objects to consider, in the viscous context: circulation, and helicity. We note that the Weber formula implies that

$$udx - vdA = -dn$$

and therefore

$$\oint_{\gamma \circ A} udx = \oint_\gamma vda \tag{4.25}$$

holds for any closed loop γ. Similarly, in view of the Cauchy formula one has that the helicity obeys

$$h = u \cdot \omega = v \cdot \tilde{\zeta} - \nabla \cdot (n\omega)$$

So, if $T \circ A$ is a vortex tube, then

$$\int_{T \circ A} h dx = \int_{T \circ A} v \cdot \tilde{\zeta} dx = \int_T v \cdot \xi da. \qquad (4.26)$$

The metric coefficients g^{ij} determine the connection coefficients, as it is well known. But they do not determine their own evolution as they change under the Navier-Stokes equations. (The evolution equation of g^{ij} involves ∇u and ∇A). It is therefore remarkable that the virtual velocity, Cauchy invariant and volume element evolve according to equations that do not involve explicitly the velocity, once one computes in a diffusive Lagrangian frame. This justifies the following terminology: we will say that a function f is *diffusively Lagrangian* under the Navier-Stokes flow if $f \circ A$ obeys an evolution equation with coefficients determined locally by the Euclidean Riemannian metric induced by the change of variables A. Thus, for instance, the metric itself is not diffusively Lagrangian. The previous calculations can be summarized thus:

Theorem 4.2. *The virtual velocity v, the Cauchy invariant ζ and the Jacobian determinant $det(\nabla A)$ associated to solutions of the Navier-Stokes equations are diffusively Lagrangian.*

5 Approximations

We will describe here approximations of the Navier-Stokes (and Euler) equations. These approximations are partial differential equations with globally smooth solutions. We'll consider a mollifier: an approximation of the identity obtained by convolution with a smooth function which decays enough at infinity, is positive and has integral equal to one. The mollified u is denoted $[u]$:

$$[u]_\delta = \delta^{-3} \int_{\mathbf{R}^3} J\left(\frac{x-y}{\delta}\right) u(y) dy = J_\delta(-i\nabla)u = J_\delta u = [u]$$

The length scale $\delta > 0$ is fixed in this section. In order to recover the original equations one must pass to the limit $\delta \to 0$. The first approximation concerns the Eulerian velocity formulation (4.1), and is due to Leray ([67]):

$$\partial_t u + [u] \cdot \nabla u - \nu \Delta u + \nabla p = 0 \qquad (5.1)$$

together with $\nabla \cdot u = 0$. The Eulerian vorticity formulation (4.3) has an approximation which corresponds to Chorin's vortex methods ([13]):

$$\partial_t \omega + [u] \cdot \omega - \nu \Delta \omega = \omega \cdot \nabla[u] \qquad (5.2)$$

with u calculated from ω using $\omega = \nabla \times u$. This relationship can be written as

$$[u] = J_\delta \left(\nabla \times (-\Delta)^{-1}\right) \omega.$$

The virtual velocity active scalar system (4.4) has an approximation

$$\begin{cases} D_\nu([u],\nabla)A = 0, \\ D_\nu([u],\nabla)v = 2\nu C\nabla v, \\ u = W[A,v]. \end{cases} \tag{5.3}$$

The relationship determining the advecting velocity is thus

$$[u] = J_\delta W[A,v]$$

with $W[A,v]$ the same Weber formula (2.6). The Cauchy invariant active scalar system (4.6) is approximated in a similar manner

$$\begin{cases} D_\nu([u],\nabla)A = 0, \\ D_\nu([u],\nabla)\zeta^q = 2\nu G_p^{qk}\partial_k\zeta^p + \nu T_p^q\zeta^p, \\ u = \nabla \times (-\Delta)^{-1}\left(\mathcal{C}[\nabla A,\zeta]\right) \end{cases} \tag{5.4}$$

The Cauchy formula (3.1) is the same, and the diffusive Lagrangian terms G_p^{qk} and T_p^q are given by the same expressions (4.7), (4.8) as in the Navier-Stokes case. In fact, the approximations of both group expansions are defined in exactly the same way: one respects the constitutive laws relating virtual velocity or Cauchy invariant to velocity, and the same resetting rules. One modifies the advecting velocity: D_ν is replaced by $D_\nu([u],\nabla)$.

All four approximations are done by mollifying, but they are not equivalent. The Leray approximate equation (5.1) has the same energy balance as the Navier-Stokes equation,

$$\frac{d}{2dt}\int |u|^2 dx + \nu \int |\nabla u|^2 dx = 0$$

but, when one sets $\nu = 0$ the circulation integrals are not conserved. The vorticity approximation conserves circulation, but has different energy structure (the integrals of $u \cdot [u]$ decay). The approximations of the active scalar systems provide convergent group expansions for the vorticity approximation, not the Leray approximation.

Theorem 5.1. *Consider $u_0 \in H^1(\mathbf{R}^3)$, divergence free, and let $T > 0$, $\nu > 0$, $\delta > 0$ be fixed. Consider a fixed smooth, normalized mollifier J. Then the group expansions of (5.3) and (5.4) converge on $[0,T]$. The Cauchy formula (3.1) gives the solution of the approximate vorticity equation (5.2). The virtual velocity, Cauchy invariant, and Jacobian determinant are diffusively Lagrangian: they solve the evolution equations (4.12, 4.13, 2.64 (or 4.21)) with coefficients determined locally from the Euclidean Riemannian metric, in coordinates $a = A(x,t)$ computed in the expansion.*

The proof starts by verifying that there exists $\tau > 0$ so that both expansions converge with resetting times $t_{i+1} - t_i \geq \tau$. This is straightforward, but rather technical. Then one considers the variable

$$w = (\nabla A)^* v.$$

It follows from (5.3) that w obeys the equation

$$D_\nu([u], \nabla)w + (\nabla[u])^* w = 0.$$

This equation is an approximation of yet another formulation of the Navier-Stokes equations ([66], [78], [82]) used in numerical simulations ([7], [8]), and related to the alpha model ([52], [59]). Taking the curl of this equation it follows that $b = \nabla \times w$ solves

$$D_\nu([u], \nabla)b = b \cdot \nabla[u]$$

The resetting conditions are such that both w and b are continuous in time. The Weber formula implies that the advecting velocity $[u]$ is given by $[u] = J_\delta \mathbf{P} w$, and hence it is continuous in time. It follows that the b solves the equation (5.2) for all $t \in [0, T]$. The initial datum for b is ω_0, so it follows that $b = \omega$ for all t. But the Cauchy formula holds for b, and that finishes the proof. The fact that the Cauchy formula holds for b is kinematic: Indeed, if A, v solve any virtual velocity system (5.3) with some smooth advecting velocity $[u]$, then the Eulerian curl b of $(\nabla A)^* v$ is related to the Lagrangian curl ζ of v by the Cauchy formula (3.1).

6 The QG Equation

Two dimensional fluid equations can be described as active scalars

$$(\partial_t + u \cdot \nabla)\theta + c\Lambda^\alpha \theta = f$$

with an incompressible velocity given in terms of a stream function

$$u = \nabla^\perp \psi = \begin{pmatrix} -\partial_2 \psi \\ \partial_1 \psi \end{pmatrix}$$

computed in terms of θ as

$$\psi = \Lambda^{-\beta}\theta.$$

The constant $c \geq 0$ and smooth time independent source term f are given. The operator $\Lambda = (-\Delta)^{\frac{1}{2}}$ is defined in the whole $\mathbf{R}^2$ or in $\mathbf{T}^2$. Interesting examples are $\beta = 2$ (usual hydrodynamical stream function) with $\alpha = 2$ (usual Laplacian dissipation), and $\beta = 1$, (surface quasigeostrophic equation, QG in short) with $\alpha = 1$ (critical dissipation). We describe briefly the quasigeostrophic equation. When $c = 0$, $f = 0$ the equation displays a number of interesting features shared with the three dimensional Euler equations ([17]). In particular, the blow up of solutions with smooth initial data is a difficult open problem ([36], [37], [40], [77]). There is additional structure: the QG equation has global weak solutions. This has been proved by Resnick in his thesis ([79]); a concise description of the idea can be found also in ([27]). The dissipative system ([9], [27], [39], [61], [91], [92]) with $c > 0$ and $\alpha \in [0, 2]$ has a maximum principle (again proved by Resnick, and again explained in ([27]). The critical dissipative QG equation ($c = 1$, $f = 0$, $\alpha = \beta = 1$) with smooth and localized initial data in $\mathbf{R}^2$ has the following properties: Weak solutions in $L^\infty(dt; L^2(dx)) \cap L^\infty(dx)) \cap L^2(dt; H^{\frac{1}{2}})$ exist for all time. The L^∞ norm is nonincreasing on solutions. The solutions are smooth (in a variety of spaces) and unique for short time. If the initial data is small in L^∞ then the solution is smooth for all time and decays. The subcritical case ($\alpha > 1$, more dissipation, less difficulty) has global unique smooth solutions. The main open problems for the QG equations are: uniqueness of weak solutions, global regularity for large data for critical and supercritical dissipation.

A nice pointwise inequality for fractional derivatives has been discovered recently by A. Cordoba and D. Cordoba ([41]. The inequality provides another proof of the maximum principle, and has independent interest. I'll explain it briefly below. One starts with the Poisson kernel

$$P(z, t) = c_n \frac{t}{(|z|^2 + t^2)^{\frac{n+1}{2}}}$$

in $\mathbf{R}^n$ for $t \geq 0$. The constant c_n is normalizing:

$$\int_{\mathbf{R}^n} P(z, t) dz = 1.$$

Convolution with $P(\cdot, t)$ is a semigroup. The semigroup identity is

$$e^{-t\Lambda} f = \int_{\mathbf{R}^n} P(z, t) \tau_z(f) dz$$

where

$$(\tau_z(f))(x) = f(x - z).$$

and Λ is the Zygmund operator,

$$\Lambda = (-\Delta)^{\frac{1}{2}}.$$

The quickest way to check the semigroup identity is by taking the Fourier transform, $\widehat{P}(\xi, t) = e^{-t|\xi|}$. If f is in the domain of the generator of a semigroup then

$$-\Lambda f = \lim_{t \to 0} t^{-1} \left(e^{-t\Lambda} - I \right) f$$

and consequently

$$-\Lambda f = \lim_{t \to 0} \int_{\mathbf{R}^n} t^{-1} P(z, t) \delta_z(f) dz$$

where

$$\delta_z(f)(x) = f(x - z) - f(x).$$

If the function f is smooth enough, then this becomes

$$-\Lambda f = c_n \int_{\mathbf{R}^n} \frac{1}{|z|^{n+1}} \delta_z(f) dz$$

and it makes sense as a singular integral. This was proved by A. Cordoba and D. Cordoba in ([41]) where they discovered and used the pointwise inequality

$$f \Lambda f \geq \frac{1}{2} \Lambda(f^2).$$

This inequality is the consequence of an identity.

Proposition 6.1. *For any two C_0^∞ functions f, g one has the pointwise identity*

$$\Lambda(fg) = f\Lambda g + g\Lambda f - I_2(f, g) \tag{6.1}$$

with I_2 defined by

$$I_2(f, g) = c_n \int_{\mathbf{R}^n} \frac{1}{|z|^{n+1}} (\delta_z(f))(\delta_z(g)) dz \tag{6.2}$$

The proof follows the calculation of Cordoba and Cordoba: Start with

$$(\delta_z(f))^2 = \delta_z(f^2) - 2f(\delta_z(f))$$

and integrate against $c_n |z|^{-n-1} dz$. One obtains

$$\Lambda(f^2) = 2f\Lambda f - I_2(f, f) \tag{6.3}$$

By polarization, i.e., by appliying (6.3) to f replaced by $f + \epsilon g$, then differentiating in ϵ and then setting $\epsilon = 0$, one deduces (6.1), and finsihes the proof of the proposition. We remark that one can also obtain higher order identities in the same manner. One starts with

$$(\delta_z f)^m = \sum_{j=1}^{m} (-1)^j \binom{m}{j} f^{m-j} \delta_z(f^j).$$

Integrating against $c_n |z|^{-n-1} dz$, one obtains the generalized identity

$$I_m(f, \ldots, f) = \sum_{j=1}^{m} (-1)^{j-1} \binom{m}{j} f^{m-j} \Lambda(f^j) \tag{6.4}$$

where the multilinear nonlocal integral is

$$I_m(f, \ldots, f) = c_n \int_{\mathbf{R}^n} \frac{1}{|z|^{n+1}} (\delta_z(f))^m dz. \tag{6.5}$$

In view of the definition of Besov spaces in terms of δ_z ([86]) it is clear that the multilinear operators I_m are well behaved in Besov spaces. Also, for m even $I_m(f,\ldots,f) \geq 0$ pointwise. Thus, for instance

$$0 \leq I_4(f,f,f,f) = 4f^3\Lambda f + 4f\Lambda f^3 - 6f^2\Lambda f^2 - \Lambda f^4$$

holds pointwise.

Let us also note here that $I_2(f,g)$ gives a quick proof of an extension ([64]) of a Moser calculus inequality of Kato and Ponce ([62]):

Proposition 6.2. *If* $1 < p < \infty$, $1 < p_i < \infty$, $i = 1, 2$ *and* $\frac{1}{p_1} + \frac{1}{p_2} = \frac{1}{p}$, *then*

$$\|\Lambda(fg) - f\Lambda g - g\Lambda f\|_{L^p(\mathbf{R}^n)} \leq$$
$$C\left(\|f\|_{L^{p_1}(\mathbf{R}^n)}\|\Lambda f\|_{L^{p_1}(\mathbf{R}^n)}\|g\|_{L^{p_2}(\mathbf{R}^n)}\|\Lambda g\|_{L^{p_2}(\mathbf{R}^n)}\right)^{\frac{1}{2}}$$

holds for all $f \in W^{1,p_1}$, $g \in W^{1,p_2}$.

The proof is straightforward: In view of (6.1) one needs to check the inequality for $I_2(f,g)$. We write

$$I_2(f,g) = J_r(f,g) + K_r(f,g)$$

with

$$J_r(f,g) = c_n \int_{|z| \leq r} |z|^{-n-1}(\delta_z(f))(\delta_z(g))dz$$

and $K_r(f,g)$ the rest. We note that

$$\|J_r(f,g)\|_{L^p(\mathbf{R}^n)} \leq Cr\|\Lambda f\|_{L^{p_1}(\mathbf{R}^n)}\|\Lambda g\|_{L^{p_2}(\mathbf{R}^n)}$$

and

$$\|K_r(f,g)\|_{L^p(\mathbf{R}^n)} \leq Cr^{-1}\|f\|_{L^{p_1}(\mathbf{R}^n)}\|g\|_{L^{p_2}(\mathbf{R}^n)}$$

with C depending on n, p_1 and p_2 only. These inequalities follow from the elementary

$$\|\delta_z f\|_{L^p(\mathbf{R}^n)} \leq |z|\|\nabla f\|_{L^p(\mathbf{R}^n)}$$

and the boundedness of Riesz transforms in L^p. Optimizing in r we obtain the desired result. It is quite obvious also that

$$\|J_r(f,g)\|_{L^p(\mathbf{R}^n)} \leq Cr^{s_1+s_2-1}\|f\|_{B_{p_1}^{s_1,\infty}}\|g\|_{B_{p_2}^{s_2,\infty}}$$

holds for any $0 \leq s_1 < 1$, $0 \leq s_2 < 1$, $s_1 + s_2 > 1$. Here we used

$$\|f\|_{B_p^{s,\infty}} = \|f\|_{L^p(\mathbf{R}^n)} + \sup |z|^{-s}\|\delta_z f\|_{L^p(\mathbf{R}^n)}$$

Consequently

$$\|I_2(f,g)\|_{L^p(\mathbf{R}^n)} \leq C\|f\|_{B_{p_1}^{s_1,\infty}}\|g\|_{B_{p_2}^{s_2,\infty}} \tag{6.6}$$

holds.

Proposition 6.3. *Consider F, a convex C^2 function of one variable. Assume that the function f is smooth and bounded. Then*

$$F'(f)\Lambda f \geq \Lambda(F(f)) \qquad (6.7)$$

holds pointwise. In particular, if $F(0) = 0$ and $f \in C_0^\infty$

$$\int_{\mathbf{R}^n} F'(f)\Lambda f\, dx \geq 0.$$

Indeed

$$F(\beta) - F(\alpha) - F'(\alpha)\,(\beta - \alpha) \geq 0$$

holds for all α, β. We substitute $\beta = f(x - z)$, $\alpha = f(x)$ and integrate against $c_n|z|^{-n-1}dz$. Note that $F(f(x))$ is bounded because F is continuous and thus bounded on the range of f. The inequality

$$D_{2m} = \int f^{2m-1}\Lambda f\, dx \geq 0$$

implies that the L^{2m} norms do not increase on solutions of the critical dissipative QG equation, and the maximum principle follows.

7 Dissipation and Spectra

Turbulence theory concerns itself with statistical properties of fluids. Some of the objects encountered in turbulence theory are the mean velocity $\langle u \rangle$, Eulerian velocity fluctuation $v = u - \langle u \rangle$, energy dissipation rate $\nu\langle|\nabla u|^2\rangle$, velocity correlation functions $\langle \delta_y v \otimes v \rangle$, velocity structure functions $\langle \delta_y v \otimes \delta_y v \rangle$, higher order structure functions, $\langle \delta_{y_1} v \otimes \cdots \delta_{y_m} v \rangle$. We use the notation $\delta_y u = u(x - y) - u(x)$. The operation $\langle \cdots \rangle$ is ensemble average, a functional integral. The Navier-Stokes equations represent the underlying dynamics. A mathematical framework related to the Navier-Stokes equations has been developed ([60], [51], [89]), but the mathematical advance has been slow. Not complexity but rather simplicity is the essence of the difficulty: turbulent flows obey nontrivial statistical laws. Among these laws, the law for wall bounded flows ([5]), and for scaling of Nusselt number with Rayleigh number in Rayleigh-Bénard turbulence in Helium ([73]) are major examples. A celebrated physical theoretical prediction concerning universality in turbulence, is due to Kolmogorov ([65],[58]). One of the simplest and most important question in turbulence is: how much energy is dissipated by the flow. This question is of major importance for engineering applications because the energy disspated by turbulence is transferred to objects immersed in it. Mathematically, the question is about the long time average of certain integrals, low order (first and second) moments of the velocity and velocity gradients in forced flows. There is a rigorous mathematical method ([25], [26], [35], [43]) to obtain bounds for these bulk quantities, and make contact with experiments in convection and shear dominated turbulence.

The Kolmogorov 2/3 law for homogeneous turbulence is

$$S_2(r) \sim (\epsilon r)^{\frac{2}{3}}$$

with

$$S_2(r) = \langle (\delta_r u)^2 \rangle$$

the second order longitudinal structure function and with

$$\epsilon = \nu \langle |\nabla u|^2 \rangle$$

the energy dissipation rate. The law is meant to hold asymptotically, for large Reynolds numbers, and for r in a range of scales, $[L, k_d^{-1}]$, called the inertial range. The energy dissipation rate ϵ per unit mass has dimension of energy per time, $cm^2 sec^{-3}$. The Kolmogorov law follows from dimensional analysis if one postulates that in the inertial range the (second order) statistics of the flow depend only on the parameter ϵ, because the typical longitudinal velocity fluctuations over a distance r should be an expression with units of $cm sec^{-1}$ and such an expression is $(\epsilon r)^{\frac{1}{3}}$. The energy spectrum for homogeneous turbulence is

$$E(k) = \frac{1}{2} \int_{|\xi|=k} \langle |\hat{u}(\xi, t)|^2 \rangle dS(\xi)$$

with $\hat{u}$ the spatial Fourier transform. The Kolmogorov-Obukhov energy spectrum law is

$$E(k) \sim \epsilon^{\frac{2}{3}} k^{-\frac{5}{3}}$$

The asymptotic equality takes place for $k \in [k_0, k_d]$ where

$$k_d = \nu^{-\frac{3}{4}} \epsilon^{\frac{1}{4}}$$

is the dissipation wave number, and k_0 is the integral scale. The interval $[k_0, k_d]$ is the inertial range. The statements about asymptotic equality refer to the high Reynolds number limit, $k^{\frac{5}{3}} \epsilon^{-\frac{2}{3}} E(k) \to C$, as $Re \to \infty$. The Reynolds number is

$$Re = \frac{UL}{\nu} = \sqrt{\langle |u|^2 \rangle}^{\frac{1}{2}} (k_0 \nu)^{-1}$$

where $U^2 = \langle |u|^2 \rangle$, $L = k_0^{-1}$. The physical intuition is the following: energy is put into the system at large scales $L = k_0^{-1}$. This energy is transferred to small scales without loss, in a statistically selfsimilar manner. At scale k (in wave number space) the only allowed external parameter is ϵ. The energy spectrum has dimension of energy per wave number, because the energy spectrum is the integrand in the one dimensional integral

$$E = \int_0^\infty E(k)dk.$$

Thus $E(k)$ is measured in $cm^3 sec^{-2}$. The wave number is measured in cm^{-1}. A time scale formed with ϵ and k is $t_k = \epsilon^{-\frac{1}{3}}k^{-\frac{2}{3}}$. This is the time of transfer of energy at wave-number scale k. Using this time scale and the length scale k^{-1} one arrives at the expression $E(k) \sim k^{-3}\epsilon^{\frac{2}{3}}k^{\frac{4}{3}}$, the Kolmogorov-Obukhov spectrum. The spectrum can be derived also from the 2/3 law.

The dissipation of the energy occurs at a dissipation scale k_d. This is an inverse length scale formed using only the kinematic viscosity (measured in $cm^2 sec^{-1}$) and ϵ.

The fact that Laplacian dissipation and no other should be used to study turbulence acquires a physical justification: the Kolmogorov dissipation scale is based on the Laplacian, and it is observed experimentally quite convincingly.

Let us describe in more mathematical terms the issues. We start with the dissipation law. The Kolmogorov theory predicts that the energy dissipation is

$$\epsilon \sim \frac{U^3}{L}.$$

The first difficulty one encounters is due to the need to solve three dimensional Navier-Stokes equations. The second difficulty concerns the ensemble average: one needs homogeneity (translation invariance of the statistics) and bounded solutions. One way out of this is to take a set of bounded solutions and perform a well defined operation M that is manifestly translation invariant, normalized and positivity preserving ([30]). This gives upper bounds for energy dissipation, and structure functions of order one and two in the whole space. Assuming scaling $M((\delta_r u)^2) \sim r^s$ we proved in ([30]) that $s \geq 2/3$. In ([30]) the body forced were assumed to be uniformly bounded, but not uniformly square integrable. If the body forces are uniformly square integrable, for instance in bounded domains, then an idea of Foias ([50]) can be used to bound the dissipation without assuming that the velocities are bounded. An upper bound on dissipation in bounded domains, exploiting this idea is given in ([44]).

I will explain the idea below in the whole space, with square integrable forces. Consider solutions of Leray regularized solutions of Navier-Stokes equations in $\mathbf{R}^3$,

$$D_\nu([u], \nabla)u + \nabla p = f, \quad \nabla \cdot u = 0,$$

with smooth, time independent, deterministic divergence-free body forces f,

$$f(x) = \int_{\mathbf{R}^3} e^{2\pi i \xi \cdot x} \widehat{f}(\xi) d\xi$$

with $\widehat{f}$ supported in $|\xi| \leq k_0$. We consider a long time average M_T.

$$\frac{1}{T}\int_0^T h(\cdot,t)dt = M_T(h)$$

We define the averaging procedure to be

$$\langle h(x,t)\rangle = k_0^3 \lim_{T\to\infty} \sup M_T \int_{\mathbf{R}^3} h(x,t)dx$$

We set

$$U^2 = \langle |u|^2\rangle,\ F^2 = \langle |f|^2\rangle,$$

and

$$L^{-1} = \|\nabla f\|_{L^\infty} F^{-1}.$$

Note that $L^{-1} \le 2\pi k_0$. The energy dissipation of the Leray regularized solution $\nu\langle|\nabla u|^2\rangle$ is bounded uniformly, independently of the regularization. The upper bound is:

$$\epsilon \le \frac{U^3}{L} + \sqrt{\epsilon}\sqrt{\nu}\frac{U}{L}$$

This implies, of course,

$$\epsilon \le \frac{U^3}{L} + \frac{\nu U^2}{4L^2} + \frac{\sqrt{\nu}U}{2L}\sqrt{\frac{\nu U^2}{L^2} + \frac{4U^3}{L}}$$

and consequently

$$\lim_{Re\to\infty} \sup \epsilon L U^{-3} \le 1.$$

For the proof one applies first M_T and obtains

$$f_i = \partial_j M_T([u]_j u_i) + \partial_i M_T p - \nu\Delta M_T u + M_T u_t$$

because $M_T f = f$. One then takes the scalar product with f.

$$\int_{\mathbf{R}^3} |f|^2 dx = -\int_{\mathbf{R}^3} (\partial_j f_i) M_T([u]_j u_i)dx$$

$$+\nu\int_{\mathbf{R}^3} \nabla f \cdot \nabla M_T(u)dx + \frac{1}{T}(\int_{\mathbf{R}^3} f\cdot(u(\cdot,T) - u(\cdot,0))dx.$$

Then we deduce

$$\|f\|_{L^2}^2 \le \|\nabla f\|_{L^\infty} M_T(\|u\|_{L^2}^2) + \nu\int_{\mathbf{R}^3} |\nabla f||\nabla M_T u|\, dx + O(\frac{1}{T}).$$

We did use the fact that the mollifier is normalized, so $\|[u]\|_{L^2}^2 \le \|u\|_{L^2}^2$. But

$$|\nabla M_T u|^2 \leq M_T |\nabla u|^2$$

Multiplying by k_0^3, dividing by F it follows that:

$$F \leq \frac{1}{L} k_0^3 M_T(|u|^2) + \frac{\nu}{L} \sqrt{k_0^3 M_T \|\nabla u\|_{L^2}^2} + O(\frac{1}{T})$$

and letting $T \to \infty$

$$F \leq \frac{U^2}{L} + \sqrt{\nu}\sqrt{\epsilon}/L + O(\frac{1}{T}).$$

But

$$\epsilon \leq FU$$

follows immediatly from the energy balance, and that concludes the proof of the upper bound. We have therefore

Theorem 7.1. *Consider solutions of Leray's approximation*

$$D_\nu([u], \nabla)u + \nabla p = f, \ \nabla \cdot u = 0$$

in $\mathbf{R}^3$ with divergence-free, time independent body forces with Fourier transform supported in $|\xi| \leq k_0$. Let

$$\epsilon = \lim_{T \to \infty} \sup \frac{1}{T} \int_0^T \nu \|\nabla u(\cdot, t)\|^2 dt,$$

$$U^2 = \lim_{T \to \infty} \sup \frac{1}{T} \int_0^T \|u(\cdot, t)\|^2 dt,$$

$$F^2 = \|f(\cdot)\|^2,$$

where $\|h\|^2 = k_0^3 \int |h|^2 dx$ is normalized L^2 norm. Let also

$$L^{-1} = \frac{\|\nabla f\|_{L^\infty}}{F}$$

define a length scale associated to the forcing and

$$Re = \frac{UL}{\nu}$$

be the Reynolds number. Then

$$\epsilon \leq \frac{U^3}{L} \left(1 + Re^{-\frac{1}{2}} + \frac{3}{4} Re^{-1}\right)$$

holds, uniformly for all mollifiers, $L^2(\mathbf{R}^3)$ initial data and f with the required properties.

The normalization k_0^3 for volume is arbitrary: any other normalization works and does not change the inequality. The uniformity with respect to mollifiers implies that the result holds for suitable weak solutions of the Navier-Stokes equations. The problem of a lower bound is outstanding, and open. When one replaces $[u]$ by the Bessel potential $(I - \alpha^2 \Delta)^{-1}u$ one obtains the Leray alpha model ([10]).

In order to describe in more detail the issues concerning the spectrum it is convenient to phrase them using a Littlewood-Paley decomposition of functions in $\mathbf{R}^d$. This employs a nonnegative, nonincreasing, radially symmetric function

$$\phi_{(0)}(k) = \phi_{(0)}(|k|)$$

with properties $\phi_{(0)}(k) = 1$, $k \leq \frac{5}{8}k_0$, $\phi_{(0)}(k) = 0$, $k \geq \frac{3}{4}k_0$. The positive number k_0 is a wavenumber unit; it allows to make dimensionally correct statements. One sets

$$\phi_{(n)}(k) = \phi(2^{-n}k), \quad \psi_{(0)}(k) = \phi_{(1)}(k) - \phi_{(0)}(k)$$
$$\psi_{(n)}(k) = \psi_{(0)}(2^{-n}k), \quad n \in \mathbf{Z}.$$

The properties

$$\psi_{(n)}(k) = 1, \text{ for } k \in 2^n k_0 [\tfrac{3}{4}, \tfrac{5}{4}],$$
$$\psi_{(n)}(k) = 0, \text{ for } k \notin 2^n k_0 [\tfrac{5}{8}, \tfrac{3}{2}]$$

follow from construction. The Littlewood-Paley operators $S^{(m)}$ and Δ_n are multiplication in Fourier representation by $\phi_{(m)}(k)$ and, respectively by $\psi_{(n)}(k)$. For any $m \in \mathbf{Z}$, the Littlewood Paley decomposition of h is

$$h = S^{(m)}h + \sum_{n \geq m} h_{(n)}$$

and for mean zero function h that decay at infinity, $S^{(m)}h \to 0$ as $m \to -\infty$ and the Littlewood-Paley decomposition is:

$$h = \sum_{n=-\infty}^{\infty} h_{(n)}$$

with

$$h_{(n)} = \Delta_n h = \int_{\mathbf{R}^d} \Psi_{(n)}(y)(\delta_y h)dy.$$
$$\Psi_{(n)}(y) = \int e^{i2\pi y \cdot \xi} \psi_{(n)}(\xi)d\xi$$
$$\widehat{\Psi}_{(n)} = \psi_{(n)}, \quad (\delta_y h)(x) = h(x-y) - h(x).$$

Δ_n is a weighted sum of finite difference operators at scale $2^{-n}k_0^{-1}$ in physical space. For each fixed $k > 0$ at most three Δ_n do not vanish in their Fourier representation at ξ with $k = |\xi|$:

$$h_{(h)}(\xi) \neq 0 \quad n \in I_k = \left\{ [-1,1] + \log_2\left(\frac{k}{k_0}\right) \right\} \cap \mathbf{Z}.$$

Let us consider a slightly large set of indices

$$J_k = \left\{ [-2,2] + \log_2\left(\frac{k}{k_0}\right) \right\} \cap \mathbf{Z}.$$

If ξ is a wave number whose magnitude $|\xi|$ is comparable to k, $\frac{k}{2} \le |\xi| \le 2k$, then, if $u(x,t)$ is an L^2 valued function of t, one has for almost every ξ

$$\widehat{u}(\xi,t) = \sum_{n \in J_k} \widehat{u}_{(n)}(\xi,t)$$

because $I_{|\xi|} \subset J_k$. Consequently, because J_k has at most five elements,

$$\frac{2}{3k} \int_{\frac{k}{2}}^{2k} d\lambda \int_{|\xi|=\lambda} |\widehat{u}(\xi,t)|^2 \, dS(\xi) \le \frac{10}{3k} \sum_{n \in J_k} \|u_{(n)}(\cdot,t)\|_{L^2(\mathbf{R}^d)}^2.$$

Viceversa, because the functions $\psi_{(n)}$, $n \in J_k$ are non-negative, bounded by 1, and supported in $[\frac{5}{32}k, 6k]$ one has also

$$\frac{1}{k} \sum_{n \in J_k} \|u_{(n)}(\cdot,t)\|_{L^2(\mathbf{R}^d)}^2 \le \frac{5}{k} \int_{\frac{5k}{32}}^{6k} \left(\int_{|\xi|=\lambda} |\widehat{u}(\xi,t)|^2 dS(\xi) \right) d\lambda$$

Most of the experimental evidence on $E(k)$ is plotted on a log-log scale. So, for the purpose of estimating exponents in a power law, and in view of the above inequalities, we found it reasonable to consider ([22], [23], [38]) an average of the spectrum, defined as follows. One defines the Littlewood-Paley spectrum of the function $u(x,t)$ to be

$$E_{LP}(k) = \frac{1}{k} \sum_{n \in J_k} \lim_{T \to \infty} \sup \frac{1}{T} \int_0^T \|u_{(n)}(\cdot,t)\|_{L^2(\mathbf{R}^d)}^2 dt.$$

From the definition and the considerations above it follows that

$$c E_{LP}(k) \le \frac{3}{2k} \int_{\frac{k}{2}}^{2k} E(\lambda) d\lambda \le C E_{LP}(k)$$

holds for all $k > 0$, with c, C positive constants depending only on the choice of Littlewood-Paley template function $\phi_{(0)}$. We will consider the Leray approximation again in $\mathbf{R}^3$. Then the components obey

$$D_\nu([u], \nabla)u_{(n)} + \nabla p_{(n)} = W_n + f_{(n)} \tag{7.1}$$

where $p_{(n)} = \Delta_n p$, are the Littlewood-Paley components of the pressure, $f_{(n)} = \Delta_n f$, are the components of the force and

$$W_n(x,t) =$$

$$\int_{\mathbf{R}^3} \Psi_{(n)}(y)\, \partial_{y_j}\, (\delta_y([u]_j)(x,t)\delta_y(u)(x,t))\, dy.$$

The 4/5 law of isotropic homogeneous turbulence

$$S_3(r) \sim \epsilon r$$

where $S_3 = \langle (\delta_r u)^3 \rangle$ is the third order longitudinal structure function. The 4/5 law would follow from the Navier-Stokes equations if assumptions of isotropic, homogeneous, stationary statistics could be applied. The natural mathematical assumption associated to the law is that

$$\widehat{\epsilon} = \lim_{T \to \infty} \sup \frac{1}{T} \int_0^T \|u(\cdot,t)\|^3_{B_3^{\frac{1}{3}},\infty}\, dt < \infty.$$

The Besov space $L^3_{loc,unif}(dt)(B_3^{\frac{1}{3},\infty})$ is thus a natural space for a Kolmogorov theory. (It is also the natural space for the Onsager conjecture, ([28]). The uniform bound assumed above is used for the inequality

$$\lim_{T \to \infty} \sup \frac{k_0^3}{T} \int_0^T \int_{\mathbf{R}^3} |\delta_y u(x,t)|^3 dx dt \le \widehat{\epsilon}|y|$$

Armed with this inequality, one obtains a bound on the energy production in the Littlewood-Paley spectrum. Indeed,

$$W_n(x,t)u_{(n)}(x,t) = -\int_{\mathbf{R}^3}\int_{\mathbf{R}^3} \partial_{y_j}\Psi_{(n)}(y)\Psi_{(n)}(z)\, \{\delta_y[u_j]\delta_y u_i \delta_z u_i\}(x,t)dydz$$

and therefore

$$\langle W_n u_{(n)} \rangle \le C_\Psi \widehat{\epsilon}$$

with

$$C_\Psi = \int_{\mathbf{R}^3}\int_{\mathbf{R}^3} |y|^{\frac{2}{3}}|z|^{\frac{1}{3}}\Psi_{(0)}(y)\Psi_{(0)}(z)dydz$$

Considering $\widehat{f}(\xi)$ supported in $|\xi| \le k_0$ then, $f_{(n)} = 0$ for $n \ge 1$ and the balance

$$\epsilon_{(n)} = \langle \nu|u_{(n)}|^2 \rangle = \langle W_n u_{(n)} \rangle$$

follows. This implies

Theorem 7.2. *Consider smooth, time independent divergence free forces with Fourier transform supported in $|\xi| \le k_0$. Consider solutions of the Leray approximation*

$$D_\nu([u], \nabla)u + \nabla p = f, \quad \nabla \cdot u = 0,$$

with square integrable initial data. Then

$$\epsilon_{(n)} \leq C_\Psi \widehat{\epsilon}$$

holds for $n \geq 1$. Also

$$E_{LP}(k) \leq \beta_a \epsilon^{\frac{2}{3}} k^{-\frac{5}{3}}$$

holds for $k \in [ak_d, k_d]$, with $\frac{k_0}{k_d} \leq a \leq 1$ and with

$$\beta_a = C_\Psi a^{-\frac{4}{3}} \widehat{\epsilon}(\epsilon)^{-1}$$

Consequently

$$\frac{1}{k} \int_{\frac{k}{2}}^{2k} E(\lambda)d\lambda \leq \gamma_a \epsilon^{\frac{2}{3}} k^{-\frac{5}{3}}$$

holds with $\gamma_a = C\beta_a$.

This result contains an unconditional statement. For each fixed mollifier J_δ, the Leray system has global smooth solutions and $\widehat{\epsilon}$ is finite. The bound on the spectrum depends on the mollifier through $\widehat{\epsilon}$. But even if we make the assumption that $\widehat{\epsilon}$ is bounded independently of δ, we still have a result only in a limited range of physical scales.

Similar results can be obtained for two dimensional turbulence. In two dimensions there is no need to consider approximations, because solutions are well behaved. But the range limitations are still there. The spectrum suggested for the direct cascade is the 2D Kraichnan spectrum:

$$E(k) = C\eta^{\frac{2}{3}} k^{-3}.$$

with $\eta = \langle \nu | \nabla \omega |^2 \rangle$, the rate of dissipation of enstrophy. The dissipative cutoff scale is the wave number k_η formed with ν and η:

$$k_\eta = \nu^{-\frac{1}{2}} \eta^{\frac{1}{6}}$$

Bounds on the spectrum for two dimensional turbulence have been addressed by Foias and collaborators ([33], [53], [54]). The result of ([22]) is

Theorem 7.3. *Consider two dimensional incompressible Navier-Stokes equations*

$$D_\nu u + \nabla p = f$$

with time independent, divergence-free forces whose Fourier transform is supported in $|\xi| \leq k_0$. For any $\frac{k_0}{k_\eta} \leq a \leq 1$ there exists a constant C_a such that the Litlewood-Paley energy spectrum of solutions of two dimensional forced Navier-Stokes equations obeys the bound

$$E_{LP}(k) \leq C_a k^{-3}$$

for $k \in [ak_\eta, k_\eta]$. Consequently

$$\frac{1}{k} \int_{\frac{k}{2}}^{2k} E(\lambda) d\lambda \leq \tilde{C}_a k^{-3}$$

holds with $\tilde{C}_a = CC_a$.

The range of physical space scales is again bounded. The mathematical reason for this limitation is the fact that one uses $-\nu\Delta$ and one has to let $\nu \to 0$. The technical tools we have at this moment do not allow us to rigorously obtain an effective nonzero eddy diffusivity to replace kinematic viscosity. When the model allows a non-vanishing positive linear operator then the scale limitation is no longer present. This is the case if one uses the forced surface quasigeostrophic model for the inverse cascade, appropriate for boundary forced rotating geophysical systems. One obtains ([23]) a spectrum consistent with the spectrum

$$E(k) \sim k^{-2}, \ k \leq k_f.$$

obtained in Swinney's lab ([4]):

Theorem 7.4. *Consider the L G equation*

$$\partial_t \theta + u \cdot \nabla \theta + c\Lambda\theta = f$$

in $\mathbf{R}^2$, with $u = c_1 R^\perp \theta$, with deterministic forcing f with Fourier transform supported in $|\xi| \leq k_f$, and with L^2 initial data. Then

$$E_{LP}(k) \leq Ck^{-2}$$

holds for weak solutions solutions of L G, for all $k < k_f$.

References

1. Arnold, V. I., Khesin, B. A., *Topological methods in Hydrodynamics*, Applied Mathematical Sciences, **125**, Springer-Verlag, New York.
2. A.V. Babin, M.I. Visik, Attractors of partial differential equations and estimates of their dimension, Uspekhi Mat. Nauk **38** (1982), 133-187.
3. A. Babin, A. Mahalov, B. Nicolaenko, Long time averaged Euler and Navier-Stokes equations for rotating fluids, in "Structure and nonlinear waves in fluids" (K. Kirchgassner and A. Mielke eds.,) World Scientific 1995, 145-157.
4. C. Baroud, B. Plapp, Z-S. She, H. Swinney, Anomalous self-similarity in rotating turbulence, Phys. Rev. Lett. **88** (2002), 114501.
5. G. I. Barenblatt, A. J. Chorin, V. M. Prostokishin, Self-similar intermediate structures in turbulent boundary layers at large Reynolds numbers, J. Fluid. Mech. **410** (2000), 263-283.

6. Y. Brenier, Topics on hydrodynamics and area preserving maps,in Handbook of Mathematical fluid dynamics (S. Friedlander, D. Serre eds.,) Elsevier (2003), 55-86.
7. T. Buttke, Lagrangian numerical methods which preserve the Hamiltonian structure of incompressible fluid flow, in "Vortex flows and related numerical methods", J.T. Beale, G. H. Cottet and S. Huberson (Eds), NATO ASI Series, Vol. 395, Kluwer, Norwell, 1993.
8. T. Buttke and A. Chorin, Turbulence calculations in magnetization variables, Appl. Num. Math. **12** (1993), 47-54.
9. D. Chae, J. Lee, Global well-posedness in the supercritical dissipative quasi-geostrophic equation, CMP **233** (2003), 297-311.
10. A. Cheskidov, D. Holm, E, Olson, E. Titi, On a Leray alpha model of turbulence, Proc. Roy. Soc. London, to appear (2004).
11. Sir S. W. Churchill, *"I cannot forecast to you the action of Russia. It is a riddle wrapped in a mystery inside an enigma"*, Radio Broadcast, Oct. 1939.
12. S. Childress, G. R. Ierley, E. A. Spiegel, W. R. Young, Blow -up of unsteady two-dimensional Euler and Navier-Stokes solutions having stagnation-point form, J. Fluid. Mech. **203** (1989), 1-22.
13. A. Chorin, Numerical study of slightly viscous flow, J. Fluid. Mech **57** (1973), 785-796.
14. A. Chorin, Vorticity and Turbulence, Apllied Mathematical Sciences **103**, Springer-Verlag.
15. P. Constantin, Note on loss of regularity for solutions of the 3D incompressible and related equations, Commun. Math. Phys. **106** (1986), 311-325.
16. P. Constantin, Navier-Stokes equations and area of interfaces, Commun. Math. Phys. **129** (1990), 241 - 266.
17. P. Constantin, Geometric and analytic studies in turbulence, in Trends and Perspectives in Appl. Math., L Sirovich ed., Appl. Math. Sciences **100**, Springer-Verlag (1994).
18. P. Constantin, The Euler Equations and Nonlocal Conservative Riccati Equations, Intern. Math. Res. Notes, **9** (2000), 455-465.
19. P. Constantin, An Eulerian-Lagrangian approach for incompressible fluids: local theory, Journal of the AMS, **14** (2001), 263-278.
20. P. Constantin, An Eulerian-Lagrangian approach to the Navier-Stokes equations, Commun. Math. Phys. **216** (2001), 663-686.
21. P. Constantin, Near identity transformations for the Navier-Stokes equations, in *Handbook of Mathematical Fluid Dynamics*, Volume 2, S. Friedlander and D. Serre Eds, Elsevier (2003), 117–141.
22. P. Constantin, The Littlewood-Paley spectrum in 2D turbulence, Theor. Comp. Fluid Dyn.**9** (1997), 183-189.
23. P. Constantin, Energy spectrum of quasi-geostrophic turbulence, Phys. Rev. Lett. **89** (October 2002) 18, 184501.
24. P. Constantin, Transport in rotating fluids. Partial differential equations and applications. Discrete Contin. Dyn. Syşt. **10** (2004), no. 1-2, 165–176.
25. P. Constantin, Ch. Doering, Variational bounds on energy dissipation in incompressible flows: II. Channel flow, Phys. Rev E **51** (1995), 3192-3198.
26. P. Constantin and Ch. Doering, Heat transfer in convective turbulence, Nonlinearity **9** (1996) 1049-1060.

27. P. Constantin, D. Cordoba and J. Wu, On the critical dissipative quasi-geostrophic equation, Indiana University Mathematics Journal, **50** (2001), 97-107.

28. P. Constantin,W. E, E. S. Titi, Onsager's conjecture on the energy conservation for solutions of Euler's equations, Commun. Math. Phys., **165**(1994), 207-209.

29. P. Constantin and C. Fefferman, Direction of vorticity and the problem of global regularity for the Navier-Stokes equations, Indiana Univ. Math. Journal, **42** (1993), 775.

30. P. Constantin, C. Fefferman, Scaling exponents in fluid turbulence: some analytic results. Nonlinearity **7** (1994), 41-57.

31. P. Constantin and C. Foias, Global Lyapunov exponents, Kaplan-Yorke formulas and the dimension of the attractors for 2D Navier-Stokes equations, Commun. Pure apll. Math. **38** (1985), 1-27.

32. P. Constantin, C. Foias, *Navier-Stokes Equations*, The University of Chicago Press, Chicago, 1988.

33. P. Constantin, C. Foias and O. Manley, Effects of the forcing function spectrum on the energy spectrum in 2-D turbulence, Phys. Fluids, **6**, (1994), 427.

34. P. Constantin, C. Foias, R. Temam, On the dimension of the attractors in two dimensional turbulence, Physica **D 30**, (1988), 284-296.

35. P. Constantin, C. Hallstrom and V. Putkaradze, Heat transport in rotating convection, Physica **D 125**, (1999), 275-284.

36. P. Constantin, A. Majda and E. Tabak, Formation of strong fronts in the 2D quasi-geostrophic thermal active scalar, Nonlinearity, **7** (1994), 1495-1533.

37. P. Constantin, Q. Nie, N. Schorghoffer, Nonsingular surface quasi-geostrophic flow, Phys. Letters A **241** (27 April 1998), 168-172.

38. P. Constantin, Q. Nie and S. Tanveer, Bounds for second order structure functions and energy spectrum in turbulence Phys. Fluids **11** (1999), 2251-2256.

39. P. Constantin and J. Wu, Behavior of solutions to 2D quasi-geostrophic equations, SIAM J. Math. Anal., **30** (1999) 937-948.

40. D. Cordoba, Nonexistence of simple hyperbolic blow up for the quasigeostrophic equation, Ann. of Math. **148** (1998), 1135-1152.

41. A. Cordoba and D. Cordoba, A maximum principle applied to quasi-geostrophic equations, Comunn. Math. Phys, accepted 2003.

42. B. I . Davydov, Dokl. Akad. Nauk. SSSR **2** (1949), 165.

43. C. Doering and P. Constantin, Energy dissipation in shear driven turbulence, Phys.Rev.Lett. **69** (1992), 1648-1651.

44. C. Doering and C. Foias, Energy dissipation in body-forced turbulence, Journal of Fluid Mechanics **467**, (2002), 289-306.

45. D. Ebin and J. Marsden, Groups of diffeomorphisms and the motion of an incompressible fluid, Ann. of Math. **92** (1970), 102-163.

46. P. Embid and A. Majda, Averaging over fast gravity waves for geophysical flows with arbitrary potential vorticity, Commun. PDE **21** (1996), 619-658.

47. L. C. Evans and R. Gariepy, Measure theory and fine properties of functions, Studies in Advanced Mathematics, CRC Press (1992).

48. G. Eyink, Energy dissipation without viscosity in the ideal hydrodynamics, I. Fourier analysis and local energy transfer, Phys. **D 3-4** (1994), 222-240.

49. M. Feigenbaum, The transition to aperiodic behavior in turbulent systems, Comm. Math. Phys. **77** (1980), 65-86.

50. Foias, private communication (1999).

51. C. Foias, Statistical study of Navier-Stokes equations, I and II, Rend. Sem. Univ. Padova, **48** (1973), 219-349; **49** (1973)

52. C. Foias, D. Holm, E. Titi, The Navier-Stokes alpha model of turbulence, Physica **D 152** (2001), 505-519.

53. C. Foias, M. Jolly, O. Manley, Kraichnan turbulence via finite time averages, preprint 2003.

54. C. Foias, O. Manley, R. Rosa, R. Temam, *Navier-Stokes equations and turbulence*, Cambridge University Press, (2001).

55. C. Foias and G. Prodi, Sur le comportement global des solutions non-stationnaires des équations de Navier-Stokes en dimension 2, Rend. Sem. Mat. Univ. Padova **39** (1967), 1-34.

56. C. Foias, R. Temam, On the Hausdorff dimension of an attractor for the two dimensional Navier-Stokes equations, Phys. Lett. A **93**, (1983), 451-454.

57. C. Foias, R. Temam, Gevrey class regularity for the solutions of the Navier-Stokes equations, J. Funct. Anal. **87** (1989), 359-369.

58. U. Fisch, *Turbulence*, (1995) Cambridge University Press.

59. D. Holm, J. Marsden, T. Ratiu, Euler-Poincare models of ideal fluids with nonlinear dispersion, Phys. Rev. Lett. **349** (1998), 4173-4177.

60. E. Hopf, Statistical Hydrodynamics and functional calculus, J. Rat. Mech. Anal., **1** (1952), 87-123.

61. Ning Ju, Existence and uniqueness of the solution to the dissipative 2D quasi-geostrophic equation in Sobolev space, CMP, to appear.

62. T. Kato, G. Ponce, Commutator estimates and Euler and Navier-Stokes equations, Commun. Pure Appl. Math **41** (1988), 891-907.

63. Lord Kelvin, Maximum and minimum energy in vortex motion, *Mathematical and Physical Papers* IV, (1910), Cambridge University Press.

64. C. Kenig, G. Ponce, L. Vega, Well-posedness of the initial value problem for the Korteweg - de Vries equation, J. Amer. Math. Soc., **4** (1991), 323-347.

65. A.N. Kolmogorov, Local structure of turbulence in an incompressible fluid at very high Reynolds number, Dokl. Akad. Nauk. SSSR **30** (1941), 299-303.

66. G. A.Kuzmin, Ideal incompressible hydrodynamics in terms of momentum density, Phys. Lett. A **96** (1983), 88-90.

67. J. Leray, Essai sur le mouvement d'un liquide visqueux emplissant l'espace, Acta Mathematics **63**(1934), 193-248.

68. E. N. Lorenz, Deterministic nonperiodic flow, J. Atm. Sci. **20** (1963), 130-141.

69. O. A. Ladyzhenskaya, On finite dimensionality of bounded invariant sets for the Navier-Stokes equations and some other dissipative systems, Zap. Nauchn. Sem. LOMI **115** (1982), 137-155.

70. A. Majda and A. Bertozzi, Vorticity and incompressible flow, Cambridge Texts in Applied Mathematics, Cambridge University Press 2002.

71. H. K. Moffatt, The degree of knottedness of tangled vortex lines, J. Fluid Mech., **35** (1969), 117-129.

72. L. Onsager, Statistical Hydrodynamics, Nuovo Cimento **6**(2) (1949), 279-287.

73. J.J. Niemela, L. Skrbek, K.R. Sreenivasan,and R.J. Donnelly, Turbulent convection at very high Rayleigh numbers, Nature, **404** (2000), 837-840.

74. K. Ohkitani, P. Constantin, Numerical study of the Eulerian-Lagrangian formulation of the Navier-Stokes equations, Phys. Fluids **15-10** (2003), 3251-3254.

75. K. Ohkitani, P. Constantin, Numerical study of the Eulerian-Lagrangian analysis of the Navier-Stokes turbulence, work in preparation.

76. K. Ohkitani and J. Gibbon, Numerical study of singularity formation in a class of Euler and Navier-Stokes flows, Phys. Fluids **12** (2000), 3181.
77. K. Ohkitani and M. Yamada, Inviscid and inviscid-limit behavior of a surface quasi-geostrophic flow, Phys. Fluids **9** (1997), 876.
78. V. I. Oseledets, On a new way of writing the Navier-Stokes equation. The Hamiltonian formalism. Commun. Moscow Math. Soc (1988), Russ. Math. Surveys **44** (1989), 210-211.
79. S. Resnick, Dynamical Problems in Nonlinear Advective Partial Differential equations, Ph.D. Thesis, University of Chicago, Chicago (1995).
80. O. Reynolds, On the dynamical theory of incompressible viscous fluids and the determination of the criterion, Phil. Trans. Roy. Soc. London **186** (1884), 123-161.
81. R. Robert, Statistical Hydrodynamics (Onsager revisited), in Handbook of Mathematical Fluid Dynamics, II, (S. Friedlander and D. Serre eds.,) Elsevier (2003), 3-54.
82. P.H. Roberts, A Hamiltonian theory for weakly interacting vortices, Mathematica **19** (1972), 169-179.
83. D. Ruelle, F. Takens, On the nature of turbulence, Comm. Math. Phys. **20** (1971), 167-192.
84. J. Serrin, On the interior regularity of weak solutions of the Navier-Stokes Equations, Arch. Ratl. Mech. Anal **9**, (1962), 187-195.
85. A. Shnirelman, Weak solutions of incompressible Euler equations, in Handbook of Mathematical Fluid Dynamics, (S. Friedlander, D. Serre eds.,) Elsevier (2003), 87–116.
86. E. Stein, *Singular integrals and differentiability properties of functions*, Princeton University Press (1970).
87. J.T. Stuart, Nonlinear Euler partial differential equations: Singularities in their solution, in *Applied Mathematics, Fluid Mechanics, Astrophysics ΚCambridge MA MΝΟP*, World Sci., Singapore 1988, 81-95.
88. Paul Valéry, Oeuvres II, Tel Quel I, Choses Tues, Bibl. de la Pléiade, NRF, Gallimard (1960), 503.
89. M.I Vishik and A.V. Fursikov, *Mathematical problems of statistical hydrodynamics*, (1988) Kluwer, Dordrecht.
90. W. Weber, Uber eine Transformation der hydrodynamischen Gleichungen, J. Reine Angew. Math. **68** (1868), 286-292.
91. J. Wu, the quasigeostophic equation and its two regularizations, Commun. PDE **27**(2002), 1161-1181.
92. J. Wu, Existence of solutions to the dissipative quasigeostrophic equations in Besov space, preprint 2003.
93. V. E. Zakharov, E. A. Kuznetsov, Variational principle and canonical variables in magnetohydrodynamics, Doklady Akademii Nauk SSSR **194** (1970), 1288-1289.

CKN Theory of Singularities of Weak Solutions of the Navier-Stokes Equations

Giovanni Gallavotti

Dipartimento di Fisica e I.N.F.N., Università di Roma "La Sapienza"
P.le Moro 2, 00185 Roma, Italia
giovanni.gallavotti romaMinfn.it

Notations

The lectures are devoted to a complete exposition of the theory of singularities of the Navier Stokes equations solution studied by Leray, in a simple geometrical setting in which the fluid is enclosed in a container Ω with periodic boundary conditions and side size L. The theory is due to the work of Scheffer, Caffarelli, Kohn, Nirenberg and is called here CKN-theory as it is inspired by the work of the last three authors which considerably improved the earlier estimates of Scheffer.

Although the theory of Leray is well known I recall it here getting at the same time a chance at establishing a few notations, [5].

(1) An underlined letter, *e.g.* $\underline{A}, \underline{A}, \ldots$, denotes a 3–dimensional vector (*i.e.* three real or complex numbers) and underlined partial derivative symbol $\underline{\partial}, \underline{\partial}, \ldots$ denotes the gradient operator $(\partial_1, \partial_2, \partial_3)$. A vector field $\underline{u}$ is a function on Ω.

(2) Repeated labels convention is used (labels are letters or other) when not ambiguous: hence $\underline{A} \cdot \underline{B}$ or $\underline{A} \cdot \underline{B}$ means sum over i of $A_i B_i$. Therefore $\underline{\partial} \cdot \underline{u}$, if $\underline{u}$ is a vector field, is the divergence of $\underline{u}$, namely $\sum_i \partial_i u_i$. $\underline{\partial} \cdot \underline{\partial} = \Delta$ is the Laplace operator.

(3) Multiple derivatives are tensors, so that $\underline{\partial}\underline{\partial} f$ is the tensor $\partial_{ij} f$. The $L_2(\Omega)$ is the space of the square integrable functions on Ω: the squared norm of $f \in L_2$ will be $\|f\|_2^2 \overset{def}{=} \int_\Omega |f(\underline{x})|^2 s\underline{x}$.

(4) The Navier-Stokes equation with regularization parameter λ is

$$\underline{\dot{u}} = \nu \Delta \underline{u} - \langle \underline{u} \rangle_\lambda \cdot \underline{\partial}\, \underline{u} - \underline{\partial} p, \qquad \underline{\partial} \cdot \underline{u} = 0, \qquad \int_\Omega \underline{u}\, d\underline{x} = \underline{0} \qquad (0.1)$$

where the unkowns are $\underline{u}(\underline{x}, t), p(\underline{x}, t)$ with zero average and, [5],

(i) $\underline{u}$ is a divergenceless field, p is a scalar field

(ii) $\langle \underline{u} \rangle_\lambda = \int_\Omega \chi_\lambda(\underline{x} - \underline{y}) \underline{u}(\underline{y}) d\underline{y}$ and χ_λ is defined in terms of a $C^\infty(\Omega)$ function $\underline{x} \to \chi(\underline{x}) \geq 0$ not vanishing in a small neighborhood of the origin and with integral $\int \chi(\underline{x}) d\underline{x} \equiv 1$: the function $\chi(\underline{x})$ can be regarded as a periodic function on Ω or as a function on R^3 with value 0 outside Ω, as we shall imagine that Ω is centered at the origin, to fix the ideas. For $\lambda \geq 1$ also the function $\chi_\lambda(\underline{x}) \overset{def}{=} \lambda^3 \chi(\lambda \underline{x})$ can be regarded as a periodic function on Ω or as a function on R^3: it is an "approximate Dirac's δ–function". Usually the NS equation contains a volume force too: here we set it equal to $\underline{0}$.[1]

(iii) The initial datum is a divergenceless velocity field $\underline{u}^0 \in L_2(\Omega)$ with $\underline{0}$ average; no initial datum for p as p is determined from $\underline{u}^0$.

(5) A *weak solution* of the NS equations with initial datum $\underline{u}_0 \in L_2(\Omega)$ is a limit on subsequences of $\lambda \to \infty$ of solutions $\underline{u}^\lambda, p^\lambda$ of (0.1). This means that the Fourier transforms of $\underline{u}^\lambda$, and p^λ exist and have components $\underline{u}_{\underline{k}}^\lambda(t) \equiv \int_\Omega e^{-i\underline{k}\underline{x}} \underline{u}^\lambda(\underline{x}) d\underline{x}$ ($\underline{k} = \frac{2\pi}{L}\underline{n}$, $\underline{n} \in Z^3$), and $p_{\underline{k}}^\lambda(t)$ which have a limit as $\lambda \to \infty$ (on subsequences) for each $\underline{k} \neq \underline{0}$ and the limit of the $\underline{u}_{\underline{k}}(t)$ is absolutely continuous. This is equivalent to the existence of the limits of the L_2 products $(\underline{u}(t), \underline{\varphi})_{L_2}$ and $(\underline{p}(t), \psi)_{L_2}$ for all $t \in (0, \infty]$ and for all test functions $\underline{f}(\underline{x}), \psi(\underline{x})$. There might be several such limits (*i.e.* the limit may depend on the subsequence) and what follows applies to any one among them.

The core of the analysis will deal with the regularized equation and the properties of its solutions, which are easily shown to be C^∞ in $\underline{x} \in \Omega$ and in $t \in (0, \infty)$ if the intial datum is $\underline{u} \in L_2(\Omega)$. The limit $\lambda \to \infty$ will be taken at the end and it is where the theory becomes non conctructive because there is need to consider the limit on subsequences.

Of course the point is to obtain bounds which are uniform in $\lambda \to \infty$ and the limit $\lambda \to \infty$ only intervenes at the end to formulate the results in a nice form.

The theory of Leray is based on the following *a priori* bounds, see section 3.2 in [5], on solutions of (0.1) with initial datum $\underline{u}_0$ with L_2 square norm E_0

$$||\underline{u}^\lambda(t)||_2^2 \leq E_0, \qquad \int_0^t d\tau ||\underline{\partial u}^\lambda(\tau)||_2^2 \leq \frac{1}{2} E_0 \nu^{-1} \qquad (0.2)$$

satisfied by the solution $\underline{u}^\lambda$.

The notes are extracted from reference [5] to which the reader is referred for details on the above results and have been made independent from [5] modulo the above results (in fact, essentially, only modulo the statements on the regularized equation (0.1)).

The proof is conceptually quite simple and is based on a few (clever) *a priori* Sobolev inequalities: the estimates are discussed in Sect. 1-3, which form an introduction.

---

[1] This is a simplicity assumption as the extension of the theory to cases with time independent smooth (*e.g.* C^∞) volume forces would be immediate and just a notational nuisance.

Their application to the analysis of (0.1) is in Sect. 4 where the main theorems are discussed and the CKN main result is reduced via the inequalities to Scheffer's theorem. The method is a kind of multiscale analysis which allows us to obtain regularity *provided* a control quantity, identified here as the "*local Reynolds number*" on various scales, is small enough. Unfortunately it is not (yet) possible to prove that the local Reynolds number is small on all small regions (physically this would mean that in such regions the flow would be laminar, hence smooth on small scale).

However the *a priori* bounds give the information that the local Reynolds number must be small near many points in Ω and, via standard techniques, an estimate of the dimension of the possibly bad points follows. The application to the fractal dimension bound is essentially an "abstract reasoning" consequence of the results of Sect. 4 and is in Sect. 5.

The proofs of the various Sobolev inequalities necessary to obtain the key Scheffer's theorem and of the new ones studied by CKN is decribed concisely but in full detail in the series of problems at the end of the text: the hints describe quickly the various steps of the proofs (however without skipping any detail, to my knowledge).

1 Leray's Solutions and Energy

The theory of space–time singularities will be partly based upon simple general *kinematic inequalities*, which therefore have little to do with the Navier–Stokes equation, and partly they will be based on the local energy conservation which follows as a consequence of the Navier–Stokes equations *but it is not equivalent to them.*

Energy conservation for the regularized equations (0.1) says that the kinetic energy variation in a given volume element Δ of the fluid, in a time interval $[t_0, t_1]$, plus the energy dissipated therein by friction, equals the sum of the kinetic energy that in the time interval $t \in [t_0, t_1]$ enters in the volume element plus the work performed by the pressure forces (on the boundary element) plus the work of the volume forces (none in our case). The analytic form of this relation is simply obtained by multiplying both sides of the first of the (0.1) by $\underline{u}$ and integrating on the volume element Δ and over the time interval $[t_0, t_1]$.

The relation that one gets can be generalized to the case in which the volume element has a time dependent shape. And an even more general relation can be obtained by multiplying both sides of (0.1) by $\varphi(\underline{x}, t)\underline{u}(\underline{x}, t)$ where φ is a $C^\infty(\Omega \times (0, s])$ function with $\varphi(\underline{x}, t)$ zero for t near 0 (here s is a positive parameter).

Energy conservation in a sharply defined volume Δ and time interval $t \in [t_0, t_1]$ can be obtained as limiting case of choices of φ in the limit in which it becomes the characteristic function of the space–time volume element $\Delta \times [t_0, t_1]$.

Making use of a regular function $\varphi(\underline{x}, t)$ is useful, particularly in the rather "desperate" situation in which we are when using the theory of Leray. The "solutions" $\underline{u}$ (obtained by removing, in (0.1), the regularization, *i.e.* letting $\lambda \to \infty$) are only "weak solutions". Therefore, the relations that are obtained in the limit $\lambda\omega\infty$ can be interpreted as valid only after suitable integrations by parts that allow us to avoid introducing derivatives of $\underline{u}$ (whose existence is not guaranteed by the theory) at the "expense" of differentiating the "test function" φ.

Performing analytically the computation of the energy balance, described above in words, in the case of the regularized equation (0.1) and via a few integrations by parts[2] we get the following relation

$$\tfrac{1}{2}\int_\Omega d\underline{\xi}\,|\underline{u}(\underline{\xi},s)|^2\varphi(\underline{\xi},s) + \nu\int_0^s dt\int_\Omega \varphi(\underline{\xi},t)|\underline{\partial}\underline{u}(\underline{\xi},t)|^2 d\underline{x} =$$
$$= \int_0^s\int_\Omega \left[\tfrac{1}{2}(\varphi_t + \nu\,\Delta\varphi)|\underline{u}|^2 + \tfrac{1}{2}|\underline{u}|^2\langle\underline{u}\rangle_\lambda\cdot\underline{\partial}\varphi + p\,\underline{u}\cdot\underline{\partial}\varphi\right] dt\,d\underline{\xi} \tag{1.1}$$

where $\varphi_t \equiv \partial_t\varphi$ and $\underline{u} = \underline{u}^\lambda$ is in fact depending also on the regularization parameter λ; here p is the pressure $p = -\sum_{ij}\Delta^{-1}\partial_i\partial_j(u_i u_j)$.

Suppose that the solution of (0.1) with fixed initial datum $\underline{u}_0$ converges (weakly in L_2), for $\lambda \to \infty$, to a "Leray solution" $\underline{u}$ possibly only over a subsequence $\lambda_n \to \infty$.

The (1.1) implies that any (in case of non uniqueness) Leray solution $\underline{u}$ verifies the *energy inequality:*

$$\tfrac{1}{2}\int_\Omega |\underline{u}(\underline{\xi},s)|^2\varphi(\underline{\xi},s)\,d\underline{\xi} + \nu\int_{t\le s}\int_\Omega \varphi(\underline{\xi},t)|\underline{\partial}\underline{u}(\underline{\xi},t)|^2 d\underline{\xi}dt \le$$
$$\le \int_{t\le s}\int_\Omega \left[\tfrac{1}{2}(\varphi_t + \nu\,\Delta\varphi)|\underline{u}|^2 + \tfrac{1}{2}|\underline{u}|^2\underline{u}\cdot\underline{\partial}\varphi + p\,\underline{u}\cdot\underline{\partial}\varphi\right] d\underline{\xi}\,dt \tag{1.2}$$

where the pressure p is given by $p = -\sum_{ij}\Delta^{-1}\partial_i\partial_j(u_i u_j) \equiv -\Delta^{-1}\underline{\partial}\,\underline{\partial}(\underline{u}\,\underline{u})$
$\equiv -\Delta^{-1}(\underline{\partial}\,\underline{u})^2$.

Remark 1.1.

(1) It is important to remark that in this relation one might expect the *equality sign*: as we shall see the fact that we cannot do better than just obtaining an inequality means that the limit necessary to reach a Leray solution can introduce a "spurious dissipation" that we are simply unable to understand on the basis of what we know (today) about the Leray solutions.

(2) The above "strange" phenomenon reflects our inability to develop a complete theory of the Navier–Stokes equation, but one can conjecture that no other dissipation can take place and that a (yet to come) complete theory of the equations could show this. Hence we should take the inequality sign in (1.2) as one more manifestation of the inadequacy of the Leray's solution.

---

[2] The solutions of (0.1) are $C^\infty(\Omega \times [0,\infty))$ so that there is no need to justify integrating by parts.

The proof of (1.2) and of the other inequalities that we shall quote and use in this section is elementary and based, *c.f.r.* problem {15} below, on a few general "kinematic inequalities" that we now list (all of them will be used in the following but only (S) and (CZ) are needed to check (1.2)).

2 Kinematic Inequalities

A first "kinematic" inequality, *i.e.* the first inequality that we shall need and that holds for any function f, is [3]

(P) *Poincaré inequality:*

$$\int_{B_r} d\underline{x} \, |f - F|^\alpha \le C_\alpha^P \, r^{3-2\alpha} \left(\int_{B_r} d\underline{x} \, |\partial f| \right)^\alpha, \qquad 1 \le \alpha \le \frac{3}{2} \qquad (2.1)$$

where F is the average of f on the ball B_r with radius r and C_α^P is a suitable constant. We shall denote (2.1) by (P).

A second kinematic inequality that we shall use is

(S) *Sobolev inequality:*

$$\int_{B_r} |\underline{u}|^q \, d\underline{x} \le C_q^S \left[\left(\int_{B_r} (\partial \underline{u})^2 \, d\underline{x} \right)^a \cdot \left(\int_{B_r} |\underline{u}|^2 \, d\underline{x} \right)^{q/2-a} + \right.$$
$$\left. + r^{-2a} \left(\int_{B_r} |\underline{u}|^2 \, d\underline{x} \right)^{q/2} \right] \qquad \text{if } 2 \le q \le 6, \quad a = \tfrac{3}{4}(q-2) \qquad (2.2)$$

where B_r is a ball of radius r and the integrals are performed with respect to $d\underline{x}$. The C_q^S is a suitable constant; the second term of the right hand side can be omitted if $\underline{u}$ has zero average over B_r. We shall denote (2.2) by (S), [9].

A third necessary kinematic inequality will be

(CZ) *Calderon–Zygmund inequality:*

$$\int_\Omega |\sum_{i,j} (\Delta^{-1}\partial_i\partial_j)(u_i u_j)|^q d\underline{\xi} \le C_q^L \int_\Omega |\underline{u}|^{2q} d\underline{\xi}, \quad 1 < q < \infty \qquad (2.3)$$

which we shall denote (CZ): here Ω is the torus of side L and C_q^L is a suitable constant, [10].

And finally

(H) *Hölder inequality:*

---

[3] The inequalities should be regarded as inequalities for C^∞ functions; they can be extended to the appropriate Sobolev spaces by continuity.

$$\left| \int f_1 f_2 \cdots f_n \right| \leq \prod_{i=1}^{n} \left(\int |f_i|^{p_i} \right)^{\frac{1}{p_i}}, \qquad \sum_{i=1}^{n} \frac{1}{p_i} = 1 \qquad (2.4)$$

which we shall denote (H): the integrals are performed over an arbitrary domain with respect to an arbitrary measure (of course the same for all integrals).

Remark 2.1. (H) are a trivial extension of the Schwartz-Hölder inequalities; while (S) and (P) (mainly in the cases, important in what follows, $q = 6$ and $\alpha = \frac{3}{2}$) and (CZ) are less elementary and we refer to the literature, [9], [10], [6], for their proofs.

An important consequence of the inequalities is

Proposition 2.1. *Let $\underline{u}$ be a Leray solution verifying (therefore) the a priori bounds in (0.2):* $\int_{\Omega} |\underline{u}(\underline{x},t)|^2 d\underline{x} \leq E_0$ *and* $\int_0^T dt \int_{\Omega} |\underline{\partial u}(\underline{x},t)|^2 d\underline{x} \leq E_0 \nu^{-1}$ *then*

$$\int_0^T dt \int_{\Omega} d\underline{x} \, |\underline{u}|^{10/3} + \int_0^T dt \int_{\Omega} d\underline{x} \, |p|^{5/3} \leq C \nu^{-1} E_0^{5/3} \qquad (2.5)$$

where C can be chosen $C_{\frac{10}{3}}^S (1 + C_{\frac{5}{3}}^L)$.

proof: Apply (S) with $q = \frac{10}{3}$ and $a = 1$:

$$\int_{\Omega} |\underline{u}|^{\frac{10}{3}} d\underline{x} \leq C_{\frac{10}{3}}^S \left(\int_{\Omega} (\underline{\partial u})^2 d\underline{x} \right)^1 \cdot \left(\int_{\Omega} \underline{u}^2 d\underline{x} \right)^{\frac{5}{3}-1} \leq C_{\frac{10}{3}}^S E_0^{\frac{2}{3}} \int_{\Omega} |\underline{\partial u}|^2 \quad (2.6)$$

hence integrating over t between 0 and T using also the second *a priori* estimate, we find

$$\int_0^T dt \int_{\Omega} |\underline{u}|^{\frac{10}{3}} d\underline{x} \leq C_{\frac{10}{3}}^S E_0^{\frac{2}{3}} \int_0^T dt \int_{\Omega} d\underline{x} \, (\underline{\partial u})^2 \leq C_{\frac{10}{3}}^S \nu^{-1} E_0^{1+\frac{2}{3}} \qquad (2.7)$$

while the (CZ) yields: $\int_{\Omega} d\underline{x} \, |p|^{\frac{5}{3}} \leq C_{\frac{5}{3}}^L \int_{\Omega} d\underline{x} \, |\underline{u}|^{\frac{10}{3}}$ which, integrated over t and combined with (2.7), gives the announced result.

3 Pseudo Navier Stokes Velocity – Pressure Pairs. Scaling Operators

As already mentioned the CKN theory will not fully use that $\underline{u}$ verifies the Navier–Stokes equation: in order to better realize this (unpleasant) property it is convenient to define separately the only properties of the Leray solutions that are really needed to develop the theory, *i.e.* to obtain an estimate of the fractal dimension of the space–time singularities set S_0. This leads to the following notion

Definition 3.1. *(pseudo NS velocity field): Let $t \to (\underline{u}(\cdot, t), p(\cdot, t))$ be a function with values in the space of zero average square integrable "velocity" and "pressure" fields on Ω. Suppose that for each $\varphi \in C^\infty(\Omega \times (0, T])$ with $\varphi(\underline{x}, t)$ vanishing for t near zero the following properties hold. For each $T < \infty$ and $s \leq T$:*

(a) $\int_\Omega \underline{u}\, d\underline{x} = \underline{0}, \qquad \underline{\partial} \cdot \underline{u} = \underline{0}, \qquad p = -\sum_{i,j} \partial_i \partial_j \Delta^{-1}(u_i u_j)$

(b) $\int_0^T dt \int_\Omega d\underline{x}\, |\underline{u}|^{10/3} + \int_0^T dt \int_\Omega d\underline{x}\, |p|^{5/3} < \infty$

(c) $\frac{1}{2} \int_\Omega d\underline{x}\, |\underline{u}(\underline{x}, s)|^2 \varphi(\underline{x}, s) + \nu \int_{t \leq s} \int_\Omega \varphi(x, t) |\underline{\partial} \underline{u}|^2 d\underline{x}\, dt \leq$ (3.1)

$$\leq \int_{t \leq s} \int_\Omega \left[\frac{1}{2}(\varphi_t + \nu\, \Delta\varphi)|\underline{u}|^2 + \frac{1}{2}|\underline{u}|^2 \underline{u} \cdot \underline{\partial}\varphi + p\, \underline{u} \cdot \underline{\partial}\varphi \right] d\underline{x}\, dt$$

Then we shall say that the pair $(\underline{u}, p)$ is a pseudo NS velocity and pressure pair. The singularity set in the time interval $[0, T]$ of $(\underline{u}, p)$ will be defined as the set S_0 of the points $(\underline{x}, t) \in \Omega \times [0, T]$ that do not admit a vicinity U where $|\underline{u}|$ is bounded.[4]

The name given to the set S_0 is justified by a general result on the theory of NS equations which shows that is a Leray's solution of the NS equations is essentially bounded in a neighborhood of a space time point then it is C^∞ near such point.

Proposition 3.1. *(velocity is unbounded near singularities): Let $\underline{u}(\underline{x}, t)$ be a Leray's solution of the NS equation in L_2. Given $t_0 > 0$ suppose that $|\underline{u}(\underline{x}, t)| \leq M$, $(\underline{x}, t) \in U_\rho(\underline{x}_0, t_0) \equiv$ sphere of radius ρ $(\rho < t_0)$ around $(\underline{x}_0, t_0)$, for some $M < \infty$: then $\underline{u} \in C^\infty(U_{\rho/2}(\underline{x}_0, t_0))$.*

Remark 3.1.
(1) This means that the only way a singularity can manifest itself, in a Leray solution of the NS equations, is through a divergence of the velocity field itself. For instance it is impossible to have a singular derivative having the velocity itself unbounded. Hence, if $d \geq 3$ velocity discontinuities are impossible (and even less so shock waves), for instance. Naturally if $\underline{u}(\underline{x}, t)$ is modified on a set of points $(\underline{x}, t)$ with zero measure it remains a weak solution (because the Fourier transform, in terms of which the notion of weak solution is defined, does not change), hence the condition $|\underline{u}(\underline{x}, t)| \leq M$ for each $(\underline{x}, t) \in U_\lambda(\underline{x}_0, t_0)$ can be replaced by the condition: *for almost all $(\underline{x}, t) \in U_\lambda(\underline{x}_0, t_0)$.*
(2) The above result is not strong enough to overcome the difficulties of a local theory of regularity of the Leray weak solutions. Therefore one looks for other results of the same type and it would be desirable to have results concerning

---

Here we mean bounded outside a set of zero measure in U or, as one says, *essentially bounded* because it is clear that, being $\underline{u}, p$ in $L_2(\Omega)$, they are defined up to a set of zero measure and it would not make sense to ask that they are bounded everywhere without specifying which realization of the functions we take.

regularity implied by *a priori* informations on the vorticity. We have already seen that bounded total vorticity implies regularity: however it is very difficult to go really beyond; hence it is interesting to note that also other properties of the vorticity may imply regularity. A striking result in this direction, although insufficient for concluding regularity (if true at all) of Leray weak solutions, is in [CF93].

(3) For a proof of the above (Serrin's) theorem see [5], proposition IV in section 3.3.

The remaining part of this section will concern the general properties of the pseudo NS pairs and their regularity at a given point $(\underline{x}, t)$: it will not have more to do with the velocity and pressure fields that solve the Navier–Stokes equations. It is indeed easy to convince oneself that the (3.1), in spite of the arbitrariness of φ, *are not equivalent*, not even formally, to the Navier–Stokes equations, and they pose far less severe on $\underline{u}, p$ restrictions. We should not be surprised, therefore, if it turned out possible to exhibit pseudo NS pairs that really have singularities on "large sets" of space–time. In a way it is already surprising that the pseudo NS fields verify the regularity properties discussed below.

The analysis of the latter properties (of pseudo NS fields) is based on the mutual relations between certain quantities that we shall call "*dimensionless operators*" relative to the space–time point $(\underline{x}_0, t_0)$

Definition 3.2. *(dimensionless "operators" for NS) Let $(\underline{x}_0, t_0) \in \Omega \times (0, \infty)$ and consider the sets* [5]

$$
\begin{aligned}
\Delta_r(t_0) &= \quad \{t \,|\, |t - t_0| < r^2 \nu^{-1}\} \\
B_r(\underline{x}_0) &= \quad \{\underline{\xi} \,|\, |\underline{\xi} - \underline{x}_0| < r\} \equiv B_r \\
Q_r(\underline{x}_0, t_0) &= \{(\underline{\xi}, \vartheta) \,|\, |\underline{\xi} - \underline{x}_0| < r, \, |\vartheta - t_0| < r^2 \nu^{-1}\} \\
Q_r(\underline{x}_0, t_0) &= \Delta_r(t_0) \times B_r(\underline{x}_0) \equiv Q_r
\end{aligned}
\tag{3.2}
$$

define:

(i) *"dimensionless kinetic energy operator" on scale r:*

$$
A(r) = \frac{1}{\nu^2 r} \sup_{|t - t_0| \leq \nu^{-1} r^2} \int_{B_r} |\underline{u}(\underline{\xi}, t)|^2 \, d\underline{\xi}
\tag{3.3}
$$

and we say that the dimension of A is 1 : this refers to the factor r^{-1} that is used to make the integral dimensionless.

(ii) *"local Reynolds number" averaged on scale r:*

$$
\delta(r) = \frac{1}{\nu r} \int_{Q_r} d\vartheta d\underline{\xi} \, |\partial \underline{u}|^2
\tag{3.4}
$$

---

[5] If $r \geq L/2$ this is interpreted as $B_r \equiv \Omega$. If $r^2 \nu^{-1} > t_0$ then $\Delta_r(t_0)$ is interpreted as $0 < t < t_0 + r^2 \nu^{-1}$

and we say that the dimension of δ is 1 : this refers in general to the power $-\alpha$ *to which r has to be raised so that an expression becomes dimensionless: in this case* $\alpha = 1$.

(iii) "dimensionless energy flux" on scale r:

$$G(r) = \frac{1}{\nu^2 r^2} \int_{Q_r} d\vartheta d\underline{\xi} \, |\underline{u}|^3 \tag{3.5}$$

The dimension of G is 2.

(iv) "dimensionless pressure power" forces on scale r:

$$J(r) = \frac{1}{\nu^2 r^2} \int_{Q_r} d\underline{\xi} d\vartheta \, |\underline{u}| \, |p| \tag{3.6}$$

The dimension of J is 2.

(v) "dimensionless non locality" on scale r:

$$K(r) = \frac{r^{-13/4}}{\nu^{3/2}} \int_{\Delta_r} d\vartheta \Big(\int_{B_r} |p| \, d\underline{\xi} \Big)^{5/4} \tag{3.7}$$

The dimension of K is 13/4.

(vi) "dimensionless intensity" on scale r:

$$S(r) = \nu^{-7/3} r^{-5/3} \int_{Q_r} (|\underline{u}|^{10/3} + |p|^{5/3}) d\vartheta d\underline{\xi} \tag{3.8}$$

where the pressure is always defined by the expression

$$p = - \sum_{i,j=1}^{3} \partial_i \partial_j \Delta^{-1} (u_i u_j).$$

The dimension of S is 5/3.

Remark 3.2.

(1) The $A(r), \ldots$ *are not* operators in the common sense of functional analysis. Their name is due to their analogy with the quantities that appear in problems that are studied with the methods of the "renormalization group" (which, also, are not operators in the common sense of the words). Perhaps a more appropriate name could be "dimensionless observables": but we shall call them operators to stress the analogy of what follows with the methods of the renormalization group.

(2) The $A(r), G(r), J(r), K(r), S(r)$ are in fact estimates of the quantities that their name evokes. We omit the qualifier "estimate" when referring to them for brevity.

(3) The interest of (i)–(iv) becomes manifest if we note that the energy inequality (3.1) can be expressed in terms of such quantities if φ is suitably chosen. Indeed let

$$\varphi \doteq \chi(\underline{x}, t) \, \frac{\exp-\left(\frac{(\underline{x}-\underline{x}_0)^2}{4(\nu(t_0-t)+2r^2)}\right)}{(4\pi\nu(t-t_0)+8\pi r^2)^{3/2}} \tag{3.9}$$

where $\chi(\underline{x}, t)$ is C^∞ and has value 1 if $(\underline{x}, t) \in Q_{r/2}$ and 0 if $(\underline{x}, t) \notin Q_r$. Then there exists a constant $C > 0$ such that

$$\begin{array}{llll} |\varphi| < \frac{C}{r^3}, & |\partial\varphi| < \frac{C}{r^4}, & |\partial_t\varphi + \nu\Delta\varphi| < \frac{C}{\nu^{-1}r^5}, & \text{everywhere} \\ |\varphi| > \frac{1}{Cr^3}, & & \text{if} \quad (\underline{x}, t) \in Q_{r/2} \end{array} \tag{3.10}$$

Hence (3.1) implies

$$\frac{\nu^2}{Cr^2}\left(A(\tfrac{r}{2}) + \delta(\tfrac{r}{2})\right) \le C\left(\frac{1}{\nu^{-1}r^5}\int_{Q_r}|\underline{u}|^2 + \frac{1}{r^4}\int_{Q_r}|\underline{u}|^3 + \frac{1}{r^4}\int_{Q_r}|\underline{u}||p|\right) \tag{3.11}$$

and, since $\int_{Q_r}|\underline{u}|^2 \le C\left(\int_{Q_r}|\underline{u}|^3\right)^{2/3}(\nu^{-1}r^5)^{1/3}$ with a suitable C, it follows that for some $\tilde{C}$

$$A(\tfrac{r}{2}) + \delta(\tfrac{r}{2}) \le \tilde{C}\left(G(r)^{2/3} + G(r) + J(r)\right) \tag{3.12}$$

(4) Note that the operator $\delta(r)$ is an average of the "local Reynolds' number" $r\int_{\Delta_r}|\partial\underline{u}|^2d\underline{\xi}$.

(5) The operator (v) appears if one tries to bound $J(\tfrac{r}{2})$ in terms of $A(r)+\delta(r)$: such an estimate is indeed possible and it will lead to the *local Scheffer theorem* discussed in the next section.

4 The Theorems of Scheffer and of Caffarelli–Kohn–Nirenberg

We can state the strongest results known (in general and to date) about the regularity of the weak solutions of Navier Stokes equations (which however hold also for the pseudo Navier Stokes velocity–pressure pairs).

Theorem 4.1 *(upper bound on the dimension of the sporadic set of singular times for NS, ("Scheffer's theorem")): There are two constants $\varepsilon_s, C > 0$ such that if $G(r) + J(r) + K(r) < \varepsilon_s$ for a certain value of r, then $\underline{u}$ is bounded in $Q_{\frac{r}{2}}(\underline{x}_0, t_0)$:*

$$|\underline{u}(\underline{x}, t)| \le C\frac{\varepsilon_s^{1/3}}{r}, \quad (\underline{x}, t) \in Q_{\frac{r}{2}}(\underline{x}_0, t_0), \quad \text{almost everywhere} \tag{4.1}$$

having set $\nu = 1$.

Remark 4.1.
(1) *c.f.r.* problems {5}–{11} for a guide to the proof.
(2) This theorem can be conveniently combined, for the purpose of checking its

hypotheses, with the inequality: $J(r) + G(r) + K(r) \leq C \left(S(r)^{9/10} + S(r)^{3/4} \right)$, which follows immediately from inequality (H) and from the definitions of the operators, with a suitable C.

(3) In other words *if the operator $S(r)$ is small enough then $(\underline{x}_0, t_0)$ is a regular point.*

(4) This will imply that the fractal dimension of the space–time singularities set is $\leq 5/3$. In fact, see section 5 below, an *a priori* estimate on the global value of an operator with dimension α implies that the Hausdorff' measure of the set of points around which the operator is large does not exceed α; here the operator $S(r)$ has dimension $5/3$ and therefore together with the *a priori* bound (2.5) it yields and estimate $< 5/3$ for the Hausdorff dimension of the singularity set. This also justifies the introduction of the operator $S(r)$.

It is easy, in terms of the just defined operators, to illustrate the strategy of the proof of the following theorem which will immediately imply, via a classical argument reproduced in section 5 below, that the fractal dimension of the space time singularities set S_0 for a pseudo NS field is ≤ 1 and that its 1–measure of Hausdorff $\mu_1(S_0)$ vanishes.

Theorem 4.2 *(sufficient condition for local regularity space-time ("CKN theorem")): There is a constant ε_{ckn} such that if $(\underline{u}, p)$ is a pseudo NS pair of velocity and pressure fields and*

$$\limsup_{r \to 0} \frac{1}{\nu r} \int_{Q_r(\underline{x}_0, t_0)} |\partial \underline{u}(\underline{x}', t')|^2 \, d\underline{x}' dt' \equiv \limsup_{r \to 0} \delta(r) < \varepsilon_{ckn} \qquad (4.2)$$

then $\underline{u}(\underline{x}', t')$, $p(\underline{x}', t')$ are C^∞ in the vicinity of $(\underline{x}_0, t_0)$.[6]

For fixed $(\underline{x}_0, t_0)$, consider the "sequence of length scales": $r_n \equiv L2^n$, with $n = 0, -1, -2, \ldots$. We shall set $\alpha_n \equiv A(r_n)$, $\kappa_n = K_n^{8/5}$, $j_n = J_n$, $g_n = G_n^{2/3}$, $\delta_n = \delta(r_n)$ which is a natural definition as it will shortly appear. And define $\underline{X}_n \equiv (\alpha_n, \kappa_n, j_n, g_n) \in R_+^4$. Then the proof of this theorem is based on a bound that allows us to estimate the size of $\underline{X}_n$, defined as the sum of its components, in terms of the size of $\underline{X}_{n+p}$ provided the Reynolds number δ_{n+p} on scale $n + p$ is $\leq \delta$.

The inequality will have the form (if $p > 0$ and $0 < \delta < 1$)

$$\underline{X}_n \leq \mathcal{B}_p(\underline{X}_{n+p}; \delta) \qquad (4.3)$$

where $\mathcal{B}_p(\cdot; \delta)$ is a map of the whole R_+^4 into itself and the inequality has to be understood "component wise", *i.e.* in the sense that each component of the l.h.s. is bounded by the corresponding component of the r.h.s. We call $|\underline{X}|$ the sum of the components of $\underline{X} \in R_+^4$.

This means that near $(\underline{x}, t)$ the functions $\underline{u}(\underline{x}, t)$, $p(\underline{x}, t)$ coincide with C^∞ functions apart form a set of zero measure (recall that the pseudo NS fields are defined as fields in $L_2(\Omega)$)

The map $\mathcal{B}_p(\cdot\,;\delta)$, which to some readers will appear as strongly related to the *"beta function"* for the "running couplings" of the "renormalization group approaches",[7] will enjoy the following property

Proposition 4.1. *Suppose that p is large enough; given $\rho > 0$ there exists $\delta_p(\rho) > 0$ such that if $\delta < \delta_p(\rho)$ then the iterates of the map $\mathcal{B}_p(\cdot\,;\delta)$ contract any given ball in R^4_+, within a finite number of iterations, into the ball of radius ρ: i.e. $|\mathcal{B}_p^k(\underline{X};\delta)| < \rho$ for all large k's.*

Assuming the above proposition the main theorem II follows:

proof: Let $\rho = \varepsilon_s$, c.f.r. theorem I, and let p be so large that the above proposition holds. We set $\varepsilon_{ckn} = \delta_p(\varepsilon_s)$ and it will be, by the assumption (4.2), that $\delta_n < \varepsilon_{ckn}$ for all $n \leq n_0$ for a suitable n_0 (recall that the scale labels n are negative).

Therefore it follows that $|\mathcal{B}_p^k(\underline{X}_{n_0};\varepsilon_{ckn})| < \varepsilon_s$ for some k. Therefore by the theorem I we conclude that $(\underline{x}_0, t_0)$ is a regularity point.

Proposition 5.1 follows immediately from the following general "Sobolev inequalities"

(1) *"Kinematic inequalities":* i.e. inequalities depending only on the fact that $\underline{u}$ is a divergence zero, average zero and is in $L_2(\Omega)$ and $p = -\Delta^{-1}(\partial\,\underline{u})^2$

$$
\begin{aligned}
J_n &\leq C\left(2^{-p/5}A_{n+p}^{1/5}G_n^{1/5}K_{n+p}^{4/5} + 2^{2p}A_{n+p}^{1/2}\delta_{n+p}\right)\\
K_n &\leq C\left(2^{-p/2}K_{n+p} + 2^{5p/4}A_{n+p}^{5/8}\delta_{n+p}^{5/8}\right)\\
G_n^{2/3} &\leq C\left(2^{-2p}A_{n+p} + 2^{2p}A_{n+p}^{1/2}\delta_{n+p}^{1/2}\right)
\end{aligned}
\tag{4.4}
$$

where C denotes a suitable constant (*independent on the particular pseudo NS field*). The proof of the inequalities (4.4) is not difficult, assuming the (S,H,CZ,P) inequalities above, and it is illustrated in the problems $\{1\}$, $\{2\}$, $\{3\}$.

(2) *"Dynamical inequality":* i.e. an inequality based on the energy inequality (c) in (3.1) which implies, quite easily, the following *"dynamic inequality"*[8]

$$
A_n \leq C\left(2^p G_{n+p}^{2/3} + 2^p A_{n+p}\delta_{n+p} + 2^p J_{n+p}\right)
\tag{4.5}
$$

whose proof is illustrated in problem $\{4\}$.

---

[7] Indeed it relates properties of operators on a scale to those on a different scale. Note, however, that the couplings on scale n, i.e. the components of $\underline{X}_n$, provide information on those of X_{n+p} rather than on those of $\underline{X}_{n-p}$ as usual in the renormalization group methods, see [1].

[8] We call it "dynamic" because it follows from the energy inequality, i.e. from the equations of motion.

Proof of proposition: Assume the above inequalities (4.4), (4.5) and setting $\alpha_n = A_n, \kappa_n = K_n^{8/5}, j_n = J_n, g_n = G_n^{2/3}, \delta_{n+p} = \delta$ and, as above, $\underline{X}_n = (\alpha_n, \kappa_n, j_n, g_n)$. The r.h.s. of the inequalities defines the map $\mathcal{B}_p(\underline{X}; \delta)$.

If one stares long enough at them one realizes that the contraction property of the proposition is an immediate consequence of

(1) The exponents to which $\varepsilon = 2^{-p}$ is raised in the various terms are either positive or not; in the latter cases the inverse power of ε is always appearing multiplied by a power of δ_{n+p} which we can take so small to compensate for the size of ε to any negative power, *except in the one case corresponding to the last term in* (4.5) where we see ε^{-1} without any compensating δ_{n+p}.

(2) Furthermore the sum of the powers of the components of $\underline{X}_n$ in each term of the inequalities is *always* ≤ 1: this means that the inequalities are "almost linear" and a linear map that "bounds" $\mathcal{B}_p$ exists and it is described by a matrix with small entries *except one off–diagonal element*. The iterates of the matrix therefore contract unless the large matrix element "ill placed" in the matrix: and one easily sees that it is not.

A formal argument can be devised in many ways: we present one in which several choices appear that are quite arbitrary and that the reader can replace with alternatives. In a way one should really try to see why a formal argument is not necessary.

The relation (4.5) can be "iterated" by using the expressions (4.4) for G_{n+p}, J_{n+p} and then the first of (4.4) to express $G_{n+p}^{1/5}$ in terms of A_{n+2p} with n replaced by $n + p$:

$$
\begin{aligned}
\alpha_n \leq C \ (&2^{-p}\alpha_{n+2p} + 2^{3p}\delta_{n+2p}^{1/2}\alpha_{n+2p}^{1/2}+ \\
&+2^{p/5}(\alpha_{n+2p}\kappa_{n+2p})^{1/2} + 2^{7p/5}\delta_{n+2p}\alpha_{n+2p}^{7/20}\kappa_{n+2p}^{1/2}+ \\
&+2^{3p}\delta_{n+2p}\alpha_{n+2p})
\end{aligned} \tag{4.6}
$$

It is convenient to take advantage of the simple inequalities $(ab)^{\frac{1}{2}} \leq za + z^{-1}b$ and $a^x \leq 1 + a$ for $a, b, z, x > 0$, $x \leq 1$.

The (4.6) can be turned into a relation between α_n and $\alpha_{n+p}, \kappa_{n+p}$ by replacing p by $\frac{1}{2}p$. Furthermore, in the relation between α_n and $\alpha_{n+p}, \kappa_{n+p}$ obtained after the latter replacement, we choose $z = 2^{-p/5}$ to disentangle $2^{p/10}(\alpha_{n+p}\kappa_{n+p})^{1/2}$ we obtain recurrent (generous) estimates for α_n, κ_n

$$
\begin{aligned}
\alpha_n &\leq C \ (2^{-p/10}\alpha_{n+p} + 2^{3p/10}\kappa_{n+p} + \xi_{n+p}^\alpha) \\
\kappa_n &\leq C \ (2^{-4p/5}\kappa_{n+p} + \xi_{n+p}) \\
\xi_{n+p}^\alpha &\stackrel{def}{=} 2^{3p}\delta_{n+p}(\alpha_{n+p} + \kappa_{n+p} + 1) \\
\xi_{n+p} &\stackrel{def}{=} 2^{3p}\delta_{n+p}\alpha_{n+p}
\end{aligned} \tag{4.7}
$$

We fix p once and for all such that $2^{-p/10}C < \frac{1}{3}$.

Then if $C2^{3p}\delta_n$ is small enough, *i.e.* if δ_n is small enough, say for $\delta_n < \overline{\delta}$ *for all* $|n| \geq \overline{n}$, the matrix $M = C \begin{pmatrix} 2^{-p/10} + 2^{3p}\delta_{n+p} & 2^{3p/10} + 2^{3p}\delta_{n+p} \\ 0 & 2^{-4p/5} + 2^{3p}\delta_{n+p} \end{pmatrix}$ will have the two eigenvalues $< \frac{1}{2}$ and iteration of (4.6) will contract any ball in the plane α, κ to the ball of radius $2\overline{\delta}$.

If α_n, κ_n are bounded by a constant $\overline{\delta}$ for all $|n|$ large enough the (4.4) show that also g_n, j_n are going to be eventually bounded proportionally to $\overline{\delta}$.

Hence by imposing that δ is so small that $|X_n| = \alpha_n + \kappa_n + j_n + g_n < \rho$ we see that proposition 3 holds (hence theorem 2 as a consequence of theorem 1).

5 Fractal Dimension of Singularities of the Navier–Stokes Equation, $d = 3$

Here we ask which could be the structure of the possible set of the singularity points of the solutions of the Navier–Stokes equation in $d = 3$. The answer is an immediate consequence of theorem II and we describe it here for completeness: the technique is a classic method (Almgren) to link *a priori* estimates to fractal dimension estimates.

It has been shown already by Leray that the set of times at which a singularity is possible has zero measure (on the time axis), see §3.4 in [5].

Obviously sets of zero measure can be quite structured and even large in other senses. One can think to the example of the Cantor set which is non denumerable and obtained from an interval I by deleting an open concentric subinterval of length $1/3$ that of I and then repeating recursively this operation on each of the remaining intervals (called n–th generation intervals after n steps); or one can think to the set of rational points which is dense.

5.1 Dimension and Measure of Hausdorff

An interesting geometric characteristic of the size of a set is given by the Hausdorff dimension and by the Hausdorff measure, *c.f.r.* [4], p.174.

Definition 5.1. *(Hausdorff α–measure): The Hausdorff α-measure of a set A contained in a metric space M is defined by considering for each $\delta > 0$ all coverings $\mathcal{C}_\delta$ of A by closed sets F with diameter $0 < d(F) \leq \delta$ and setting*

$$\mu_\alpha(A) = \lim_{\delta \to 0} \inf_{\mathcal{C}_\delta} \sum_{F \in \mathcal{C}_\delta} d(F)^\alpha \tag{5.1}$$

Remark 5.1.
(1) The limit over δ exists because the quantity $\inf_{\mathcal{C}_\delta} \dots$ is monotonic nondecreasing.
(2) It is possible to show that the function defined on the sets A of M by

$A \to \mu_\alpha(A)$ is completely additive on the smallest family of sets containing all closed sets and invariant with respect to the operations of complementation and countable union (which is called the σ-*algebra* / of the Borel sets of M), *c.f.r.* [4].

One checks immediately that given $A \in /$ there is α_c such that

$$\mu_\alpha(A) = \infty \text{ if } \alpha < \alpha_c, \qquad \mu_\alpha(A) = 0 \text{ if } \alpha > \alpha_c \qquad (5.2)$$

and it is therefore natural to set up the following definition

Definition 5.2. *(Hausdorff measure and Hausdorff dimension): Given a set $A \subset R^d$ the quantity α_c, (5.2), is called Hausdorff dimension of A, while $\mu_\alpha(A)$ is called Hausdorff measure of A.*

It is not difficult to check that

(1) Denumerable sets in $[0,1]$ have zero Hausdorff dimension and measure.

(2) Hausdorff dimension of n-dimensional regular surfaces in R^d is n and, furthermore, the Hausdorff measure of their Borel subsets defines on the surface a measure μ_α that is equivalent to the area measure μ: namely there is a $\rho(x)$ such that $\mu_\alpha(dx) = \rho(x)\mu(dx)$.

(3) The Cantor set, defined also as the set of all numbers in $[0,1]$ which in the representation in base 3 do not contain the digit 1, has

$$\alpha_c = \log_3 2 \qquad (5.3)$$

as Hausdorff dimension.[9]

5.2 Hausdorff Dimension of Singular Times in the Navier–Stokes Solutions $(d = 3)$

We now attempt to estimate the Hausdorff dimension of the sets of times $t \leq T < \infty$ at which appear singularities of a given weak solution of Leray,

---

[+] Indeed with 2^n disjoint segments with size 3^{-n}, uniquely determined (the n-th generation segments), one covers the whole set C; hence

$$\mu_{\alpha\,\delta} \text{ r } \inf_{C_\delta} \sum_{C_\delta} d(F)^\alpha \quad 1 \quad \text{ if } \quad \alpha \text{ r } \alpha_0 \text{ r } \log_3 2$$

and $\mu_{\alpha_0}(C)$ 1: *i.e.* $\mu_\alpha(C)$ r 0 if $\alpha > \alpha_0$. Furthermore, *c.f.r.* problem 16 below, if $\alpha < \alpha_0$ one checks that the covering   0 realizing the smallest value of $\sum_{C_\delta} d(F)^\alpha$ with r 3^{-n} is precisely the just considered one consisting in the 2^n intervals of length 3^{-n} of the n-th generation and the value of the sum on such covering diverges for $n \to \infty$. Hence $\mu_\alpha(C)$ r ∞ if $\alpha < \alpha_0$ so that $\alpha_0 \equiv \alpha$ and $\mu_\alpha(C)$ r 1.

i.e. a solution of the type discussed in (0.1). Here T is an *a priori* arbitrarily prefixed time.

We need a key property of Leray's solutions, namely that if at time t_0 it is $J_1(t_0) = L^{-1} \int (\partial \underline{u})^2 d\underline{x} < \infty$, *i.e.* if the Reynolds number $R(t_0) = J_1(t_0)^{1/2}/V_c \equiv V_1/V_c$ with $V_c \stackrel{def}{=} \nu L^{-1}$ is $< +\infty$, then the solution stays regular in a time interval $(t_0, t_0 + \tau]$ with (see proposition II in §3.3 of [5], eq. (3.3.34)):

$$\tau = \min F \frac{T_c}{R(t_0)^4}, T_c \qquad T_c = \frac{L^2}{\nu} \tag{5.4}$$

From this it will follow, see below, that there are $A > 0, \gamma > 0$ such that if

$$\liminf_{\sigma \to 0} \left(\frac{\sigma}{T_c} \right)^\gamma \int_{t-\acute{\sigma}}^t \frac{d\vartheta}{\sigma} R^2(\vartheta) < A \tag{5.5}$$

then $\tau > \sigma$ and the solution is regular in an interval that contains t so that the instant t is an instant at which the solution is regular. Here, as in the following, we could fix $\gamma = 1/2$: but γ is left arbitrary in order to make clearer why the choice $\gamma = 1/2$ is the "best".

We first show that, indeed, from (5.5) we deduce the existence of a sequence $\sigma_i \to 0$ such that

$$\int_{t-\sigma_i}^t \frac{d\vartheta}{\sigma_i} R^2(\vartheta) < A \left(\frac{\sigma_i}{T_c} \right)^{-\gamma} \tag{5.6}$$

therefore, the l.h.s. being a time average, there must exist $\vartheta_{0i} \in (t - \sigma_i, t)$ such that

$$R^2(\vartheta_{0i}) < A \left(\frac{\sigma_i}{T_c} \right)^{-\gamma} \tag{5.7}$$

and then the solution is regular in the interval $(\vartheta_{0i}, \vartheta_{0i} + \tau_i)$ with length τ_i at least

$$\tau_i = FT_c \frac{(\sigma_i/T_c)^{2\gamma}}{A^2} > \sigma_i \tag{5.8}$$

provided $\gamma \leq 1/2$, and σ_i is small enough and if A is small enough (if $\gamma = \frac{1}{2}$ then this means $2A^2 < F$). Under these conditions the size of the regularity interval is longer than σ_i and *therefore it contains t itself.*

It follows that, if t is in the set S of the times at which a singularity is present, it must be

$$\liminf_{\sigma \to 0} \left(\frac{\sigma}{T_c} \right)^\gamma \int_{t-\sigma}^t \frac{d\vartheta}{\sigma} R^2(\vartheta) \geq A \qquad \text{if } t \in S \tag{5.9}$$

i.e. every singularity point is covered by a family of infinitely many intervals F with diameters σ *arbitrarily small* and satisfying

$$\int_{t-\sigma}^t d\vartheta \, R^2(\vartheta) \geq \frac{A}{2} \sigma \left(\frac{\sigma}{T_c} \right)^{-\gamma} \tag{5.10}$$

From Vitali's covering theorem (*c.f.r.* problem {19}) it follows that, given $\delta > 0$, one can find a denumerable family of intervals $F_1, F_2, \ldots$, with $F_i = (t_i - \sigma_i, t_i)$, pairwise disjoint and verifying the (5.10) and $\sigma_i < \delta/4$, such that the intervals $5F_i \overset{def}{=} (t_i - 7\sigma_i/2, t_i + 5\sigma_i/2)$ (obtained by dilating the intervals F_i by a factor 5 about their center) cover S

$$S \subset {}_i 5F_i \tag{5.11}$$

Consider therefore the covering $\mathcal{C}$ generated by the sets $5F_i$ and compute the sum in (5.1) with $\alpha = 1 - \gamma$:

$$\begin{aligned} \sum_i (5\sigma_i)\left(\tfrac{5\sigma_i}{T}\right)^{-\gamma} &= 5^{1-\gamma} \sum_i \sigma_i \left(\tfrac{\sigma_i}{T}\right)^{-\gamma} < \\ &< \tfrac{25^{1-}}{A\sqrt{T}} \sum_i \int_{F_i} d\vartheta \, R^2(\vartheta) \le \tfrac{25^{1-}}{A\sqrt{T}} \int_0^T d\vartheta \, R^2(\vartheta) < \infty \end{aligned} \tag{5.12}$$

where we have made use of the *a priori* estimates on vorticity (0.2) and we must recall that $\gamma \le 1/2$ is a necessary condition in order that what has been derived be valid (*c.f.r.* comment to (5.8)).

Hence it is clear that for each $\alpha \ge 1/2$ it is $\mu_\alpha(S) < \infty$ (pick, in fact, $\alpha = 1-\gamma$, with $\gamma \le 1/2$) hence the Hausdorff dimension of S is $\alpha_c \le 1/2$. Obviously the choice that gives the best regularity result (with the informations that we gathered) is precisely $\gamma = 1/2$.

Moreover one can check that $\mu_{1/2}(S) = 0$: indeed we know that S has zero measure, hence there is an open set $G \quad S$ with measure smaller than a prefixed ε. And we can choose the intervals F_i considered above so that they also verify $F_i \subset G$: hence we can replace the integral in the right hand side of (5.12) with the integral over G hence, since the integrand is summable, we shall find that the value of the integral can be supposed as small as wished, so that $\mu_{1/2}(S) = 0$.

5.3 Hausdorff Dimension in Space–Time of the Solutions of NS, ($d = 3$)

The problem of which is the Hausdorff dimension of the points $(\underline{x}, t) \in \Omega \times [0, T]$ which are singularity points for the Leray's solutions is quite different.

Indeed, *a priori*, it could even happen that, at one of the times $t \in S$ where the solution has a singularity as a function of time, *all* points $(\underline{x}, t)$, with $\underline{x} \in \Omega$, are singularity points and therefore the set S_0 of the singularity points thought of as a set in space–time could have dimension 3 (and perhaps even 3.5 if we take into account the dimension of the singular times discussed in (B) above).

From theorem II we know that if

$$\limsup_{r \to 0} r^{-1} \int_{t-r^2/2\nu}^{t+r^2/2\nu} \int_{S(\underline{x},r)} \frac{d\vartheta}{\nu} \, d\underline{\xi} \, (\partial \underline{u})^2 < \varepsilon \tag{5.13}$$

then regularity at the point $(\underline{x}, t)$ follows.

It follows that the set S_0 of the singularity points in space–time can be covered by sets $C_r = S(\underline{x}, r) \times (t - r^2\nu^{-1}, t + \frac{1}{2}r^2\nu^{-1}]$ with r arbitrarily small and such that

$$\frac{1}{r\nu} \int_{t-\frac{r^2}{2\nu}}^{t+\frac{r^2}{2\nu}} d\vartheta \int_{S(\underline{x},r)} d\underline{x} \, (\partial \underline{u})^2 > \varepsilon \tag{5.14}$$

which is the negation of the property in (5.13).

Again by a covering theorem of Vitali (*c.f.r.* problems {16}, {17}), we can find a family F_i of sets of the form $F_i = S(\underline{x}_i, r_i) \times (t_i - \frac{r_i^2}{2\nu}, t_i + \frac{r_i^2}{2\nu}]$ pairwise disjoint and such that the sets $6F_i =$ set of points $(\underline{x}', t')$ at distance $\leq 6r_i$ from the points of F_i covers the singularity set S_0.[10] One can then estimate the sum in (5.1) for such a covering, by using that the sets F_i are pairwise disjoint and that $5F_i$ has diameter, if $\max r_i$ is small enough, not larger than $18r_i$:

$$\sum_i (18r_i) \leq \frac{36}{\nu\varepsilon} \sum_i \int_{F_i} (\partial \underline{u})^2 d\underline{\xi} dt \leq \frac{36}{\nu\varepsilon} \int_0^T \int_\Omega (\partial \underline{u})^2 d\underline{\xi} dt < \infty \tag{5.15}$$

i.e. the 1-measure of Hausdorff $\mu_1(S_0)$ would be $< \infty$ hence the Hausdorff dimension of S_0 would be ≤ 1.

Since S_0 has zero measure, being contained in $\Omega \times S$ where S is the set of times at which a singularity occurs somewhere, see (5.9), it follows (still from the covering theorems) that in fact it is possible to choose the sets F_i so that their union U is contained into an open set G which differs from S_0 by a set of measure that exceeds by as little as desired that of S_0 (which is zero); one follows the same method used above in the analysis of the time–singularity. Hence we can replace the last integral in (5.15) with an integral extended to the union U of the $F_i's$: the latter integral can be made as small as wished by letting the measure of G to 0. It follows that not only the Hausdorff dimension of S_0 is ≤ 1, but also the $\mu_1(S_0) = 0$.

Remark 5.2.
(1) In this way we exclude that the set S_0 of the space–time singularities contains a regular curve: singularities, *if existent*, cannot move along trajectories (like flow lines) otherwise the dimension of S would be $1 > 1/2$) nor they can distribute, at fixed time, along lines and, hence, in a sense they must appear isolates and immediately disappear (always assuming their real existence).
(2) A conjecture (much debated and that I favor) that is behind all our discussions is that *if the initial datum $\underline{u}^0$ is in $C^\infty(\Omega)$ then there exists a solution*

---

[10] Here the constant 5, as well as the other numerical constants that we meet below like $5, 6, 18$ have no importance for our purposes and are just simple constants for which the estimates work.

to the Navier Stokes equation that is of class C^∞ in $(\underline{x}, t)$", i.e. $S_0 = $! The problem is, still, open: counterexamples to the conjecture are not known (*i.e.* singular Leray's solutions with initial data and external force of class C^∞) but the matter is much debated and different alternative conjectures are possible (*c.f.r.* [7]).

(3) In this respect one should keep in mind that if $d \geq 4$ it is possible to show that *not all* smooth initial data evolve into regular solutions: counterexamples to smoothness can indeed be constructed, *c.f.r.* [8].

Acknowledgements: The notes are extracted from reference [5].

6 Problems. The Dimensional Bounds of the CKN Theory

In the following problems we shall set $\nu = 1$, with no loss of generality, thus fixing the units so that time is a square length. The symbols $(\underline{u}, p)$ will denote a pseudo NS field, according to definition 1. Moreover, for notational simplicity, we shall set $A_\rho \equiv A(\rho)$, $G_\rho \equiv G(\rho), \ldots$, and sometimes we shall write $A_{r_n}, G_{r_n} \ldots$ as $A_n, \ldots$ with an abuse that should not generate ambiguities. The validity of the (3.1) for Leray's solution is checked in problem {15}, at the end of the problems section, to stress that the theorems of Scheffer and CKN concern pseudo NS velocity–pressure fields: however it is independent of the first 14 problems. There will many constants that we generically denote C: they are not the same but one should think that they increase monotnically by a finite amount at each inequality. The integration elements like $d\underline{x}$ and dt are often omitted to simplify the notations and they should be easily understood from the integration domains.

(1) Let $\rho = r_{n+p}$ and $r = r_n$, with $r_n = L2^n$, *c.f.r.* lines following (4.2), and apply (S),(2.2), with $q = 3$ and $a = \frac{3}{4}$, to the field $\underline{u}$, at t fixed in Δ_r and using definition 2 deduce

$$\int_{B_r} |\underline{u}|^3 d\underline{x} \leq C_3^S \left[\left(\int_{B_r} |\partial \underline{u}|^2 \, d\underline{x} \right)^{\frac{3}{4}} \left(\int_{B_r} |\underline{u}|^2 \, d\underline{x} \right)^{\frac{3}{4}} + r^{-3/2} \left(\int_{B_r} |\underline{u}|^2 \right)^{3/2} \right] \leq$$
$$\leq C_3^S \left[\rho^{3/4} A_\rho^{3/4} \left(\int_{B_r} |\partial \underline{u}|^2 d\underline{x} \right)^{3/4} + r^{-3/2} \left(\int_{B_r} |\underline{u}|^2 \right)^{3/2} \right]$$

Infer from the above the third of (4.4). (*Idea* Let $\overline{|\underline{u}|^2_\rho}$ be the average of $\underline{u}^2$ on the ball B_ρ; apply the inequality (P), with $\alpha = 1$, to show that there is $C > 0$ such that

$$\int_{B_r} d\underline{x} \, |\underline{u}|^2 \leq \left(\int_{B_\rho} d\underline{x} \left| |\underline{u}|^2 - \overline{|\underline{u}|^2_\rho} \right| \right) + \overline{|\underline{u}|^2_\rho} \int_{B_r} d\underline{x} \leq$$
$$\leq C\rho \int_{B_\rho} d\underline{x} \, |\underline{u}| |\partial \underline{u}| + C \left(\frac{r}{\rho} \right)^3 \int_{B_\rho} d\underline{x} \, |\underline{u}|^2 \leq C\rho^{3/2} A_\rho^{1/2} \left(\int_{B_\rho} d\underline{x} \, |\partial \underline{u}|^2 \right)^{1/2} +$$
$$+ C \left(\frac{r}{\rho} \right)^3 \rho A_\rho$$

where the dependence from $t \in \Delta_r$ is not explicitly indicated; hence

$$\int_{B_r} d\underline{x}\,|\underline{u}|^3 \le C\,(r\rho^{-1})^3 A_\rho^{3/2} + C\,(\rho^{3/4} + \rho^{9/4} r^{-3/2}) A_\rho^{3/4} \left(\int_{B_\rho} d\underline{x}\,|\partial\underline{u}|^2 \right)^{3/4}$$

then integrate both sides with respect to $t \in \Delta_r$ and apply (H) and definition 2.)

(2) Let $\varphi \le 1$ be a non negative C^∞ function with value 1 if $|\underline{x}| \le 3\rho/4$ and 0 if $|\underline{x}| > 4\rho/5$; we suppose that it has the "scaling" form $\varphi = \varphi_1(\underline{x}/\rho$ with $\varphi_1 \ge 0$ a C^∞ function fixed once and for all. Let B_ρ be the ball centered at $\underline{x}$ with radius ρ; and note that, if $\rho = r_{n+p}$ and $r = r_n$, pressure can be written, at each time (without explicitly exhibiting the time dependence), as $p(\underline{x}) = p'(\underline{x}) + p''(\underline{x})$ with

$$p'(\underline{x}) = \tfrac{1}{4\pi} \int_{B_\rho} \tfrac{1}{|\underline{x}-\underline{y}|} p(\underline{y}) \Delta\varphi(\underline{y})\,d\underline{y} + \tfrac{1}{2\pi} \int_{B_\rho} \tfrac{\underline{x}-\underline{y}}{|\underline{x}-\underline{y}|^3} \cdot \partial\varphi(\underline{y})\,p(\underline{y})\,d\underline{y}$$
$$p''(\underline{x}) = \tfrac{1}{4\pi} \int_{B_\rho} \tfrac{1}{|\underline{x}-\underline{y}|} \varphi(\underline{y})\,(\partial\underline{u}(\underline{y})) \cdot (\partial\underline{u}(\underline{y}))\,d\underline{y}$$

if $|\underline{x}| < 3\rho/4$; and also $|p'(\underline{x})| \le C\rho^{-3} \int_{B_\rho} d\underline{y}\,|p(\underline{y})|$ and all functions are evaluated at a fixed $t \in \Delta_r$. Deduce from this remark the first of the (4.4). (*Idea* First note the identity $p = -(4\pi)^{-1} \int_{B_\rho} |\underline{x} - \underline{y}|^{-1} \Delta\,(\varphi p)$ for $\underline{x} \in B_r$ because if $\underline{x} \in B_{3\rho/4}$ it is $\varphi p \equiv p$. Then note the identity $\Delta\,(\varphi p) = p\,\Delta\varphi + 2\partial p \cdot \partial\varphi + \varphi\,\Delta p$ and since $\Delta p = -\partial \cdot (\underline{u} \cdot \partial\,\underline{u}) = -(\partial\underline{u}) \cdot (\partial\,\underline{u})$:

the second of the latter relations generates p'' while $p\Delta\varphi$ combines with the contribution from $2\partial p \cdot \partial\varphi$, after integrating the latter by parts, and generates the two contributions to p'.
From the expression for p'' we see that

$$\int_{B_r} d\underline{x}\,|p''(\underline{x})|^2 \le \int_{B_\rho \times B_\rho} d\underline{y}\,d\underline{y}'\,|\partial\underline{u}(\underline{y})|^2 |\partial\underline{u}(\underline{y}')|^2 \int_{B_r} d\underline{x} \tfrac{1}{|\underline{x}-\underline{y}||\underline{x}-\underline{y}'|} \le$$
$$\le C\rho (\int_{B_\rho} d\underline{y}\,|\partial\underline{u}(\underline{y})|^2)^2 \tag{6.1}$$

The part with p' is more interesting: since its expression above contains inside the integral kernels apparently singular at $\underline{x} = \underline{y}$ like $|\underline{x} - \underline{y}|^{-1} \Delta\varphi$ and $|\underline{x} - \underline{y}|^{-1} \partial\varphi$ one remarks that, in fact, there is no singularity because the derivatives of φ vanish if $\underline{y} \in B_{3\rho/4}$ (where $\varphi \equiv 1$). Hence $|\underline{x} - \underline{y}|^{-k}$ can be bounded "dimensionally" by ρ^{-k} in the whole region $B_\rho/B_{3\rho/4}$ for all $k \ge 0$ (this remark also motivates why one should think p as sum of p' and p'').

Thus replacing the (apparently) singular kernels with their dimensional bounds we get

$$\int_{B_r} d\underline{x}\,|\underline{u}||p'| \le \frac{C}{\rho^3} \left(\int_{B_r} d\underline{x}\,|\underline{u}| \right) \cdot \left(\int_{B_\rho} d\underline{x}\,|p| \right)$$

which can be bounded by using inequality (H) as

$$\leq \frac{C}{\rho^3} \left(\int_{B_r} d\underline{x} \, |\underline{u}|^{2/5} \cdot |\underline{u}|^{3/5} \cdot 1 \right) \cdot \left(\int_{B_\rho} d\underline{x} \, |p| \right) \leq$$

$$\leq \frac{C}{\rho^3} \left(\int_{B_r} d\underline{x} \, |\underline{u}|^2 \right)^{1/5} \cdot \left(\int_{B_r} d\underline{x} \, |\underline{u}|^3 \right)^{1/5} (r^3)^{3/5} \cdot \int_{B_\rho} d\underline{x} \, |p| \leq$$

$$\leq \frac{C}{\rho^3} (\rho A_\rho)^{1/5} \left(\int_{B_r} d\underline{x} \, |\underline{u}|^3 \right)^{1/5} \cdot \left(\int_{B_\rho} d\underline{x} \, |p| \right)$$

where all functions depend on $\underline{x}$ (and of course on t) and then, integrating over $t \in \Delta_r$ and dividing by r^2 one finds, for a suitable $C > 0$:

$$\frac{1}{r^2} \int_{Q_r} dt d\underline{x} \, |\underline{u}||p'| \leq C \left(\frac{r}{\rho}\right)^{1/5} G_r^{1/5} K_\rho^{4/5} A_\rho^{1/5}$$

that is combined with $\int_{B_r} d\underline{x}|\underline{u}||p''| \leq (\int_{B_r} d\underline{x} \, |\underline{u}|^2)^{1/2}(\int_{B_r} d\underline{x} \, |p''|^2)^{1/2}$ which, integrating over time, dividing by ρ^2 and using inequality (6.1) for $\int_{B_r} d\underline{x} \, |p''|^2$ yields: $r^{-2} \int_{Q_r} dt d\underline{x} \, |\underline{u}||p''| \leq C(\rho r^{-1})^2 A_\rho^{1/2} \delta_\rho)$.

(3) In the context of the hint and notations for p of the preceding problem check that $\int_{B_r} d\underline{x} \, |p'| \leq C(r\rho^{-1})^3 \int_{B_\rho} d\underline{x}|p|$. Integrate over t the power $5/4$ of this inequality, rendered adimensional by dividing it by $r^{13/4}$; one gets: $r^{-13/4} \int_{\Delta_r} (\int |p'|)^{5/4} \leq C(r\rho^{-1})^{1/2} K_\rho$, which yields the first term of the second inequality in (4.4). Complete the derivation of the second of (4.4). (*Idea* Note that $p''(\underline{x}, t)$ can be written, in the interior of B_r, as $p'' = \tilde{p} + \hat{p}$ with:

$$\tilde{p}(\underline{x}) = -\frac{1}{4\pi} \int_{B_\rho} \frac{\underline{x} - \underline{y}}{|\underline{x} - \underline{y}|^3} \varphi(\underline{y})\underline{u} \cdot \underline{\partial} \, \underline{u} \, dy, \qquad \hat{p}(\underline{x}) = -\frac{1}{4\pi} \int_{B_\rho} \frac{\partial \varphi(\underline{y}) \cdot (\underline{u} \cdot \underline{\partial})\underline{u}}{|\underline{x} - \underline{y}|} dy$$

(always at fixed t and not declaring explicitly the t–dependence). Hence by using $|\underline{x} - \underline{y}| > \rho/4$, for $\underline{x} \in B_r$ and $\underline{y} \in B_\rho/B_{3\rho/4}$, *i.e.* for $\underline{y}$ in the part of B_ρ where $\underline{\partial}\varphi \neq \underline{0}$) we find

$$\int_{B_r} |\tilde{p}| \, d\underline{x} \leq C \int_{B_\rho} d\underline{y} \left(\int_{B_r} \frac{d\underline{x}}{|\underline{x}-\underline{y}|^2} |\underline{u}(\underline{y})| |\underline{\partial} \, \underline{u}(\underline{y})| \right) \leq$$

$$\leq \quad Cr \left(\int_{B_\rho} |\underline{u}|^2 \right)^{1/2} \left(\int_{B_\rho} |\underline{\partial} \underline{u}|^2 \right)^{1/2} \leq Cr\rho^{1/2} A_\rho^{1/2} \left(\int_{B_\rho} |\underline{\partial} \underline{u}|^2 \right)^{1/2}$$

$$\int_{B_r} |\hat{p}| \, d\underline{x} \leq C \frac{r^3}{\rho^2} \int_{B_\rho} |\underline{u}||\underline{\partial} \underline{u}| \leq C \, r\rho^{1/2} A_\rho^{1/2} \left(\int_{B_\rho} |\underline{\partial} \underline{u}|^2 \right)^{1/2}$$

and $\left(\int_{B_r} |p''| \right)^{5/4}$ is bounded by raising the right hand sides of the last inequalities to the power $5/4$ and integrating over t, and finally applying inequality (H) to generate the integral $\left(\int_{Q_\rho} |\underline{\partial} \underline{u}|^2 \right)^{5/8}$).

(4) Deduce that (4.5) holds for a pseudo–NS field $(\underline{u}, p)$, *c.f.r.* definition 2.1. (*Idea* Let $\varphi(\underline{x}, t)$ be a C^∞ function which is 1 on $Q_{\rho/2}$ and 0 outside Q_ρ; it is: $0 \leq \varphi(\underline{x}, t) \leq 1$, $|\partial\varphi| \leq \frac{C}{\rho}$, $|\Delta\varphi + \partial_t\varphi| \leq \frac{C}{\rho^2}$, if we suppose that φ has the form $\varphi(\underline{x},t) = \varphi_2(\frac{\underline{x}}{\rho}, \frac{t}{\rho^2}) \geq 0$ for some φ_2 suitably fixed and smooth. Then, by applying the third of (3.1) and using the notations of the preceding problems, if $\bar{t} \in \Delta_{\rho/2}(t_0)$, it is

$$\int_{B_r \times \{\bar{t}\}} |\underline{u}(\underline{x},t)|^2 d\underline{x} \leq \frac{C}{\rho^2} \int_{Q_\rho} dt d\underline{x} \, |\underline{u}|^2 + \int_{Q_\rho} dt d\underline{x} \, (|\underline{u}|^2 + 2p) \underline{u} \cdot \partial\varphi \leq$$

$$\leq \frac{C}{\rho^2} \int_{Q_\rho} dt d\underline{x} \, |\underline{u}|^2 + \left| \int_{Q_\rho} dt d\underline{x} \, (|\underline{u}|^2 - \overline{|\underline{u}|_\rho^2}) \underline{u} \cdot \partial\varphi \right| + 2 \int_{Q_\rho} dt d\underline{x} \, p \, \underline{u} \cdot \partial\varphi \leq$$

$$\leq \frac{C}{\rho^{1/3}} \left(\int_{Q_\rho} dt d\underline{x} \, |\underline{u}|^3 \right)^{2/3} + \left| \int_{Q_\rho} dt d\underline{x} \left(|\underline{u}|^2 - \overline{|\underline{u}|_\rho^2} \right) \underline{u} \cdot \partial\varphi \right| + \frac{2C}{\rho} \int_{B_\rho} dt d\underline{x} \, |p| |\underline{u}| \leq$$

$$\leq C\rho G_\rho^{2/3} + C\rho J_\rho + \rho \left| \frac{1}{\rho} \int_{Q_\rho} dt d\underline{x} \, (|\underline{u}|^2 - \overline{|\underline{u}|_\rho^2}) \underline{u} \cdot \partial\varphi \right| \qquad (*)$$

We now use the following inequality, at t fixed and with the integrals over $d\underline{x}$

$$\frac{1}{\rho} \left| \int_{B_\rho} d\underline{x} \, (|\underline{u}|^2 - \overline{|\underline{u}|_\rho^2}) \underline{u} \cdot \partial\varphi \right| \leq \frac{C}{\rho^2} \int_{B_\rho} d\underline{x} \, |\underline{u}| \left| |\underline{u}|^2 - \overline{|\underline{u}|_\rho^2} \right| \leq$$

$$\leq \frac{C}{\rho^2} \left(\int_{B_\rho} d\underline{x} \, |\underline{u}|^3 \right)^{1/3} \left(\int_{B_\rho} \left| \underline{u}^2 - \overline{|\underline{u}|_\rho^2} \right|^{3/2} \right)^{2/3}$$

and we also take into account inequality (P) with $f = \underline{u}^2$ and $\alpha = 3/2$ which yields (always at t fixed and with integrals over $d\underline{x}$):

$$\left(\int_{B_\rho} \left| \underline{u}^2 - \overline{|\underline{u}|_\rho^2} \right|^{3/2} \right)^{2/3} \leq C \left(\int_{B_\rho} |\underline{u}| |\partial \underline{u}| \right)$$

then we see that

$$\int_{B_\rho} \left| |\underline{u}|^2 - \overline{|\underline{u}|_\rho^2} \right| |\underline{u}| \, |\partial\varphi| \leq \frac{C}{\rho} \left(\int_{B_\rho} |\underline{u}|^3 \right)^{1/3} \left(\int_{B_\rho} |\underline{u}| |\partial \underline{u}| \right) \leq$$

$$\leq \frac{C}{\rho} \left(\int_{B_\rho} |\underline{u}|^3 \right)^{1/3} \left(\int_{B_\rho} |\underline{u}|^2 \right)^{1/2} \left(\int_{B_\rho} |\partial \underline{u}|^2 \right)^{1/2} \leq$$

$$\leq \frac{C}{\rho} \rho^{1/2} A_\rho^{1/2} \left(\int_{B_\rho} |\underline{u}|^3 \right)^{1/3} \cdot \left(\int_{B_\rho} |\partial \underline{u}|^2 \right)^{1/2} \cdot 1$$

Integrating over t and applying (H) with exponents $3, 2, 6$, respectively, on the last three factors of the right hand side we get

$$\frac{1}{\rho^2} \int_{Q_\rho} |\underline{u}| \left| |\underline{u}|^2 - \overline{|\underline{u}|_\rho^2} \right| \leq C A_\rho^{1/2} G_\rho^{1/3} \delta_\rho^{1/2} \leq C \left(G_\rho^{2/3} + A_\rho \delta_\rho \right)$$

and placing this in the first of the preceding inequalities (*) we obtain the desired result).

The following problems provide a guide to the proof of theorem II. Below we replace, unless explicitly stated the sets B_r, Q_r, Δ_r introduced in definition 2, in (C) above, and employed in the previous problems with B_r^0, Δ_r^0, Q_r^0 with $B_r^0 = \{\underline{x} | |\underline{x} - \underline{x}_0| < r\}$, $\Delta_r^0 = \{t | t_0 > t > t_0 - r^2\}$, $Q_r^0 = \{(\underline{x}, t) | |\underline{x} - \underline{x}_0| < r, t_0 > t > t_0 - r^2\} = B_r^0 \times \Delta_r^0$. Likewise we shall set $B_{r_n}^0 = B_n^0, \Delta_{r_n}^0 = \Delta_n^0, Q_{r_n}^0 = Q_n^0$ and we shall define new operators A, δ, G, J, K, S by the same *expressions in (3.2)%(3.8) in (C) above but with the just defined new meaning of the integration domains. However, to avoid confusion, we shall call them* $A^0, \delta^0, \ldots$ *with a superscript 0 added.*

(5) With the above conventions check the following inequalities

$$A_n^0 \leq C A_{n+1}^0, \qquad G_n^0 \leq C G_{n+1}^0, \qquad G_n^0 \leq C \left(A_n^{0\,3/2} + A_n^{0\,3/4} \delta_n^{0\,3/4} \right)$$

(*Idea* The first two are trivial consequences of the fact that the integration domains of the right hand sides are larger than those of the left hand sides, and the radii of the balls differ only by a factor 2 so that C can be chosen 2 in the first inequality and 4 in the second. The third inequality follows from (S) with $a = \frac{3}{4}$, $q = 3$:

$$\int_{B_r^0} |\underline{u}|^3 \leq C \left[\left(\int_{B_r^0} |\partial \underline{u}|^2 \right)^{3/4} \left(\int_{B_r^0} |\underline{u}|^2 \right)^{3/4} + r^{-3/2} \left(\int_{B_r^0} |\underline{u}|^2 \right)^{3/2} \right] \leq$$

$$\leq C \left[r^{3/4} A_r^{0\,3/4} \left(\int_{B_r^0} |\partial \underline{u}|^2 \right)^{3/4} + A_r^{0\,3/2} \right]$$

where the integrals are over $d\underline{x}$ at t fixed; and integrating over t we estimate G_r^0 by applying (H) to the last integral over t.)

(6) Let $n_0 = n+p$ and $Q_n^0 = \{(\underline{x},t) | \, |\underline{x}-\underline{x}_0| < r_n, \, t_0 > t > t-r_n^2\} \overset{def}{=} B_n^0 \times \Delta_n^0$ consider the function:

$$\varphi_n(\underline{x},t) = \frac{\exp(-(\underline{x}-\underline{x}_0)^2/4(r_n^2 + t_0 - t))}{(4\pi(r_n^2 + t_0 - t))^{3/2}}, \qquad (\underline{x},t) \in Q_{n_0}^0$$

and a function $\chi_{n_0}(\underline{x},t) = 1$ on $Q_{n_0-1}^0$ and 0 outside $Q_{n_0}^0$, for instance choosing, a function which has the form $\chi_{n_0}(\underline{x},t) = \tilde{\varphi}(r_{n_0}^{-1}\underline{x}, r_{n_0}^{-1/2}t) \geq 0$, with $\tilde{\varphi}$ a C^∞ function fixed once and for all. Then write (3.1) using $\varphi = \varphi_n \chi_{n_0}$ and deduce the inequality

$$\frac{A_n^0 + \delta_n^0}{r_n^2} \leq C \left[r_{n+p}^{-2} G_{n+p}^{0\,2/3} + \sum_{k=n+1}^{n+p} r_k^{-2} G_k^0 + r_{n+p}^{-2} J_{n+p}^0 + \sum_{k=n+1}^{n+p-1} r_k^{-2} L_k \right] \quad (6.2)$$

where $L_k = r_k^{-2} \int_{Q_k^0} d\underline{x}\, dt\, |\underline{u}| |p - \overline{p^k}|$ with $\overline{p^k}$ equals the average of p on the ball B_k^0; for each $p > 0$. (*Idea* Consider the function φ and note that $\varphi \geq (Cr_n^3)^{-1}$ in Q_n^0, which allows us to estimate *from below* the left hand side term in (3.1), with $(Cr_n^2)^{-1}(A_n^0 + \delta^0{}_n)$. Moreover one checks that

$$|\varphi| \leq \frac{C}{r_m^3}, \qquad |\partial \varphi| \leq \frac{C}{r_m^4}, \qquad n \leq m \leq n+p \equiv n_0, \quad \text{in} \quad Q_{m+1}^0/Q_m^0$$

$$|\partial_t \varphi + \Delta \varphi| \leq \frac{C}{r_{n_0}^5} \qquad\qquad\qquad \text{in} \quad Q_{n_0}^0$$

and the second relation follows from $\partial_t \varphi + \Delta \varphi \equiv 0$ in the "dangerous places", *i.e.* $\chi = 1$, because φ is a solution of the heat equation (backward in time).

Hence the first term in the right hand side of (3.1) can be bounded *from above* by

$$\int_{Q_{n_0}^0} |\underline{u}|^2 |\partial_t \varphi_n + \Delta\varphi_n| \le \frac{C}{r_{n_0}^5} \int_{Q_{n_0}^0} |\underline{u}|^2 \le \frac{C}{r_{n_0}^5} \Big(\int_{Q_{n_0}^0} |\underline{u}|^3\Big)^{2/3} r_{n_0}^{5/3} \le \frac{C}{r_{n_0}^2} G^{0\,2/3}_{n_0}$$

getting the first term in the r.h.s. of (6.2).

Using here the scaling properties of the function φ the second term is bounded by

$$\int_{Q_{n_0}^0} |\underline{u}|^3 |\partial\varphi_n| \le \frac{C}{r_n^4} \int_{Q_{n+1}^0} |\underline{u}|^3 + \sum_{k=n+2}^{n_0} \frac{C}{r_k^4} \int_{Q_k^0/Q_{k-1}^0} |\underline{u}|^3 \le$$
$$\le \sum_{k=n+1}^{n_0} \frac{C}{r_k^4} \int_{Q_k^0} |\underline{u}|^3 \le C \sum_{k=n+1}^{n_0} \frac{G_k^0}{r_k^2}$$

Calling the third term (*c.f.r.* (1.1)) Z we see that it is bounded by

$$Z \le \Big|\int_{Q_{n_0}^0} p\,\underline{u}\cdot\underline{\partial}\chi_{n_0}\varphi_n\Big| \le \Big|\int_{Q_{n+1}^0} p\,\underline{u}\cdot\underline{\partial}\chi_{n+1}\varphi_n\Big| +$$
$$+\sum_{k=n+2}^{n_0} \Big|\int_{Q_k^0} p\,\underline{u}\cdot\underline{\partial}(\chi_k - \chi_{k-1})\varphi_n\Big| \le \Big|\int_{Q_{n+1}^0}(p - \overline{p^{n+1}})\,\underline{u}\cdot\underline{\partial}\chi_{n+1}\varphi_n\Big| +$$
$$+\sum_{k=n+2}^{n_0-1} \Big|\int_{Q_k^0}(p - \overline{p^k})\,\underline{u}\cdot\underline{\partial}(\chi_k - \chi_{k-1})\varphi_n\Big| + \int_{Q_{n_0}^0} |\underline{u}|\,|p|\,|\underline{\partial}(\chi_{n_0} - \chi_{n_0-1})\varphi_n|$$

where $\overline{p^m}$ denotes the average of p over B_m^0 (which only depends on t): the possibility of replacing p by $p - \overline{p}$ in the integrals is simply due to the fact that the 0 divergence of $\underline{u}$ allows us to add to p *any* constant because, by integration by parts, it will contribute 0 to the value of the integral.

From the last inequality it follows

$$Z \le \sum_{k=n+1}^{n_0-1} \frac{C}{r_k^4} \int_{Q_k^0} |p - \overline{p^k}|\,|\underline{u}| + J_{n_0}^0 r_{n_0}^{-2} = \sum_{k=n+1}^{n_0-1} \frac{C}{r_k^2} L_k + J_{n_0}^0 r_{n_0}^{-2}$$

then sum the above estimates.)

(7) If $\underline{x}_0$ is the center of Ω the function $\chi_{n_0}p$ can be regarded, if $n_0 < -1$, as defined on the whole R^3 and zero outside the torus Ω. Then if Δ is the Laplace operator *on the whole* R^3 note that the expression of p in terms of $\underline{u}$ (*c.f.r.* (a) of (3.1)) implies that in $Q_{n_0}^0$:

$$\chi_{n_0}p = \Delta^{-1}\Delta\chi_{n_0}p \equiv \Delta^{-1}\Big(p\Delta\chi_{n_0} + 2(\underline{\partial}\chi_{n_0})\cdot(\underline{\partial}p) - \chi_{n_0}\underline{\partial}\underline{\partial}\cdot(\underline{u}\,\underline{u})\Big)$$

Show that this expression can be rewritten, for $n < n_0$, as

$$\chi_{n_0}p = -\Delta^{-1}(\chi_{n_0}\underline{\partial}\underline{\partial}(\underline{u}\,\underline{u})) + [\Delta^{-1}(p\Delta\chi_{n_0}) + 2(\underline{\partial}\Delta^{-1})((\underline{\partial}\chi_{n_0})p)] =$$
$$= [-(\underline{\partial}\underline{\partial}\Delta^{-1})(\chi_{n_0}\underline{u}\,\underline{u})] + [2(\underline{\partial}\Delta^{-1})(\underline{\partial}\chi_{n_0}\underline{u}\,\underline{u}) - \Delta^{-1}((\underline{\partial}\underline{\partial}\chi_{n_0})\underline{u}\,\underline{u})] +$$
$$+[\Delta^{-1}(p\Delta\chi_{n_0}) + 2(\underline{\partial}\Delta^{-1})((\underline{\partial}\chi_{n_0})p)] \stackrel{def}{=} p_1 + p_2 + p_3 + p_4$$

with $p_1 = -(\partial\underline{\partial}\Delta^{-1})(\chi_{n_0}\vartheta_{n+1}\underline{u}\underline{u})$ and $p_2 = -(\partial\underline{\partial}\Delta^{-1})(\chi_{n_0}(1-\vartheta_{n+1})\underline{u}\underline{u})$

where ϑ_k is the characteristic function of $B^0{}_k$ and p_3, p_4 are the last two terms in square brackets. (*Idea* Use, for $\underline{x}, t \in Q^0_{n_0}$, Poisson formula

$$\chi_{n_0}(\underline{x},t)p(\underline{x},t) = \frac{-1}{4\pi}\int_{B^0_{n_0}}\frac{\Delta((\ _{n_0}p)(\underline{y},t))}{|\underline{x}-\underline{y}|}d\underline{y} =$$

$$= \frac{-1}{4\pi}\int_{B^0_{n_0}}\frac{p\Delta\ _{n_0}+2\underline{\partial}\ _{n_0}\cdot\underline{\partial}p-\ _{n_0}\partial\underline{\partial}\cdot(\underline{u}\underline{u})}{|\underline{x}-\underline{y}|}d\underline{y}$$

and suitably integrate by parts).

(8) In the context of the previous problem check that the formulae derived there can be written more explicitly as

$$p_1 = -(\partial\underline{\partial}\Delta^{-1})\cdot(\chi_{n_0}\vartheta_{n+1}\underline{u}\,\underline{u}), \qquad p_2 = -\frac{1}{4\pi}\int_{B^0_{n_0}/B^0_{n+1}}\left(\partial\underline{\partial}\frac{1}{|\underline{x}-\underline{y}|}\right)\cdot\chi_{n_0}\underline{u}\,\underline{u}$$

$$p_3 = \frac{1}{2\pi}\int_{B^0_{n_0}}\frac{\underline{x}-\underline{y}}{|\underline{x}-\underline{y}|^3}(\underline{\partial}\chi_{n_0})\underline{u}\,\underline{u} + \frac{1}{4\pi}\int_{B^0_{n_0}}\frac{1}{|\underline{x}-\underline{y}|}(\partial\underline{\partial}\chi_{n_0})\underline{u}\,\underline{u}$$

$$p_4 = -\frac{1}{4\pi}\int\frac{1}{|\underline{x}-\underline{y}|}p(\underline{y})\Delta\chi_{n_0} + \frac{2}{4\pi}\int p(\underline{y})\frac{\underline{x}-\underline{y}}{|\underline{x}-\underline{y}|^3}\cdot\underline{\partial}\chi_{n_0}$$

where $n < n_0$ and the integrals are over $\underline{y}$ at t fixed, and the functions in the left hand side are evaluated in $\underline{x}, t$.

(9) Consider the quantity L_n, introduced in (6),

$$L_n \overset{def}{=} r_n^{-2}\int_{Q_n}|\underline{u}|\,|p-\overline{p}_n(\vartheta)|\,d\xi d\vartheta$$

and show that, setting $n_0 = n+p$, $p > 0$, it is

$$L_n \le C\Big[\left(\frac{r_{n+1}}{r_{n_0}}\right)^{7/5}A^{01/5}_{n+1}G^{01/5}_{n+1}K^{04/5}_{n_0} + \left(\frac{r_{n+1}}{r_{n_0}}\right)^{5/3}G^{01/3}_{n+1}G^{02/3}_{n_0} +$$

$$+G^0_{n+1} + r^3_{n+1}G^{01/3}_{n+1}\sum_{k=n+1}^{n_0}r_k^{-3}A^0_k\Big]$$

(*Idea* Refer to (8) to bound L_n by: $\sum_{i=1}^4 r_n^{-2}\int_{Q^0_n}|\underline{u}||p_i - \overline{p_i^n}|$ where $\overline{p_i^n}$ is the average of p_i over B^0_n; and estimate separately the four terms. For the first it is not necessary to subtract the average and the difference $|p_1 - \overline{p_1^n}|$ can be divided into the sum of the absolute values each of which contributes equally to the final estimate which is obtained via the (CZ), and the (H)

$$\int_{B^0_{n+1}}|p_1 - \overline{p_1}||\underline{u}| \le 2\left(\int_{B^0_{n+1}}|p_1|^{3/2}\right)^{2/3}\left(\int_{B^0_{n+1}}|\underline{u}|^3\right)^{1/3} \le C\int_{B^0_{n+1}}|\underline{u}|^3$$

and the contribution of p_1 at L_n is bounded, therefore, by CG^0_{n+1}: note that this would not be true with p instead of p_1 because in the right hand side there would be $\int_\Omega|p|^{3/2}$ rather than $\int_{B^0_{n+1}}|p|^{3/2}$, because the (CZ) is a "nonlocal" inequality. The term with p_2 is bounded as

$$\int_{\Delta_n^0} \int_{B_n^0} |p_2 - \overline{p_2^n}||\underline{u}| \le \int_{\Delta_n^0} \int_{B_n^0} |\underline{u}| \, r_n \max |\partial p_2| \le$$

$$\le r_n \Big(\int_{Q_n^0} \frac{|\underline{u}|^3}{r_n^2} \Big)^{1/3} r_n^{2/3} r_n^{10/3} \max_{Q_n^0} |\partial p_2| \le$$

$$\le r_n^5 G_n^{0\,1/3} \sum_{m=n+1}^{n_0-1} \max_{t \in \Delta_m^0} \int_{B_{m+1}^0/B_m^0} \frac{|\underline{u}|^2}{r_m^4} = r_n^5 G_n^{0\,1/3} \sum_{m=n+1}^{n_0-1} \frac{A_m^0}{r_m^3}$$

Analogously the term with p_3 is bounded by using $|\partial p_3| \le C r_{n_0}^{-4} \int_{B_{n_0}^0} |\underline{u}|^2$ which is majorized by $C r_{n_0}^{-3} (\int_{B_{n_0}^0} |\underline{u}|^3)^{2/3}$ obtaining

$$\frac{1}{r_n^2} \int_{Q_n^0} |\underline{u}||p_3 - \overline{p_3^n}| \le \frac{C}{r_n^2} r_{n_0}^{-3} \int_{\Delta_n^0} [(\int_{B_{n_0}^0} |\underline{u}|^3)^{2/3} r_n \int_{B_n^0} |\underline{u}|] \le$$
$$\le \frac{C}{r_n^2} r_n^3 r_{n_0}^{-3} \int_{\Delta_n^0} (\int_{B_{n_0}^0} |\underline{u}|^3)^{2/3} (\int_{B_n^0} |\underline{u}|^3)^{1/3} \le$$
$$\le \frac{C}{r_n^2} (\frac{r_n}{r_{n_0}})^3 r_{n_0}^{4/3} r_n^{2/3} G_{n_0}^{0\,2/3} G_n^{0\,1/3} = C (\frac{r_n}{r_{n_0}})^{5/3} G_{n_0}^{0\,2/3} G_n^{0\,1/3}$$

Finally the term with p_4 is bounded (taking into account that the derivatives $\Delta \chi_n, \partial \chi_n$ vanish where the kernels become bigger than what suggested by their dimension) by noting that

$$\int_{B_n^0} |p_4 - \overline{p_4^n}||\underline{u}| \le C r_n \int_{B_n^0} |\underline{u}| \max_{B_n^0} |\partial p_4| \le C r_n \Big(\int_{B_n^0} |\underline{u}| \Big) \Big(\int_{B_{n_0}^0} \frac{|\underline{p}|}{r_{n_0}^4} \Big)$$

Denoting with $\tilde{K}_{n_0}^0$ the operator $K_{n_0}^0$ without the factor $r_{n_0}^{-13/4}$ which makes it dimensionless, and introducing, similarly, $\tilde{A}_n^0, \tilde{G}_n^0$ we obtain the following chain of inequalities, using repeatedly (H)

$$\frac{1}{r_n^2} \int_{Q_n^0} |p_4 - \overline{p_4^n}||\underline{u}| \le \frac{C}{r_n^2} r_n \Big(\int_{\Delta_n^0} \Big(\int_{B_{n_0}^0} \frac{|\underline{p}|}{r_{n_0}^4} \Big)^{5/4} \Big)^{4/5} \Big(\int_{\Delta_n^0} \Big(\int_{B_n^0} |\underline{u}| \Big)^5 \Big)^{1/5} \le$$
$$\le \frac{C}{r_n^2} \frac{r_n}{r_{n_0}^4} \tilde{K}_{n_0}^{0\,4/5} \Big(\int_{\Delta_n^0} \Big(\int_{B_n^0} |\underline{u}|^{2/5} |\underline{u}|^{3/5} \cdot 1 \Big) 5 \Big)^{1/5} \le$$
$$\le \frac{C}{r_n^2} \frac{r_n}{r_{n_0}^4} \tilde{K}_{n_0}^{0\,4/5} \Big(\int_{B_n^0} |\underline{u}|^2 \Big)^{1/5} \Big(\int_{Q_n^0} |\underline{u}|^3 \Big)^{1/5} r_n^{9/5} \le$$
$$\le \frac{C}{r_n^2} \frac{r_n}{r_{n_0}^4} r_n^{12/5} \tilde{K}_{n_0}^{0\,4/5} \tilde{A}_n^{0\,1/5} \tilde{G}_n^{0\,1/5} \le C \Big(\frac{r_n}{r_{n_0}} \Big)^{7/5} A_n^{0\,1/5} G_n^{0\,1/5} K_{n_0}^{0\,4/5}$$

Finally use the inequalities of (5) and combine the estimates above on the terms p_j, $j = 1, .., 4$.)

(10) Let $T_n = (A_n^0 + \delta_n^0)$; combine inequalities of (6) and (9), and (5) to deduce

$$T_n \le 2^{2n} \Big(2^{-2n_0} \varepsilon + \sum_{k=n+1}^{n_0-1} 2^{-2k} T_k^{3/2} + 2^{-2n_0} \varepsilon +$$
$$+ 2^{-7n_0/5} \varepsilon \sum_{k=n+1}^{n_0-1} 2^{-3k/5} T_k^{1/2} + \varepsilon 2^{-5n_0/3} \sum_{k=n_0+2}^{n_0-1} 2^{-k/3} T_k^{1/2} +$$
$$\sum_{k=n+1}^{n_0-1} 2^{-2k} T_k^{3/2} + \sum_{k=n+1}^{n_0-1} 2^k T_k^{1/2} \sum_{p=k}^{n_0} 2^{-3p} T_p \Big)$$
$$\varepsilon \equiv C \max(G_{n_0}^{0\,2/3}, K_{n_0}^{0\,4/5}, J_{n_0}^0)$$

and show that, by induction, if ε is small enough then $r_n^{-2} T_n \le \varepsilon^{2/3} r_{n_0}^{-2}$.

(11) If $G(r_0) + J(r_0) + K(r_0) < \varepsilon_s$ with ε_s small enough, then given $(\underline{x}', t') \in Q_{r_0/4}(\underline{x}_0, t_0)$, show that if one calls $G_r^0, J_r^0, K_r^0, A_r^0, \delta_r^0$ the operators associated with $Q_r^0(\underline{x}', t')$ then

$$\limsup_{n \to \infty} \frac{1}{r_n^2} A_n^0 \leq C \frac{\varepsilon_s^{2/3}}{r_0^2}$$

for a suitable constant C. (*Idea* Note that $Q_{r_0/4}^0(\underline{x}', t') \subset Q_{r_0}(\underline{x}_0, t_0)$ hence $G_{r_0/4}^0, J_{r_0/4}^0, \ldots$ are bounded by a constant, $(\leq 4^2)$, times $G(r_0), J(r_0) \ldots$ respectively. Then apply the result of (10)).

(12) Check that the result of (11) implies theorem II. (*Idea* Indeed

$$\frac{1}{r_n^2} A_n^0 \geq \frac{1}{r_n^3} \int_{B_n^0} |\underline{u}(\underline{x}, t')|^2 d\underline{x} \xrightarrow[n \to -\infty]{} \frac{4\pi}{3} |\underline{u}(\underline{x}', t')|^2$$

where B_n^0 is the ball centered at $\underline{x}'$, *for almost all* the points $(\underline{x},' , t') \in Q_{r_0/4}^0$; hence $|\underline{u}(\underline{x}', t')|$ is bounded in $Q_{r_0/4}^0$ and one can apply proposition 2).

(13) Let f be a function with zero average over B_r^0. Since $f(\underline{x}) = f(\underline{y}) + \int_0^1 ds \, \underline{\partial} f(\underline{y} + (\underline{x} - \underline{y})s) \cdot (\underline{x} - \underline{y})$ for each $\underline{y} \in B_r^0$, averaging this identity over $\underline{y}$ one gets

$$f(\underline{x}) = \int_{B_r^0} \frac{d\underline{y}}{|B_r^0|} \int_0^1 ds \, \underline{\partial} f(\underline{y} + (\underline{x} - \underline{y})s) \cdot (\underline{x} - \underline{y})$$

Assuming $\alpha = 1$ prove (P). (*Idea* Change variables as $\underline{y} \to \underline{z} = \underline{y} + (\underline{x} - \underline{y})s$ so that for α integer

$$\int_{B_r^0} |f(\underline{x})|^\alpha \frac{d\underline{x}}{|B_r^0|} \equiv \int_{B_r^0} \frac{d\underline{x}}{|B_r^0|} \Big| \int_0^1 \int_{B_r^0} \frac{d\underline{z}}{|B_r^0|} \frac{ds}{(1-s)^3} \underline{\partial} f(\underline{z}) \cdot (\underline{z} - \underline{x}) \Big|^\alpha$$

where the integration domain of $\underline{z}$ depends from $\underline{x}$ and s, and it is contained in the ball with radius $2(1 - s)r$ around $\underline{x}$. The integral can then be bounded by

$$\int \frac{d\underline{z}_1}{|B_r^0|} \frac{ds_1}{1 - s_1} \cdots \frac{d\underline{z}_\alpha}{|B_r^0|} \frac{ds_\alpha}{1 - s_\alpha} (2r)^\alpha |\underline{\partial} f(\underline{z}_1)| \ldots |\underline{\partial} f(\underline{z}_\alpha)| \int \frac{d\underline{x}}{|B_r^0|}$$

where $\underline{x}$ varies in a domain with $|\underline{x} - \underline{z}_i| \leq 2(1 - s_i)r$ for each i. Hence the integral over $\frac{d\underline{x}}{|B_r^0|}$ is bounded by $8(1 - s_i)^3$ for each i. Performing a geometric average of such bounds (over α terms)

$$\int_{B_r^0} |f(\underline{x})|^\alpha \frac{d\underline{x}}{|B_r^0|} \leq 2^{\alpha+3} r^\alpha \prod_{i=1}^\alpha \int \frac{d\underline{z}_i \, ds_i}{|B_r^0|(1-s_i)} ||\underline{\partial} f(\underline{z}_i)| (1 - s_i)^{3/\alpha} \leq$$
$$\leq 2^{\alpha+3} r^\alpha \Big(\int_{B_r^0} |\underline{\partial} f(\underline{z})| \frac{d\underline{z}}{|B_r^0|} \Big)^\alpha \cdot \Big(\int_0^1 \frac{ds}{(1-s)^{3-3/\alpha}} \Big)^\alpha$$

getting (P) and an explicit estimate of the constant C_α^P only for $\alpha = 1$ and a hint that (P) should hold for $\alpha < \frac{3}{2}$ at least.)

(14) Differentiate twice with respect to α^{-1} and check the convexity of $\alpha^{-1} \to ||f||_\alpha \equiv (\int |f(\underline{x})|^\alpha \, d\underline{x}/|B_r^0|)^{1/\alpha}$. Use this to get (P) for each $1 \le \alpha < \alpha_0$ if it holds for $\alpha = \alpha_0$. (*Idea* Since (P) can be written $||f||_\alpha \le C_\alpha (\int |\partial f| \, d\underline{x}/r^2)$ then if $\alpha^{-1} = \vartheta \alpha_0^{-1} + (1-\vartheta)$ it follows that C_α can be taken $C_\alpha = \vartheta C_{\alpha_0} + (1-\vartheta) C_1$).

(15) Consider a sequence $\underline{u}^\lambda$ of solutions of the Leray regularized equations which converges *weakly* (*i.e.* for each Fourier component) to a Leray solution. By construction the $\underline{u}^\lambda, \underline{u}$ verify the *a priori* bounds in (0.2) and (hence) (2.5). Deduce that $\underline{u}$ verifies the (3.1). (*Idea* Only (c) has to be proved. Note that if $\underline{u}^\lambda \to \underline{u}^0$ weakly, then the left hand side of (1.1) is semi continuous hence the value computed with $\underline{u}^0$ is not larger than the limit of the right hand side in (1.1). On the other hand the right hand side of (1.1) is *continuous* in the limit $\lambda \to \infty$. Indeed given $N > 0$ weak convergence implies

$$\lim\nolimits_{\lambda\to\infty} \int_0^{T_0} dt \int_\Omega |\underline{u}^\lambda - \underline{u}^0|^2 \, d\underline{x} \equiv \lim\nolimits_{\lambda\to\infty} \int_0^{T_0} dt \sum_{0<|\underline{k}|} |\gamma_{\underline{k}}^\lambda(t) - \gamma_{\underline{k}}^0(t)|^2 \le$$

$$\le \lim\nolimits_{\lambda\to\infty} \left(\sum_{0<|\underline{k}|<N} \int_0^{T_0} dt |\gamma_{\underline{k}}^\lambda(t) - \gamma_{\underline{k}}^0(t)|^2 + \sum_{|\underline{k}|\ge N} \int_0^{T_0} dt \frac{|\underline{k}|^2}{N^2} |\gamma_{\underline{k}}^\lambda(t) - \gamma_{\underline{k}}^0(t)|^2 \right) \le$$

$$\le \lim\nolimits_{\lambda\to\infty} \left(\sum_{0<|\underline{k}|<N} \int_0^{T_0} dt |\gamma_{\underline{k}}^\lambda(t) - \gamma_{\underline{k}}^0(t)|^2 + \frac{1}{N^2} \int_0^{T_0} dt \int_\Omega |\partial(\underline{u}^\lambda - \underline{u}^0)|^2 \right) =$$

$$= \lim\nolimits_{\lambda\to\infty} \frac{1}{N^2} \int_0^{T_0} dt \int_\Omega |\partial(\underline{u}^\lambda - \underline{u}^0)|^2 \le \frac{2E_0\nu^{-1}}{N^2}$$

using the *a priori* bound in (0.2) (with zero force) and componentwise convergence of the Fourier transform $\gamma_{\underline{k}}(t)$ of $\underline{u}(t)$ to the Fourier transform $\gamma_{\underline{k}}^0(t)$ of $\underline{u}^0$. Hence $\int_0^{T_0} \int_\Omega |\underline{u}^\lambda - \underline{u}^0|^2 \to 0$ showing the convergence of the first two terms of the right hand side of (1.1) to the corresponding terms of (c) in (3.1).

Apply, next, the inequality (S), (2.2), with $q = 3$, $a = \frac{3}{4}$, $\frac{q}{2} - a = \frac{3}{4}$, and again by the *a priori* bounds in (0.2) we get

$$\int_0^{T_0} dt \int_\Omega |\underline{u}^\lambda - \underline{u}^0|^3 \, d\underline{x} \le C \int_0^{T_0} dt \, ||\partial(\underline{u}^\lambda - \underline{u}^0)||_2^{3/2} \, ||\underline{u}^\lambda - \underline{u}^0||_2^{3/2} \le$$

$$\le C \left(\int_0^{T_0} dt \, ||\partial(\underline{u}^\lambda - \underline{u}^0)||_2^2 \right)^{3/4} \left(\int_0^{T_0} dt \, ||\underline{u}^\lambda - \underline{u}^0||_2^6 \right)^{1/4} \le$$

$$\le C(2E_0\nu^{-1})^{3/4}(2\sqrt{E_0}) \int_0^{T_0} dt \, ||\underline{u}^\lambda - \underline{u}^0||_2^2 \xrightarrow[\lambda\to\infty]{} 0$$

showing continuity of the third term in the second member of (3.1). Finally, and analogously, if we recall that $p^\lambda = -\Delta^{-1} \sum_{ij} \partial_i \partial_j (u_i^\lambda u_j^\lambda)$ and if we apply the inequalities (CZ) and (H), we get

$$\int_0^{T_0} dt \int_\Omega d\underline{x} |p^\lambda \underline{u}^\lambda - p^0 \underline{u}^0| \le \int \int |p^\lambda - p^0| \, |\underline{u}^\lambda| + \int \int |p^0| \, |\underline{u}^\lambda - \underline{u}^0| \le$$

$$\left(\int \int |p^\lambda - p^0|^{3/2} \right)^{2/3} \left(\int \int |\underline{u}^\lambda|^3 \right)^{1/3} + \left(\int \int |p^0|^{3/2} \right)^{2/3} \left(\int \int |\underline{u}^\lambda - \underline{u}^0|^3 \right)^{1/3}$$

where the last integral tends to zero by the previous relation while the first, via (CZ), will be such that $\int_0^{T_0} \int_\Omega |p^\lambda - p^0|^{3/2} \le \left(\int \int |\underline{u}^\lambda - \underline{u}^0|^3 \right)^{2/3} \xrightarrow[\lambda\to\infty]{} 0$ proving the continuity of the fourth term in the right hand side of (c) in (3.1). Hence the right hand side is continuous in the considered limit).

(16) *(covering theorem,* (Vitali)) Let S be an arbitrary set inside a sphere of R^n. Consider a *covering* of S with little open balls with the *Vitali property:* *i.e.* such that every point of S is contained in a family of open balls of the covering whose radii have a zero greatest lower bound. Given $\eta > 0$ show that if $\lambda > 1$ is large enough it is possible to find a denumerable family $F_1, F_2, \ldots$ of pairwise disjoint balls of the covering with diameter $< \eta$ such that $\cup_i \lambda F_i \supset S$ where λF_i denotes the ball with the same center of F_i and radius λ times longer. Furthermore λ can be chosen independent of S, see also problem {17}. *(Idea* Let $\mathcal{F}$ be the covering and let $a = max_{\mathcal{F}} \, diam(F)$. Define $a_k = a2^{-k}$ and let $\mathcal{F}_1$ be a *maximal* family of *pairwise disjoint* ball of $\mathcal{F}$ with radii $\geq a2^{-1}$ and $< a$. Likewise let $\mathcal{F}_2$ be a maximal set of balls of $\mathcal{F}$ with radii between $a2^{-2}$ and $a2^{-1}$ pairwise disjoint between themselves and with the ones of the family $\mathcal{F}_1$. Inductively we define $\mathcal{F}_1, \ldots, \mathcal{F}_k, \ldots$. It is now important to note that if $x \notin \cup_k \mathcal{F}_k$ it must be: $distance(x, \mathcal{F}_k) < \lambda a2^{-k}$ for some k, if λ is large enough. If indeed δ is the radius of a ball S_δ containing x and if $a2^{-k_0} \leq \delta < a2^{-k_0+1}$ then the point of S_δ farthest away from x is at most at distance $\leq 2\delta < 4a2^{-k_0}$; and if, therefore, it was $d(x, \mathcal{F}_{k_0}) \geq 4a2^{-k_0}$ we would find that the set $\mathcal{F}_{k_0}$ could be made larger by adding to it S_δ, against the maximality supposed for $\mathcal{F}_{k_0}$. Note that $\lambda = 5$ is a possible choice.)

(17) Show that if the balls in problem {16} are replaced by the *parabolic cylinders* which are Cartesian products of a radius r ball in the first k coordinates and one of radius r^α, with $\alpha \geq 1$ in the $n - k$ remaining ones, then the result still holds if one replaces $5F_i$ with λF_i where λ is a suitable homothety factor (with respect to the center of F_i). Show that if $\alpha = 1, 2$ then $\lambda = 5$ is enough (and, in general, $\lambda = (4^2 + 2^{2(1+\alpha)/\alpha})^{1/2}$ is enough).

(18) Check that the Hausdorff dimension of the Cantor set C is $\log_3 2$, *c.f.r.* (5.3). *(Idea* It remains to see, given the equation in footnote[9], that if $\alpha < \alpha_0$ then $\mu_\alpha(C) = \infty$. If $\delta = 3^{-n}$ the covering $\mathcal{C}_n$ of C with the n–th generation intervals is "the best" among those with sets of diameter $\leq 3^{-n}$ because another covering could be refined by deleting from each if its intervals the points that are out of the n–th generation intervals. Furthermore the inequality $1 < 23^{-\alpha}$ for $\alpha < \log_3 2$ shows that it will not be convenient to further subdivide the intervals of $\mathcal{C}_n$ for the purpose of diminishing the sum $\sum |F_i|^\alpha$. Hence for $\delta = 3^{-n}$ the minimum value of the sum is $2^n 3^{-n\alpha} \xrightarrow[n\to\infty]{} \infty$.)

References

1. Benfatto, G., Gallavotti, G.: *enormali ation roup*, p. 1–144, Princeton University Press, 1995.
2. Caffarelli, L., Kohn, R., Nirenberg, L : *Partial regularity of suitable weak solutions of the Navier- Stokes equations*, Communications on pure and applied mathematics, 35, 771–831, 1982.

3. Constantin, P., Foias, C.: *Navier Stokes Equations*, Chicago Lectures in Mathematics series, University of Chicago Press, 1988.
4. Dunford, N., Schwartz, I.: *Linear operators*, Interscience, 1960.
5. Gallavotti, G.: *Foundations of Fluid Mechanics*, p. 1–513, Springer-Verlag, Berlin, 2003.
6. Lieb, E., Loss, M.: *Analysis*, American Mathematical Society, Providence, 2001.
7. Pumir, A., Siggia, E.: *Vortex dynamics and the existence of solutions to the Navier Stokes equations*, Physics of Fluids, 30, 1606–1626, 1987.
8. Scheffer, V.: *Hausdorff dimension and the Navier Stokes equations*, Communications in Mathematical Physics, 55, 97–112, 1977. And *Boundary regularity for the Navier Sokes equation in half space*, Communications in Mathematical Physics, 85, 275–299, 1982.
9. Sobolev, S.L.: *Applications of Functional analysis in Mathematical Physics*, Translations of the American Mathematical Society, vol 7, 1963, Providence.
10. Stein, E.: *Harmonic analysis*, Princeton University press, 1993.

Approximation of Weak Limits and Related Problems

Alexandre V. Kazhikhov

Lavrentyev Institute of Hydrodynamics
630090, Novosibirsk, Russia
kazhikhov hydro.nsc.ru

Mathematical Subject Classification (1991): 35Q30, 35Q35
Key words: Weak convergence, strong approximation, Orlicz function spaces, compactness, compensated compactness.

Preface

This lecture course is concerned with some mathematical problems originated from the theory of compressible Navier-Stokes equations (cf.[9],[15],[16]).

The lecture notes consist of three sections. We discuss the problem of strong approximation of weak limits in section I and prove, firstly, that weak limit of some sequence of functions in Orlicz space can be approximated in strong sense (in norm) by the subsequence of averaged functions if the radius of averaging tends to zero slowly enough. This result allows us to control the order of approximation by weakly converging sequence. In particular, it justifies the smoothing approach near singularities in computation of non-smooth solutions to partial differential equations. Secondly, we consider the weak converging sequence of approximate solutions to averaged Navier-Stokes equations for incompressible fluids and obtain strong convergence to the solution of the limiting equations.

The section II contents the recent results [23],[25] in theory of transport equations in Orlicz spaces. We introduce special class of convex functions (Young functions) and corresponding Orlicz spaces. It allows us to obtain exact well-posedness (existence and uniqueness) results for linear transport equations in Orlicz spaces and describe the optimal conditions for coefficients. It is worthy to be mentioned that this class of Young functions (of fast growth at infinity, greater than any polynomial) is connected with Gronwall-type inequality and Osgood's uniqueness theorem (cf.[26]) for Cauchy problem in ordinary differential equations theory. Namely, we obtain optimal conditions for coefficients in Gronwall inequality. At the same time it gives the relations between Orlicz-Sobolev spaces and Osgood's condition in uniqueness problem.

The section III concerns with some problems related to compactness arguments. Besides the classical compactness (cf.[1],[2],[3]) so called compensated compactness ([4]-[7]) is also under consideration. We expose new version of classic compactness theorem which, in fact, is a particular case of compensated compactness. On the other side, we give a new and very simple proof of "div-curl" lemma (the mostly important tool in applications of compensated compactness theory to nonlinear P.D.E's.) which reduces the compensated compactness to the current one. Finally, a new viewpoint on the general compensated compactness theorem (Theorem of L.Tartar) is suggested: vanishing of quadratic form on the kernel of operator instead of algebraic conditions. Such approach, perhaps, can be more convenient in some cases when it is possible to describe the kernel of differential operator.

The lecture course was given to participants of Summer School "Mathematical Foundations of Turbulent Viscous Flows" organized by C.I.M.E. Author would like to express a deep gratitude to School Scientific Directors Prof. Marco Cannone and Prof. Tetsuro Miyakawa as well as C.I.M.E. officials Direttore Prof. Pietro Zecca, Segretario Prof. Elvira Mascolo and School Secretary Dr. Veronika Sustik.

1 Strong Approximation of Weak Limits by Averagings

1.1 Notations and Basic Notions from Orlicz Function Spaces Theory

Let $\Omega \subset \mathbb{R}^n$ be bounded domain with smooth boundary Γ, and $\mathbf{x} = (x_1, \ldots, x_n)$ are the points of Ω. By $L^1(\Omega)$ we denote the space of integrable functions on Ω. $L^\infty(\Omega)$ is the space of essentially bounded functions, $L^p(\Omega)$, $1 < p < \infty$ – the Lebesgue space of functions which are integrable in power p. We shall use also the Orlicz function spaces, and remind the basic notions (cf. [18]).

Let m(r) be defined on $[0, \infty)$ function, continuous from the right, non-negative, non-decreasing and such that

$$m(0) = 0, \quad m(r) \to \infty \text{ as } r \to \infty. \tag{1.1}$$

The convex function (Young function)

$$M(t) = \int_0^t m(r)dr \tag{1.2}$$

produces Orlicz class $K_M(\Omega)$ containing the functions $f(\mathbf{x}) \in L^1(\Omega)$ such that $M(|f(\mathbf{x})|)$ belong to $L^1(\Omega)$, too. The liner span of $K_M(\Omega)$ endowed with the norm

$$\|f\|_{L_M(\Omega)} = \inf \left\{ \lambda > 0 \,\middle|\, \int_\Omega M\left(\frac{|f(\mathbf{x})|}{\lambda}\right) d\mathbf{x} \leq 1 \right\} \tag{1.3}$$

is called as Orlicz space $L_M(\Omega)$ associated with Young function $M(t)$. The closure of $L^\infty(\Omega)$ in the norm (1.3) yields, in general, another Orlicz space $E_M(\Omega)$, and the inclusions take place

$$E_M(\Omega) \quad K_M(\Omega) \quad L_M(\Omega). \tag{1.4}$$

One says function $M(t)$ satisfies $\Delta_2-condition$ (cf.[18]) if there exist constants $C > 0$ and $t_0 > 0$ such that

$$M(2t) \leq C\,M(t) \quad \text{for } \forall t \geq t_0. \tag{1.5}$$

Three sets E_M, K_M and L_M coincide if and only if $M(t)$ satisfies Δ_2-condition. As rule, it's possible to compare two Young functions $M_1(t)$ and $M_2(t)$, namely, $M_2(t)$ dominates $M_1(t)$ if there exist constants $C > 0$ and $\alpha > 0$ such that

$$M_1(\alpha t) \leq M_2(t) \quad \text{for } \forall\, t \geq t_0 = t_0(\alpha, C). \tag{1.6}$$

In this case

$$E_{M_2} \quad E_{M_1} \text{ and } L_{M_2} \quad L_{M_1}.$$

If each function M_k, $k = 1, 2$, dominates the other one, then M_1 and M_2 are equivalent, $M_1 \cong M_2$, and corresponding Orlicz spaces are the same. Further, one says function M_2 dominates M_1 essentially if

$$\lim_{t \to \infty} \frac{M_1(\beta t)}{M_2(t)} = 0, \quad \forall \beta = \text{const} > 0. \tag{1.7}$$

In this case the strong embeddings

$$E_{M_2} \subset E_{M_1}, \quad L_{M_2} \subset L_{M_1} \tag{1.8}$$

are valid.

Denote by

$$n(r) = m^{-1}(r) \tag{1.9}$$

the inverse function to $m(r)$ and introduce Young function

$$N(t) = \int_0^t n(r)dr. \tag{1.10}$$

This function is called as complementary convex function to $M(t)$, and it's equivalent to the next one:

$$N(t) \cong \sup_{r>0} \{tr - M(r)\}. \tag{1.11}$$

Two Orlicz spaces L_M and L_N are supplementary, and for any $f(\mathbf{x}) \in L_M(\Omega)$, $g(\mathbf{x}) \in L_N(\Omega)$ there exists the integral

$$< f, g > \equiv \int_\Omega f(\mathbf{x})g(\mathbf{x})d\mathbf{x} \tag{1.12}$$

which defines the linear continuous functional on $E_N(\Omega)$ with fixed $f \in L_M(\Omega)$ and any $g \in E_M(\Omega)$. It gives the notion of weak convergence (actually, weak* convergence) in $L_M(\Omega)$: the sequence $\{f_n(\mathbf{x})\}$converges weakly to $f(\mathbf{x}) \in L_M(\Omega)$, $f_n \rightharpoonup f$ if

$$< f_n, g > \to < f, g > \quad \text{for each } g \in E_N(\Omega). \tag{1.13}$$

At the same time it's possible to define mean convergence in $L_M(\Omega)$:

$$f_n \to f \quad \text{in mean value, if}$$

$$\int_\Omega M(|f_n - f|)d\mathbf{x} \to 0 \quad \text{as } n \to \infty. \tag{1.14}$$

If $M(t)$ satisfies Δ_2-condition, mean 2convergence is equivalent to the strong-convergence, i.e. in norm of $L_M(\Omega)$. Otherwise, mean convergence is stronger than weak, but weaker than strong one. As an important and interesting examples we shall use three Young functions: $M_1(t) = t^p, 1 < p < \infty$, $M_2(t) = e^t - t - 1$ and $M_3(t) = (1 + t)\ln(1 + t) - t$. The first function $M_1(t)$ yields the Lebesgue space $L^p(\Omega)$, the second one $M_2(t)$ produces two Orlicz spaces L_{M_2} and E_{M_2} because $M_2(t)$ doesn't satisfy Δ_2-condition, and $M_3(t)$ is slowly increasing function which is essentially dominated by$M_1(t)$. The Orlicz space $L_{M_3}(\Omega)$ is located between $L^1(\Omega)$ and any $L^p(\Omega)$, $p > 1$.

Finally, for any $f(\mathbf{x}) \in L^1(\Omega)$ and $h > 0$, let us denote

$$f_h(\mathbf{x}) = \frac{1}{2}h^N \int_\Omega f(\mathbf{y})\omega\left(\frac{\mathbf{x} - \mathbf{y}}{h}\right) d\mathbf{y} \tag{1.15}$$

an averaging of $f(\mathbf{x})$ where $\omega(\mathbf{z})$ is the kernel of averaging:

$$\omega(\mathbf{z}) \in C_0^\infty(\mathbb{R}^N), \quad \omega(\mathbf{z}) \geq 0, \quad \int_{R^N} \omega(\mathbf{z})d\mathbf{z} = 1.$$

Formula (1.15) often is called as the mollification of function f and, in fact, it is the convolution of f with the mollifier $\frac{1}{h^N}\omega(\mathbf{z}/h)$.
It's the well-known fact that

$$\|f_h - f\|_{L_M(\Omega)} \to 0 \quad \text{as } h \to 0 \tag{1.16}$$

if $M(t)$ satisfies Δ_2-condition, and

$$\int_\Omega M(|f_h - f|)dx \to \quad \text{as } h \to 0 \tag{1.17}$$

for any $M(t)$, i.e. the sequence of $\{f_h\}$ approximates f in the sense of mean convergence.

1.2 Strong Approximation of eak Limits

Let us consider some sequence of functions $\{f_n(\mathbf{x})\}$, $n = 1, 2, \ldots$, from Orlicz space $L_M(\Omega)$ such that $f_n \rightharpoonup f$ weakly in $L_M(\Omega)$, as $n \to \infty$. For each $f_n(\mathbf{x})$ we construct the family of averaged functions $(f_n)_h(\mathbf{x})$.

Theorem 1.1 If $M(t)$ satisfies Δ_2-condition then there exists subsequence $(f_m)_{h_m}$ such that

$$(f_m)_{h_m} \to f \quad \text{strongly in } L_M(\Omega)$$

$$\text{as } m \to \infty, \ h \to 0.$$

For any $M(t)$ there exists subsequence $(f_m)_{h_m}$ converging to f in mean value, i.e. in the sense (1.14).

Proof. Step 1. Simple example.

In order to understand the problem we consider firstly, one very simple example of the sequence $f_n(x) = \sin(nx)$, $n = 1, 2, \ldots$, $x \in \mathbb{R}^1$, $\Omega = (0, 2\pi)$, of periodic functions. In this case we have

$$f_n(x) \rightharpoonup 0 \text{ weakly in } L^2(0, 2\pi).$$

Let us take the Steklov averaging

$$(f_n)_h(x) = \frac{1}{2h} \int_{x-h}^{x+h} f_n(\xi)d\xi. \tag{1.18}$$

It's easy to calculate

$$(f_n)_h(x) = \frac{\sin nh}{nh} \sin(nx). \tag{1.19}$$

and thereby to obtain:

  a) If $h = h_n \to 0$ as $n \to \infty$, but $nh_n \to \infty$ (for example, $h_n = n^{-\alpha}$, $0 < \alpha < 1$) then

$$(f_n)_{h_n} \to 0 \text{ as } n \to \infty,$$

  i.e. $(f_n)_{h_n}(x)$ tends to 0 strongly when $h_n \to 0$ slowly enough.
  b) If $nh_n \to const \neq 0$ (or are bounded), then $(f_n)_{h_n} \rightharpoonup 0$ weakly only.

Step 2. One-dimensional case, Steklov averaging.

Now consider the case of $N = 1$, $x \in \mathbb{R}^1$, and $f_n(x) \rightharpoonup f(x)$ weakly in $L_M(\Omega)$, where $M(t)$ satisfies Δ_2-condition.

For the sake of simplicity we assume $f_n(x)$, $f(x)$ to be T-periodic, $T = $ const > 0, and, moreover, without loss of generality, one can admit $f(x) \equiv 0$, since it is possible to consider the sequence of differences $f_n - f$ instead of f_n, i.e.

$$f_n(x) \rightharpoonup 0 \text{ weakly in } L_M(0, T). \tag{1.20}$$

Let us construct the family of functions

$$(f_n)_h(x) = \frac{1}{2h} \int\limits_{x-h}^{x+h} f_n(\xi)\xi \tag{1.21}$$

and the sequence

$$F_n(x) = \int\limits_{0}^{x} f_n(\xi)d\xi \tag{1.22}$$

which doesn't depend on h.

It means

$$(f_n)_h(x) = \frac{1}{2h}(F_n(x + h) - F_n(x - h)). \tag{1.23}$$

The sequence $\{F_n(x)\}$ possesses the estimates

$$\sup_x |F_n(x)| \le C, \quad \|F_n'()x\|_{L_M(0,T)} \le C \tag{1.24}$$

with constant C independent on n.

Compactness theorem yields $F_n(x) \to 0$ strongly in $L_M(0, T)$, i.e.

$$\|F_n(x)\|_{L_M(0,T)} \le C_n, \quad C_n \to 0 \text{ as } n \to \infty. \tag{1.25}$$

If we take $F_n(x + h)$ or $F_n(x - h)$ (displacements of $F_n(x)$) then

$$\|F_n(x + h)\|_{L_M} \le C_n \cdot C, \quad \|F_n(x - h)\|_{L_M} \le C_n \cdot C \tag{1.26}$$

with C independent on $h, h \in (01]$.

It gives

$$\|(f_n)_h\|_{L_M(0,T)} \le C \cdot \frac{C_n}{h}. \tag{1.27}$$

So, if we take $h = h_n$ such that $C_n \cdot h_n^{-1} \to 0$ as $n \to \infty$, for example, $h_n = C_n^\beta, 0 < \beta = const < 1$, then

$$(f_n)_{h_n} \to 0 \text{ strongly in } L_M(0, T) \tag{1.28}$$

It proves the theorem 1 in 1-dimensional case.

Step 3. The general case.

Let the sequence $\{f_n(\mathbf{x})\}$, $\mathbf{x} \in \Omega \subset \mathbb{R}^N$, be weakly converging in $L_M(\Omega)$ to $f(\mathbf{x}) \equiv 0$, where $M(t)$ satisfies Δ_2-condition.

We extend $f_n(\mathbf{x})$ by 0 outside of Ω and consider the family $(f_n)_h(\mathbf{x})$ given by the formula (1.15)

$$(f_n)_h(\mathbf{x}) = \frac{1}{h^n} \int\limits_{\mathbb{R}^N} f_n(\mathbf{y})\omega\left(\frac{\mathbf{x}-\mathbf{y}}{h}\right) d\mathbf{y} \qquad (1.29)$$

with arbitrary kernel of averaging $\omega(\mathbf{z})$.

At the beginning we fix some $h = h_0$, for example $h_0 = 1$, and consider the sequence

$$F_n(\mathbf{x}) = \int\limits_{\mathbb{R}^N} f_n(\mathbf{y})\omega(\mathbf{x}-\mathbf{y})d\mathbf{y} \equiv \int\limits_{\mathbb{R}^N} f_n(\mathbf{x}-\mathbf{z})\omega(\mathbf{z})d\mathbf{z}$$

which doesn't depend on h.

As in one-dimensional case we conclude by the compactness theorem

$$F_m(\mathbf{x}) \to 0 \text{ strongly in } L_M(\Omega),$$

i.e. $\|F_m\|_{L_M(\Omega)} \leq C_m$, $C_m \to 0$ as $m \to \infty$. If we take the family

$$F_{m\,h}(\mathbf{x}) \equiv \int\limits_{\mathbb{R}^N} f_m(\mathbf{y})\omega\left(\frac{\mathbf{x}-\mathbf{y}}{h}\right) d\mathbf{y}$$

then for $h \leq h_0 = 1$

$$\|F_{mh}\|_{L_M(\Omega)} \leq C\|F_m\|_{L_M(\Omega)}$$

with constant C independent on h, i.e.

$$\|F_{mh}\|_{L_M(\Omega)} \leq C \cdot C_m.$$

In view of $(f_n)_h = h^{-N}F_{nh}$ it means that if we choose $h = h_m$ such that

$$C_m h_m^{-N} \to 0 \text{ as } m \to \infty, \; h_m \to 0, \qquad (1.30)$$

for instance, $h_m^N = C_m^\beta$, $0 < \beta < 1$, then $(f_m)_{h_m} \to 0$ strongly in $L_M(\Omega)$.

The theorem 1.1 is proved for the case of Young function $M(t)$ satisfying Δ_2-condition. For any $M(t)$ the same proof is acceptable if one considers the mean convergence (1.14) instead of strong or strong convergence in any Banach function space where the set of smooth functions is dense, for example, in the space $E_M(\Omega)$.

Remark 1.1

The main significance of theorem 1.1 seems to be useful for justification of the smoothing approach in computation of non-smooth solutions to P.D.E's. Big oscillations occur near singularities. The appearance of oscillations can be connected with weak convergence of approximate solutions to the exact one. The procedure of "smoothing" means, in fact, an averaging, and the theorem 1.1 indicates that the radius of averaging must be big enough according to assumption (1.30).

1.3 Applications to Navier-Stokes Equations

It this section we illustrate the theorem 1.1 by one simple example of Navier-Stokes equations for viscous incompressible fluid. We consider the sequence $\{\mathbf{u}_n\}$ of solutions to Navier-Stokes equations [20]:

$$\frac{\partial \mathbf{u}_n}{\partial t} + (\mathbf{u}_n \cdot \nabla)\mathbf{u}_n + \nabla p_n = \nu \Delta \mathbf{u}_n + \mathbf{f}_n,$$
$$\operatorname{div}\mathbf{u}_n = 0, \quad (\mathbf{x}, t) \in Q = \Omega \times (0, T), \quad \Omega \subset \mathbb{R}^3, \quad \Gamma = \partial \Omega. \tag{1.31}$$

which are complemented with the initial and boundary data

$$\mathbf{u}_n\bigg|_{t=0} = \mathbf{u}_n^0(\mathbf{x}), \quad \mathbf{x} \in \Omega, \; \mathbf{u}_n\bigg|_{\Gamma} = 0, \quad t \in (0, T) \tag{1.32}$$

Let us suppose

$$\mathbf{u}_n^0(\mathbf{x}) \rightharpoonup \mathbf{u}^0(\mathbf{x}) \text{ weakly in } L^2(\Omega),$$

$$\mathbf{f}_n \rightharpoonup \mathbf{f} \text{ weakly in } \; L^2(0, T; L^2(\Omega)) \quad as \quad n \to \infty.$$

In view of well-known a priori estimate

$$\sup_{0 < t < T} \|\mathbf{u}_n(t)\|_{L^2(\Omega)} + \int_0^T \|\nabla \mathbf{u}_n(t)\|_{L^2(\Omega)}^2 dt \le C \tag{1.33}$$

with constant C independent on n, we may admit

$$\mathbf{u}_n \rightharpoonup \mathbf{u} \text{ weakly in } L^2(0, T; W^{1,2}(\Omega)) \cap L^p(0, T; L^2(\Omega)) \tag{1.34}$$

with any p, $1 \le p < \infty$.

According to the theorem 1.1 it's possible to extract subsequence $\{(\mathbf{u}_m)_{h_m}\}$ of averaged functions (with respect all independent variables or to spatial variables only) such that

$$(\mathbf{u}_m)_{h_m} \to \mathbf{u} \quad \text{strongly in } L^2(0, T; W^{1,2}(\Omega)) \cap L^p(0, T; L^2(\Omega)),$$

$$(\mathbf{f}_m)_{h_m} \to \mathbf{f} \text{ strongly in} L^2(0, T; L^2(\Omega)).$$

And the question arises here: are $\{(\mathbf{u}_m)_{h_m}\}$ approximate solutionsto the limiting equations

$$\frac{\partial \mathbf{u}}{\partial t} + (\mathbf{u} \cdot \nabla)\mathbf{u} + \nabla p = \nu \Delta \mathbf{u} + \mathbf{f}, \quad \mathrm{div}\,\mathbf{u} = 0 \qquad (1.35)$$

in some strong sense?

To give answer to this question we apply the operator of averaging to equations (1.31) and obtain the system (1.35) for $(\mathbf{u})_{h_m}$ with new right part

$$\mathbf{F}_m = (\mathbf{f}_m)_{h_m} + [((\mathbf{u}_m)_{h_m} \cdot \nabla)(\mathbf{u}_m)_{h_m} - ((\mathbf{u}_m \cdot \nabla)\mathbf{u}_m)_{h_m}] \equiv$$

$$(\mathbf{f}_m)_{h_m} + \varphi_m.$$

Theorem 1.2 There exists subsequence $(\mathbf{u}_m)_{h_m}$ such that the difference φ_m tends to zero in the norm of space $L^p(0,T;L^q(\Omega))$ with exponents (p,q), $p \in [1,2]$, $q \in [1,3/2]$, $1/p + 3/2q > 2$.

The proof follows from theorem 1.1 because the sequence $\{(\mathbf{u}_n \cdot \nabla)\mathbf{u}_n\}$ is bounded in $L^p(0,T;L^q(\Omega))$ with such exponents (cf.[20]).

2 Transport Equations in Orlicz Spaces

2.1 Statement of Problem

Here we are concerned with the spatially periodic Cauchy problem for the linear first-order equation (such as the well-known continuity equation in fluid mechanics)

$$\frac{\partial \rho}{\partial t} + \mathrm{div}(\rho \mathbf{u}) = h, \qquad \mathbf{x} = (x_1, \dots, x_n) \in \mathbb{R}^n, \quad t \in (0,T) \qquad (2.1)$$

complemented with the initial data

$$\rho|_{t=0} = \rho_0(\mathbf{x}), \qquad \mathbf{x} \in \mathbb{R}^n. \qquad (2.2)$$

Besides, we consider the adjoint problem

$$\frac{\partial \zeta}{\partial t} + \mathbf{u} \cdot \nabla \zeta = g, \qquad \mathbf{x} \in \mathbb{R}^n, \quad t \in (0,T) \qquad (2.3)$$

$$\zeta|_{t=T} = \zeta_T(\mathbf{x}), \qquad \mathbf{x} \in \mathbb{R}^n \qquad (2.4)$$

Here $\rho(\mathbf{x},t)$ and $\zeta(\mathbf{x},t)$ are unknown functions, while $\mathbf{u} = (u_1, \dots, u_n)$, h, g, ρ_0, and ζ_T are given ones, and we assume, for the sake of simplicity, all functions being periodic with respect to all spatial variables x_k, $k = 1, 2, \dots, n$.

We denote by Ω the period set, $\Omega = \prod_{k=1}^{n}(0, T_k)$, $0 < T_k < \infty$, and by $Q = \Omega \times (0, T)$ the domain of functions in the space of independent variables $(\mathbf{x}, t)$.

Our main goal is to search for the minimal, as it is possible, conditions for the smoothness of the coefficients $\mathbf{u} = (u_1, \ldots, u_n)$ (forming the velocity vector in mechanics) to provide the existence and uniqueness of generalized solutions to the problems (2.1),(2.2) and (2.3),(2.4). There is a particular interest in compressible Navier-Stokes equations concerned with conditions namely for the divergency of the velocity vector $(\mathrm{div}\mathbf{u} = \sum_{i=1}^{n}\frac{\partial u_i}{\partial x_i})$ as far as this quantity and the density play the important role in the estimating of solutions (cf.[9],[15],[16]).

For the case of solutions ρ, ζ from Lebesgue spaces $L^p(\Omega)$, $1 \leqslant p \leqslant \infty$, such conditions for $\mathrm{div}\mathbf{u}$ were obtained by R.J.DiPerna and P.L.Lions [23], namely

$$\mathrm{div}\mathbf{u} \in L^1(0, T, L^{\infty}(\Omega)) \tag{2.5}$$

and these conditions seem to be optimal except the case of $p = 1$ for the problem (2.1), (2.2) and $p = \infty$ for the problem (2.3), (2.4), respectively.

We use Orlicz function spaces associated with Young functions of low and fast growth at the infinity instead of $L^p(\Omega)$ for construction of generalized solutions to the problems (2.1),(2.2) and (2.3),(2.4), and replace the condition (2.5) with the assumption of integrability in certain Orlicz spaces. We remind the Hölder inequality

$$\int_D f(\mathbf{x})g(\mathbf{x})d\mathbf{x} \leqslant C\|f\|_{L_M(D)} \cdot \|g\|_{L_N(D)} \tag{2.6}$$

for conjugate Orlicz spaces. We use Orlicz spaces associated with Young function of rather fast increase. Namely, let us introduce the class $\mathcal{K}$ of convex functions M such that (cf.[16],[25])

$$\mathcal{K} = \left\{ M(r) \ \Big| \ \int^{\infty} \frac{\ln M(r)}{r^2}dr = \infty \right\} \tag{2.7}$$

where the last condition means the integral to be diverging at the infinity. The class $\mathcal{K}$ contains functions increasing at the infinity faster than polinomials and non slower than the exponent. Roughly speaking class $\mathcal{K}$ consists of Young functions like that

$$M(t) = exp\{ \frac{t}{ln^{\alpha_1}t \cdot ln^{\alpha_2}lnt \cdots \underbrace{ln^{\alpha_m}ln \cdots lnt}_{m\ times}} \}$$

If all $\alpha_i \leq 1$ then $M(t) \in \mathcal{K}$, but if at least one $\alpha_j > 1$ then $\mathrm{M(t)}$ does not belong to $\mathcal{K}$. Two another (but equivalent) definitions of class $\mathcal{K}$ in terms of inverse function $M^{-1}(t)$ and supplementary function $N(t)$ are the following ones (cf.[25])

$$\mathcal{K} = \left\{ M(r) \;\bigg|\; \int\limits^{\infty} \frac{1}{rM^{-1}(r)} dr = \infty \right\} = \left\{ M(r) \;\bigg|\; \int\limits^{\infty} \frac{1}{N(r)} dr = \infty \right\}$$

Remark that, given function H, the composition $H \circ M \in \mathcal{K}$ if and only if the inverse function H^{-1} increases at the infinity non faster than polinomials.

2.2 Existence and Uniqueness Theorems

Let us explain the main idea to obtain a priori estimates for the solution $\rho(\mathbf{x}, t)$ to the problem (2.1),(2.2). If ρ is a classic solution to the equation (2.1) and $\Phi(\rho)$ is an arbitrary smooth function of ρ then the equality

$$\frac{\partial \Phi(\rho)}{\partial t} + \operatorname{div}(\Phi(\rho)\mathbf{u}) + [\rho\Phi'(\rho) - \Phi(\rho)]\operatorname{div}\mathbf{u} = \Phi'(\rho) \cdot h \tag{2.8}$$

holds besides of (2.1); here $\Phi'(\rho) = \dfrac{d\Phi}{d\rho}$. So, if the function $\Phi(\rho)$ is such that

$$r\Phi'(r) - \Phi(r) \qquad \Phi(r) \tag{2.9}$$

then there exists Young function $M(s)$ to provide the relation

$$M(r\Phi'(r) - \Phi(r)) \simeq \Phi(r) \tag{2.10}$$

and one can apply the Hölder inequality (2.6) to the third term in equation (2.8). Then the complementary function $N(r)$ gives the corresponding Orlicz space for $\operatorname{div}\mathbf{u}$ to be from. The remarkable fact is that (2.9) and (2.10) are possible if and only if $N(r)$ belongs to the class $\mathcal{K}$ and $\Phi(r)$ is a function of slow growth (less than any power r^{1+} with positive ε).

Proposition 2.1. *Let $N(r) \in \mathcal{K}$, and Φ satisfies the condition (2.9). Then a priori estimates*

$$\|\rho\|_{L^\infty(0,T,L_{\ (\Omega)})} \leqslant C\left(1 + \int\limits_Q N(|\operatorname{div}\mathbf{u}|)d\mathbf{x}dt\right) \cdot \left[\|\rho_0\|_{L_{\ (\Omega)}} + \|h\|_{L^1(0,T,L_{\ (\Omega)})}\right]$$
$$\tag{2.11}$$

$$\|\zeta\|_{L^\infty(0,T,L_{\ (\Omega)})} \leqslant C\left(1 + \int\limits_Q N(|\operatorname{div}\mathbf{u}|)d\mathbf{x}dt\right) \cdot \left[\|\zeta_T\|_{L_{\ (\Omega)}} + \|g\|_{L^1(0,T,L_{\ (\Omega)})}\right]$$
$$\tag{2.12}$$

hold for the solutions of the problems (2.1),(2.2) and (2.3),(2.4) respectively, where Ψ is the complementary function to Φ.

A priori estimates (2.11) and (2.12) allow us to prove the existence of solutions under the assumption on div$\mathbf{u}$ to be from the Orlicz class $K_N(Q)$ (or from Orlicz space $E_N(Q)$) associated with some $N \in \mathcal{K}$ and $\mathbf{u}$ being from $L^1(0, T, L_\Phi(\Omega))$ or $L^1(0, T, L_\Psi(\Omega))$ for the problems (2.1),(2.2) and (2.3),(2.4) respectively.

Theorem 2.1. *If* $\rho_0 \in L_\Phi(\Omega)$, $\zeta_T \in L_\Psi(\Omega)$,

$$h \in L^1(0, T, L_\Phi(\Omega)), \qquad g \in L^1(0, T, L_\Psi(\Omega)),$$

$$\mathbf{u} \in L^1(0, T, L_\Phi(\Omega)) \quad or \quad \mathbf{u} \in L^1(0, T, L_\Psi(\Omega))$$

and div$\mathbf{u} \in K_N(Q)$ *with* $N \in \mathcal{K}$, *then there exist solutions to the problems (2.1),(2.2) or (2.3),(2.4) respectively.*

Scetch of proof. We approximate all prescribed functions with the sequences of smooth functions and construct the sequence of classic solutions which contents weakly converging subsequence in view of (2.11),(2.12). Passing to the limit yields the existence of weak solutions. (More details one can find in [25].)

To provide the uniqueness of solution it is nesessary to complete the above conditions with additional smoothness of $\mathbf{u}$.

Theorem 2.2 *If the conditions of Theorem 2.1 are fulfilled and*

$$\mathbf{u} \in L^1(0, T, W^{1,1}(\Omega))$$

then the generalized solutions of the problems (2.1),(2.2) and (2.3),(2.4) are unique.

The proof relies upon the following results given below.

2.3 Gronwall-type Inequality and Osgood Uniqueness Theorem

To prove the uniqueness in Theorem 2.2 we reduce the problem to the inequality of Gronwall type

$$\int_\Omega |\psi|d\mathbf{x} \leqslant C \int_0^t \int_\Omega |\psi| \cdot |\text{div}\mathbf{u}|d\mathbf{x}ds \tag{2.13}$$

for the difference ψ of the solutions.

This inequality yields $\psi \equiv 0$ if and only if $\text{div} \mathbf{u} \in K_N(Q)$ with some $N \in \mathcal{K}$.

Indeed, denote $\alpha(t) = \int_\Omega \psi(x,t)dx$, $\quad A(t) = \int_0^t \alpha(s)ds \quad$ and

$$\beta(t) = \begin{cases} 0, & \text{if } \alpha(t) = 0, \\ \frac{1}{\alpha(t)} \int_\Omega f(x,t)\psi(x,t)dx, & \text{if } \alpha(t) \neq 0 \end{cases}$$

Then (2.13) takes the form

$$\alpha(t) \leq \int_0^t \alpha(s)\beta(s)ds \tag{2.14}$$

It gives after changing of variables:

$$N(\frac{\alpha(t)}{A(t)}) \leq N(\frac{1}{A(t)} \int_0^{A(t)} \beta(A^{-1}(\tau))d\tau)$$

$$\leq \frac{1}{A(t)} \int_0^{A(t)} N(\beta(A^{-1}(\tau)))d\tau = \frac{1}{A(t)} \int_0^t N(\beta(s))\alpha(s)ds \tag{2.15}$$

Here we have used Jensen's inequality for convex functions

$$M(\frac{1}{mes\Omega} \int_\Omega f(\mathbf{x})dx) \leq \frac{1}{mes\Omega} \int_\Omega M(f(\mathbf{x}))dx$$

$$\forall f \in L_M(\Omega), \forall M(t) - convex.$$

Inequalities (2.14),(2.15) lead to differential inequality

$$\frac{dA}{dt} \leq A(t)N^{-1}(\frac{1}{A(t)}), \quad A(0) = 0 \tag{2.16}$$

By Osgood uniqueness theorem (2.16) implies $A(t) \equiv 0$ and then $\psi = 0$ a.e. In this connection let us consider Cauchy problem for the system of ordinary differential equations

$$\frac{d\mathbf{x}}{dt} = \mathbf{u}(\mathbf{x},t), \qquad \mathbf{x}|_{t=0} = \mathbf{x}_0 \in \mathbb{R}^n \tag{2.17}$$

We suppose $\mathbf{u}(\mathbf{x},t) \in L^1(0,T,W^1L_N(\Omega))$ with some $N \in \mathcal{K}$ (that means: gradient of $\mathbf{u}$ with respect to $\mathbf{x}$ belongs to $L^1(0,T,L_N(\Omega))$).

By the embedding theorem (cf.[24]) the field $\mathbf{u}$ possesses the continuity modulus (in the generalized Hölder sense)

$$\sigma(s) = \int_{s^{-n}}^\infty \frac{N^{-1}(\xi)}{\xi^{1+1/n}}d\xi \tag{2.18}$$

By Osgood uniqueness theorem the solution of (2.17) is unique if and only if (cf.[26], Ch.3, Corollary 6.2)

$$\int\limits_0 \frac{ds}{\sigma(s)} = \infty \tag{2.19}$$

In view of (2.18) this condition is equivalent to $N(r)$ being from $\mathcal{K}$.

Proposition 2.2 *If* $\dfrac{\partial u_i}{\partial x_j} \in L^1(0,T,L_N(\Omega))$, $i,j = 1,2,\ldots,n$, *then the uniqueness of solution to Cauchy problem (2.17) is equivalent to the assumption* $N(r) \in \mathcal{K}$.

2.4 Conclusive Remarks

The nesessity of conditions for the coefficients in well-posedness theorems we illustrate by the following example.

Example 2.1 Let $n = 2$, and $\gamma(s)$ is continuous odd function such that $\gamma(0) = 0$, and $\gamma(s) \to \infty$ as $s \to \infty$, and $\displaystyle\int\limits_0 \frac{ds}{\gamma(s)} < \infty$. We construct a vector field on the plane $\mathbb{R}^2$ by the formula

$$\mathbf{u} = \mathbf{u}(x,y) = (\gamma(x), -y\gamma'(x)) \tag{2.20}$$

Then the system of equations for the trajectories

$$\frac{dx}{dt} = \gamma(x), \qquad \frac{dy}{dt} = -y\gamma'(x) \tag{2.21}$$

yields the solutions

$$x(t) = \pm\Gamma^{-1}(t), \qquad y(t) = C\exp\left(-\int\limits_1^t \gamma'(\Gamma^{-1}(s))ds\right) \tag{2.22}$$

(here $\Gamma(z) = \displaystyle\int\limits_0^z \frac{ds}{\gamma(s)}$), forming a double surface containing the axis

$\{(x,y,t) \mid x = t = 0\}$. In particular, taking $\gamma(x) = 2\sqrt{x}$, we obtain the family of solutions

$$x = t^2, \qquad y = \frac{C}{t} \tag{2.23}$$

Thus, the function

$$\zeta(x,y,t) = \begin{cases} 1, & -\Gamma^{-1}(t) < x < \Gamma^{-1}(t) \\ 0, & \text{otherwise} \end{cases} \tag{2.24}$$

is a nontrivial solution to the transport equation (2.3) vith zero initial data.

Such a non-uniqueness example appeared due to the non-embeddings

$$\mathbf{u} \notin L^1(0, T, W^{1,1}_{A9}(\mathbb{R}^2)) \quad \text{and} \quad \mathrm{div}\,\mathbf{u} \in K_M(Q) \text{ with } M \notin \mathcal{K}$$

Example 2.2 Set

$$\zeta = \chi_{\mathcal{C}}, \quad \rho = \beta^{-n}(t)\chi_{\mathcal{C}}, \quad \mathbf{u} = \frac{\beta'(t)}{\beta(t)}\mathbf{x} \tag{2.25}$$

with $\chi_{\mathcal{C}}$ standing for the charasteristic function of $\mathcal{C}$, and $\mathcal{C}$, in its turn, is the cone $\mathcal{C} = \{\,(\mathbf{x}, t)\,|\,|\mathbf{x}| \leqslant \beta(t)\,\}$ with a positive function $\beta(t)$ vanishing as $t = T$. Since

$$\text{meas supp } \rho(t, \cdot) = \text{meas supp } \zeta(t, \cdot) \to 0 \quad \text{as } t \to T$$

then the formula (2.25) defines the example of nontrivial solution to the homogeneous problem (2.3),(2.4) and decaying solution to the problem (2.1),(2.2). Such an example is possible due to the non-embedding

$$\mathrm{div}\,\mathbf{u} \in L^1(0, T; L_{M,A9}(\mathbb{R}^n)) \quad \text{with} \quad M \in \mathcal{K}$$

Remark on "Div-condition". Here we consider the connection appearing between the functions $\Phi(r)$ satisfying (2.9) and corresponding Young functions $M(r)$ (or their complementaries $N(s)$) from (2.10) which describe the function space for $\mathrm{div}\,\mathbf{u}$. Given function $M(r)$ produces the class of functions $\Phi(r)$ with the property (2.9) which have the same growth with respect to the linear function, with the power-type difference only. Namely, two functions $\Phi_1(r)$ and $\Phi_2(r)$ satisfy (2.10) with the same $M(r)$ if and only if there exists $q = \text{const} > 0$ such that

$$\frac{\Phi_1(r)}{r} = \left(\frac{\Phi_2(r)}{r}\right)^q, \quad r > 0 \tag{2.26}$$

This explains, in particular, the condition (2.5) for the case of Lebesgue spaces L^p.

3 Some Remarks on Compensated Compactness Theory

3.1 Introduction

Compactness method (as it's named by J.-L.Lions in his famous book [1]) concerns with the solving of some boundary value problem

$$L(u) = 0 \tag{3. 1}$$

by the construction of sequence of "approximate" solutions $\{u_n\}$

$$L_n(u_n) = 0, \quad n = 1, 2, \ldots \tag{3. 2}$$

and passing to the limit as $n \to \infty$.

As rule, the sequence $\{u_n\}$ converges in certain sense to some element u, but it's not so evident for u to be the solution of (3.1), in particular, in the case of nonlinear problem (3.1).

The simple example

$$u_n(x) = sin(nx) \quad on \quad [0, 2\pi]$$

illustrates the typical situation:

$$u_n(x) \rightharpoonup 0 \ \ weakly \ in \ L^2(0, 2\pi) \quad as \quad n \to \infty$$

but $u_n^2(x) \rightharpoonup \frac{1}{2}$.

Very often one has quadratic nonlinearity

$$u_n \rightharpoonup u, \quad v_n \rightharpoonup v \quad weakly \ \ in \ L^2(\Omega)$$

Then

$$u_n \cdot v_n \rightharpoonup \chi \ \ in \ sense \ of \ distributions \quad D'(\Omega)$$

And the question is wheather equality

$$\chi(\mathbf{x}) = u(\mathbf{x}) \cdot v(\mathbf{x}) \quad in \quad D'(\Omega)$$

holds or not?

This question possesses the positive answer if one of two sequences $\{u_n\}$ or $\{v_n\}$ converges in strong sense, i.e. if

$$u_n \to u \quad or \quad v_n \to v \ \ in \ norm \ of \ L^2(\Omega)$$

So, the main goal of compactness method is to obtain a strong convergence at least for one sequence, and the principal tool here is based on the Aubin- Simon compactness theorem (cf.[1],Ch.2, theorem 5.1, [2],[3]), which is elucidated in the next section for the convenience of reader.

As a one of very useful tool in solving of non-linear P.D.E.'s is so called theory of compensated compactness (cf.[4]-[7]). Roughly speaking, this theory allows us to pass to the limit in weakly converging sequences under minimal conditions which don't provide in general case strong convergence of any sequence either $\{u_n\}$ or $\{v_n'\}$.

The mostly important for the applications to nonlinear P.D.E.'s theory is one particular version of compensated compactness, namely, so called "div-curl" lemma [4],[5],[7]. Numerous interesting results were obtained on the base of this lemma (cf. [8]-[17]).

3.2 Classical Compactness (Aubin-Simon Theorem)

Now we recall here one well-known and widely-used compactness argument (cf. [1], Ch. 2, Theorem 5.1). Let B_0, B and B_1 be three Banach spaces such that

$$B_0 \hookrightarrow\hookrightarrow B \hookrightarrow B_1$$

Here $\hookrightarrow$ means continuous embedding while $\hookrightarrow\hookrightarrow$ is continuous and compact one.

If the sequence$\{u_n(t)\}$ is bounded in $L^{p_0}(0, T; B_0)$, $1 \le p_0 \le \infty$, and the sequence of derivatives $\{du_n/dt\}$ is bounded in $L^{p_1}(0, T; B_1)$, $1 \le p_1 \le \infty$, then $\{u_n(t)\}$ is compact in $L^p(0, T; B)$ with $1 \le p \le p_0$ in the case of $p_0 < \infty$ and $1 \le p < \infty$ in the case of $p_0 = \infty$, $p_1 = 1$.

This theorem was proved firstly by J.P.Aubin [2] in the case $1 < p_0, p_1 < \infty$ and for the limiting cases by J.Simon [3].

The same result takes the place if the sequence of time-derivatives of the other sequence $\{v_n(t)\}$ instead of $\{u_n(t)\}$ is bounded (cf.[16], lemma 6).

Let$V$ and W be two Banach function spaces defined on the bounded domain $\Omega \in \mathbb{R}^N$ and such that $D(\Omega) \hookrightarrow V \hookrightarrow\hookrightarrow W \hookrightarrow D'(\Omega)$ We suppose, of course, for any Banach space$B$ located between $D(\Omega)$and $D'(\Omega)$ the product with infinitely differentiable functions is defined:

$$\forall u \in B, \forall \phi \in D(\Omega) \quad \exists (u \cdot \phi) \in B.$$

In particular, it's valid for the spaces V and W as well as for their conjugate spaces V' and W' which satisfy the embedding relations

$$D(\Omega) \hookrightarrow W' \hookrightarrow\hookrightarrow V' \hookrightarrow D'(\Omega)$$

Finally, let V_1' be some Banach space, arbitrary wide

$$V' \hookrightarrow V_1' \hookrightarrow D'(\Omega).$$

Now we suppose that there exist two sequences $\{u_n(t)\}$ and $\{v_n(t)\}$ such that

$$u_n(t) \to u(t) \quad weakly\ in\ L^p(0, T; W), 1 < 0 \le \infty$$

$$v_n(t) \to v(t) \quad weakly\ in\ L^q(0, T; W'), q \ge \frac{p}{p-1}$$

Then one can define the sequence of products $\{u_n(t) \cdot v_n(t)\}$ as a sequence in the space $L^1(0, T; D'(\Omega))$ by the rule

$$< u_n \cdot v_n, \phi >=< v_n, u_n \cdot \phi >, \quad \forall \phi \in D(\Omega),$$

where $< g, \phi >$ designates the value of distribution g on the test function ϕ. We assume also the sequence $\{u_n \cdot v_n\}$ to be converging in $D'(0, T; D'(\Omega))$

$$u_n \cdot v_n \to \chi \quad in\ D'(0, T; D'(\Omega))$$

If we suppose additionally $\{u_n(t)\}$ is bounded in $L^p(0,T;V)$ and $\{v'_n(t)\}$ is bounded in $L^{p_1}(0,T;V'_1)$, $1 \le p_1 \le \infty$ then the equality $\chi = u \cdot v$ holds in sense of $D'(0,T;D'(\Omega))$. This proposition is a particular case of compensated compactness theory (see subsection 4 below) but it's proof can be reduced to Aubin - Simon theorem if we take $B_0 = W', B = V'$ and $B_1 = V'_1$.

To apply this version of compactness theorem to P.D.E. theory we define, firstly, the sequence $\{v_n\}$ which admits a priori estimates for time-derivatives. The other factor in non-linear term gives the sequence $\{u_n\}$. It indicates the spaces V and W'. Then we have to check the compactness of embedding V into W or, if it's easier, W' into V'. (See, for example, application to compressible Navier-Stokes equations in [16], subsection 5.2, and also recent paper [31])

3.3 Compensated Compactness – "div-curl" Lemma

Let $\Omega \subset \mathbb{R}^N$ be bounded domain with boundary Γ, and let $\{\mathbf{w}_n^1\}$ and $\{\mathbf{w}_n^2\}$, $n = 1, 2, \ldots$, be two sequences of vector fields on Ω such that

$$\mathbf{w}_n^1 \rightharpoonup \mathbf{w}^1, \quad \mathbf{w}_n^2 \rightharpoonup \mathbf{w}^2 \ weakly \ in \ L^2(\Omega) \quad as \ n \to \infty. \qquad (3.3)$$

Then the sequence of scalar products $Q(\mathbf{w}_n^1, \mathbf{w}_n^2) = \sum_{i=1}^N (w_n^1)_i (w_n^2)_i$ is bounded in $L^1(\Omega)$, so

$$Q(\mathbf{w}_n^1(\mathbf{x}), \mathbf{w}_n^2(\mathbf{x})) \rightharpoonup \chi(\mathbf{x}) \quad in \ D'(\Omega) \qquad (3.4)$$

And the question is: wheather equality

$$\chi(\mathbf{x}) = Q(\mathbf{w}^1(\mathbf{x}), \mathbf{w}^2(\mathbf{x})) \quad in \ D'\Omega \qquad (3.5)$$

holds or not?

Let us introduce two operators of vector analysis:

$$div\mathbf{a} = \sum_{i=1}^N \frac{\partial a_i}{\partial x_i}, \quad (curl\mathbf{a})_{ij} = \frac{\partial a_i}{\partial x_j} - \frac{\partial a_j}{\partial x_i}, \quad i,j = 1, 2, \ldots, N. \quad (3.6)$$

Proposition 3.1 (F.Murat[4],[5], see also [7])

If the sequence $\{div\mathbf{w}_n^1\}$ is compact in $H_{loc}^{-1}(\Omega)$ and $\{curl\mathbf{w}_n^2\}$ is compact in $(H_{loc}^{-1}(\Omega))^m$, $m = N(N-1)/2$, then the equality (3.5) is valid.

Remark 3.1 If div and curl of one sequence $\{\mathbf{w}_n^1\}$ or $\{\mathbf{w}_n^2\}$ are compact, then this sequence is strongly compact in $L^2(\Omega)$ as a solution to elliptic problem for Laplace operator (see decomposition (3.7) below).

Proof. Introduce two function spaces:

$$J(\Omega) = \{\mathbf{u} \in L^2(\Omega) \mid div\mathbf{u} = 0\}$$

$$G(\Omega) = \{\mathbf{v} \in L^2(\Omega) \mid \mathbf{v} = \nabla\psi, \quad \psi \in \overset{\circ}{H}{}^1(\Omega)\}$$

which make Helmholtz-Weyl decomposition (cf.[19]-[21])

$$L^2(\Omega) = J(\Omega) \oplus G(\Omega) \tag{3.7}$$

(To find ψ for given $\mathbf{w}$ one has to solve Dirichlet problem for Poisson equation with right hand part $div\mathbf{w}$. Such operator is continious map from $H^{-1}(\Omega)$ into $\overset{\circ}{H}{}^1(\Omega)(cf.[1],[13])$. In particular, if the sequence $\{div\mathbf{w}_n\}$ is compact in $H^{-1}(\Omega)$ then the corresponding sequence $\{\mathbf{v}_n\}$ in decomposition (3.7) is compact in $L^2(\Omega)$.)

Now one can represent

$$\mathbf{w}_n^1 = \mathbf{u}_n^1 + \mathbf{v}_n^1, \quad \mathbf{w}_n^2 = \mathbf{u}_n^2 + \mathbf{v}_n^2$$

where $\mathbf{u}_n^k \in J(\Omega)$, $\mathbf{v}_n^k \in G(\Omega)$, $k = 1, 2..$ Then one has

$$\{div\mathbf{v}_n^1 = div\mathbf{w}_n^1\}$$

-compact set in $H_{loc}^{-1}(\Omega)$, according to assumption, and $curl\mathbf{v}_n^1 \equiv 0$. It means the set $\{\mathbf{v}_n^1\}$ is compact in $L^2(\Omega)$.

By the same way

$$\{curl\mathbf{u}_n^2 = curl\mathbf{w}_n^2\}$$

is compact set in $(H_{loc}^{-1}(\Omega))^m$, $m = 1/2N(N-1)$ and $div\mathbf{u}_n^2 \equiv 0$.

It implies compactness of $\{\mathbf{u}_n^2\}$ in $L^2(\Omega)$. So, the scalar product can be rewritten as follows

$$Q(\mathbf{w}_n^1, \mathbf{w}_n^2) = Q(\mathbf{u}_n^1, \mathbf{u}_n^2) + Q(\mathbf{v}_n^1, \mathbf{v}_n^2) + (\mathbf{u}_n^1 \cdot \mathbf{v}_n^2) + (\mathbf{u}_n^2 \cdot \mathbf{v}_n^1)$$

After integrating over Ω two last terms vanish since $J(\Omega)$ and $G(\Omega)$ are orthogonal subspaces. And two first terms content strongly converging sequences $\{\mathbf{u}_n^2\}$ and $\{\mathbf{v}_n^1\}$ what allows to pass to the limit in (3.4),(3.5) and to prove proposition 3.1. In addition we remark that test function $\phi \in D(\Omega)$ can be included in any sequence $\{\mathbf{w}_n^1\}$ or $\{\mathbf{w}_n^2\}$. It is also the reason why the smoothness of boundary Γ is not so important.

As it has been noted, our proof relies upon decomposition (3.7) which itself is based on the optimal estimates for the solutions of Dirichlet problem. We are able to prove "div-curl" lemma for the wide class of Orlicz spaces instead of $L^2(\Omega)$ (cf.[27],[28].) Indeed, let $\{\mathbf{w}_n^1\}$ and $\{\mathbf{w}_n^2\}$be two sequences such that

$$\mathbf{w}_n^1 \rightharpoonup \mathbf{w}^1 \quad weakly \ in \ L_M(\Omega), \quad \mathbf{w}_n^2 \rightharpoonup \mathbf{w}^2 \quad weakly \ in \ L_N(\Omega)$$

where $M(t)$ and $N(t)$ are mutually complementary Young functions.

Then we have the convergence (3.4), and to prove (3.5) it is necessary to provide decomposition (3.7) and strong compactness of $\{\mathbf{v}_n^1\}$ and $\{\mathbf{u}_n^2\}$ in $L_M(\Omega)$ and $L_N(\Omega)$, respectively. We assume additionally the set $\{div \ \mathbf{w}_n^1\}$

to be compact in Orlicz-Sobolev space $W^{-1}L_M(\Omega)$ which is conjugate for $\overset{\circ}{W}{}^1 E_N(\Omega)$, and the set $\{curl\ \mathbf{w}_n^2\}$ to be compact in $W^{-1}L_N(\Omega)$ (conjugate for $\overset{\circ}{W}{}^1 E_M(\Omega)$), respectively.

One can use decomposition (3.7) and corresponding optimal estimates if Young function M(t) or N(t) satisfies additional condition (cf.[29],[30])

$$1 < p_1 = const \le \frac{tM'(t)}{M(t)} \le p_2 = const < \infty,$$

or

$$1 < q_1 = const \le \frac{tN'(t)}{N(t)} \le q_2 = const < \infty$$

This means that the spaces $L_M(\Omega)$ and $L_N(\Omega)$ are located between Lebesgue spaces $(L^{p_1}(\Omega),\ L^{p_2}(\Omega))$ and $(L^{q_1}(\Omega),\ L^{q_2}(\Omega))$, respectively. Optimal estimates are obtained in [29],[30] by interpolation of estimates for Lebesgue spaces. For the limiting cases $p_1 = 1$ or $q_1 = 1$ and $p_2 = \infty$ or $q_2 = \infty$ the optimal results are not proved still and it remains as an interesting open problem.

3.4 Compensated Compactness-theorem of L. Tartar

Let

$$Q(\mathbf{u}) = \sum_{i,j=1}^{p} b_{ij}\, u_i\, u_j \tag{3.8}$$

be arbitrary quadratic form with constant coefficients, $(b_{ij} = const)$ on $\mathbb{R}^p$.

Let $\{\mathbf{u}_n(\mathbf{x})\}$, $\mathbf{x} \in \Omega \subset \mathbb{R}^N$ be some sequence such that

$$\mathbf{u}_n(\mathbf{x}) \rightharpoonup \mathbf{u}(\mathbf{x}) \quad weakly\ in\ (L^2(\Omega))^p \quad as\ n \to \infty \tag{3.9}$$

Then

$$Q(\mathbf{u}_n(\mathbf{x})) \rightharpoonup \chi(\mathbf{x}) \quad in\ D'(\Omega) \tag{3.10}$$

and the question is

$$\chi(\mathbf{x}) = Q(\mathbf{u}(\mathbf{x})) \quad in\ D'(\Omega)\,? \tag{3.11}$$

Suppose, some additional information is known, namely, let

$$A:\ (L^2(\Omega))^p \to (H^{-1}(\Omega))^q$$

be linear bounded operator of the form

$$A_k(\mathbf{u}) = \sum_{i=1}^{p}\sum_{j=1}^{N} a_{kij}\frac{\partial u_i}{\partial x_j},\quad k = 1,2,\ldots,q \tag{3.12}$$

where, for simplicity, a_{kij} are real constants.

Finally, let us introduce the set

$$\Lambda = \{\lambda \in \mathbb{R}^p | \exists \xi \in \mathbb{R}^N, \xi \neq 0, \sum_{i=1}^{p} \sum_{j=1}^{N} a_{kij}\lambda_i\xi_j = 0, \forall k = 1, 2, \ldots, q\}$$

(3. 13)

Proposition 3.2 (L.Tartar[6])
Assume (3.9), (3.10) and additionally

$$\{A(\mathbf{u}_n)\} \text{ is compact in } (H_{loc}^{-1}(\Omega))^q$$

(3. 14)

and

$$Q(\lambda) = 0, \quad \forall \lambda \in \Lambda$$

(3. 15)

Then the equality (3.11) holds.

Remark 3.2. "div-curl" lemma is a particular case of this proposition.

Proof. To make a proof looking like the proof of "div-curl" lemma we assume at the beginning operator A to be closed. It means the set ImA^* is closed, where $A^* : (\overset{\circ}{H^1}(\Omega))^q \to (L^2(\Omega))^p$ is an adjoint operator, or, by other words, second order operator $A \circ A^* : (\overset{\circ}{H^1}(\Omega))^q \to (H^{-1}(\Omega))^q$ is strongly elliptic.

We shall use decomposition (cf:[22])

$$L^2(\Omega) = kerA \oplus ImA^*$$

(3. 16)

instead of (3.7). Then for $\forall n = 1, 2, \ldots$ one has

$$\mathbf{u}_n = \mathbf{v}_n + \mathbf{w}_n, \quad \mathbf{v}_n \in kerA, \quad \mathbf{w}_n \in ImA^*$$

The set $\{\mathbf{w}_n\}$ is compact in $(L^2(\Omega))^p$ since operator A is invertible on ImA^*, and $\{A(\mathbf{w}_n)\} = A(\mathbf{u}_n)\}$ is compact in $(H^{-1}(\Omega))^q$ according to assumption.

Rewrite quadratic form Q:

$$Q(\mathbf{u}_n) = Q(\mathbf{v}_n) + Q(\mathbf{w}_n) + \tilde{Q}(\mathbf{v}_n, \mathbf{w}_n) + \tilde{Q}(\mathbf{w}_n, \mathbf{v}_n)$$

(3. 17)

where

$$\tilde{Q}(\mu, \nu) = \sum_{i,j=1}^{p} b_{ij}\mu_i\nu_j$$

is bilinear form on $\mathbb{R}^p \times \mathbb{R}^p$ such that $Q(\mathbf{u}) = \tilde{Q}(\mathbf{u}, \mathbf{u})$.

Formula (3.17) means that only one question is in limiting passing, namely, does $Q(\mathbf{v}_n)$ converge to $Q(\mathbf{v})$ if $\mathbf{v}_n \in kerA$?

Fourier transformation applied to equalities

$$A_k(\mathbf{v}_n) = 0 \quad k = 1, 2, \ldots, q$$

yields

$$\sum_{i=1}^{p}\sum_{j=1}^{N} a_{kij} \, (\hat{v}_n)_i \xi_j \; = \; 0$$

where $\hat{\mathbf{v}}_n$ is Fourier image of $\mathbf{v}_n$ extended by 0 on the whole $\mathbb{R}^N$ and $\xi = (\xi_1, \xi_2, \ldots, \xi_N)$ - parameters of Fourier transformation

It means both $Re(\hat{\mathbf{v}}_n)$ and $Im(\hat{\mathbf{v}}_n)$ belong to the set Λ, and according to assumption (15)

$$\tilde{Q}(\hat{\mathbf{v}}_n, \bar{\hat{\mathbf{v}}}_n) \; = \; 0.$$

It gives by Plancherel-Parseval identity

$$\int_{\mathbb{R}^N} Q(\mathbf{v}_n) dx \; = \; C_N \int_{\mathbb{R}^N} \tilde{Q}(\hat{\mathbf{v}}_n, \bar{\hat{\mathbf{v}}}_n) d\xi \; = \; 0$$

Passing to the limit in (3.17) we obtain the result of proposition 3.2 in the case of closed set ImA^* in $(L^2(\Omega))^p$. Note that construction of sequence $\mathbf{w}_n$ does not depend explicitly on the norm of inverse operator A^{-1}. In general case we can approximate each $\mathbf{w}_n$ from closure of ImA^* by sequence $\{\mathbf{w}_{nm}\}$ from ImA^* and pass to the limit by standard diagonal procedure. In conclusion of this section we underline once again that proposition 3.2 is not a new mathematical result, but the sense of condition (3.15) as vanishing of form Q on subspace $kerA$ is a new viewpoint which can be usefull in some cases when $kerA$ admits the simple description.

3.5 Generalizations and Examples

New approach to compensated compactness theory based on decomposition (3.16) allows us to give clear and short proofs of main known theorems formulated in propositions 3.1 and 3.2. At the same time we can give some natural generalizations of theory on the other cases of compensated compactness arguments.

a). We shall start from generalization of "div-curl" lemma as a most important for the theory of nonlinear partial differential equations.

Consider two operators of vector analysis (3.6):

$$A_1(\mathbf{u}) \; = \; div\mathbf{u}, \qquad A_2(\mathbf{u}) \; = \; curl\mathbf{u}$$

as linear bounded operators from $L^2(\Omega)$ onto $H^{-1}(\Omega)$. The kernel of operator A_1 is the subspace $J(\Omega)$ in Helmholtz-Weyl decomposition (3.7). It's easy to calculate the adjoint operator for A_1, namely, operator-gradient: $A_1^* = -\nabla$, which acts as bounded linear one from $\overset{\circ}{H}{}^1(\Omega)$ onto $L^2(\Omega)$ and $ImA_1^* = G(\Omega)$ - the second subspace in decomposition (7). Moreover, $G(\Omega) = kerA_2$, so (3.7) is a particular case of decomposition (3.16):

$$L^2(\Omega) \; = \; J(\Omega) \oplus G(\Omega) \; = \; kerA_1 \oplus ImA_1^* \; = \; ImA_2^* \oplus kerA_2$$

Now let us consider two arbitrary linear bounded operators instead of (3.6):

$$A_1 : (L^2(\Omega))^N \to (H^{-1}(\Omega))^p, \quad A_2 : (L^2(\Omega))^N \to (H^{-1}(\Omega))^q \qquad (3.18)$$

Then the following assertion takes place.

Proposition 3.3 Assume (3.3) and (3.4), and additionally let the sets

$$\{A_1(\mathbf{w}_n^1)\} \quad and \quad \{A_2(\mathbf{w}_n^2)\}$$

are compact in $(H_{loc}^{-1}(\Omega))^p$ and $(H_{loc}^{-1}(\Omega)^q$, respectively.
Then equality (3.5) holds, if operators A_1 and A_2 satisfy the condition

$$ker A_i \quad Im A_j^*, \quad i \neq j \qquad (3.19)$$

Proof. At the first step we use decomposition (3.16) with operator A_1 and represent $\mathbf{w}_n^1$ as follows

$$\mathbf{w}_n^1 = \mathbf{u}_n^1 + \mathbf{v}_n^1, \quad \mathbf{u}_n^1 \in ker A_1, \quad \mathbf{v}_n^1 \in Im A_1^* \qquad (3.20)$$

Then $\mathbf{v}_n^1 \to \mathbf{v}^1$ strongly in $L^2(\Omega)$, and in the scalar product

$$(\mathbf{w}_n^1 \cdot \mathbf{w}_n^2) = (\mathbf{u}_n^1 \cdot \mathbf{w}_n^2) + (\mathbf{v}_n^1 \cdot \mathbf{w}_n^2)$$

one can pass to the limit in second term. In the first term we use decomposition (3.16) with operator A_2 and represent $\mathbf{w}_n^2$:

$$\mathbf{w}_n^2 = \mathbf{u}_n^2 + \mathbf{v}_n^2, \quad \mathbf{u}_n^2 \in ker A_2, \quad \mathbf{v}_n^2 \in Im A_2^*$$

where the sequence $\{\mathbf{v}_n^2\}$ is compact (strongly) in $L^2(\Omega)$. It means that

$$(\mathbf{u}_n^1 \cdot \mathbf{w}_n^2) = (\mathbf{u}_n^1 \cdot \mathbf{u}_n^2) + (\mathbf{u}_n^1 \cdot \mathbf{v}_n^2)$$

and we can pass to the limit in the last term, while the first one vanishes after integrating over Ω since $\mathbf{u}_n^1$ and $\mathbf{u}_n^2$ are orthogonal in view of condition (3.19). It proves the proposition 3.3

If assumption (3.19) doesn't take place and subspaces $ker A_1$ and $ker A_2$ have nontrivial intersection, i.e.

$$S = ker A_1 \bigvee ker A_2 \neq \{0\} \qquad (3.21)$$

then (3.5) doesn't take place, in general. In this case some additional conditions are needed, but on the subspace S only. For example, another operator

$$A_3 : (L^2(\Omega))^N \to (H^{-1}(\Omega))^r$$

is compact on S. If $ker A_3 \bigcap S \neq \{0\}$ then one has to continue the procedure adding new operators.

Example 3.1

Consider the simple case of N=2 and $A_1 = A_2 = div$, i.e.

$$A_1(\mathbf{w}) = A_2(\mathbf{w}) = \frac{\partial w_1}{\partial x_1} + \frac{\partial w_2}{\partial x_2}$$

Then $S = ker A_1 \bigcap ker A_2 = J(\Omega) \neq \{0\}$. Taking operator A_3 of the form

$$A_3(\mathbf{w}) = \frac{\partial w_2}{\partial x_1} + a\frac{\partial w_1}{\partial x_2} + b\frac{\partial w_2}{\partial x_2}$$

with arbitrary real parameters a and b , we obtain equality (5), in assumption of compactness of the set $\{A_3(\mathbf{w}_n^k)\}$ in $H_{loc}^{-1}(\Omega)$, if $b^2 + 4a < 0$. If $b^2 + 4a \geq 0$, then another conditions are required on $ker A_3 \bigcap J(\Omega)$.

b). Most part of applications to nonlinear P.D.E. problems concerns with the case when operator A in (3.12) is defined by a priori estimates connected with conservation laws, and quadratic form (3.8) is related to non-linearity of the system of equations, so these two given objects are independent, in some sense, each on other, and don't satisfy the crucial assumption (3.15). Such situation is needed for some additional information on $ker A$, e.g. another operator $B : L^2(\Omega) \rightarrow (H^{-1}(\Omega))^r$ has locally compact image.

Example 3.2   Let $N = 2$, $p = 2$, $q = 1$, $A = div$ and

$$Q(\mathbf{u}) = u_1^2 + u_1 u_2.$$

Then
$$\Lambda = \{\lambda \in \mathbb{R}^2 \mid \exists \xi \in \mathbb{R}^2, \quad \xi \neq 0, \quad \lambda_1 \xi_1 + \lambda_2 \xi_2 = 0\}$$

which means $\Lambda = \mathbb{R}^2$, while

$$Q(\lambda) = \lambda_1(\lambda_1 + \lambda_2), \quad and \quad Q(\lambda) \neq 0 \ \forall \lambda \in \Lambda = \mathbb{R}^2$$

Since $ker A = J(\Omega)$ one can add any operator

$$B(\mathbf{u}) = \sum_{i=1}^{2} a_i \frac{\partial u_1}{\partial x_i} + \sum_{i=1}^{2} b_i \frac{\partial u_2}{\partial x_i}$$

to be compact in $H_{loc}^{-1}(\Omega)$ on the sequence $\{\mathbf{u}_n\}$ from subspace $J(\Omega)$ under condition

$$(a_1 - b_2)^2 + 4a_2 b_1 < 0$$

which provides operator B to be elliptic on $J(\Omega)$.

References

1. J.-L. Lions, Quelques methodes de resolution des problemes aux limites nonlineaires, Dunod, Paris, 1969.

2. J.P. Aubin, Un theoreme de compacite, C.R. Ac.Sc.Paris, 256(1963), 5042-5044.
3. J. Simon, Compact sets in $L^p(0,T;B)$, Ann.Mat.Pura Appl., VCXLVI (1987), 65-96.
4. F. Murat, Compacite par compensation, Ann.Sc.Norm.Sup. Pisa, 5(1978), 489-507.
5. F. Murat, A survey on compensated compactness, Contributions to Modern Calculus of Variations, Research Notes in Mathematics 148, Pitman, (1987), 145-183.
6. L. Tartar, Compensated compactness and applications to P.D.E.- Nonlinear Analysis and Mechanics, Heriot-Watt Symposium IV, Research Notes in Mathematics 39, Pitman, (1979), 136-212.
7. R. Coifman, P.-L. Lions, Y. Meyer, Semmes S. Compensated compactness and Hardy spaces, J. Math. Pures Appl., 72 (1993), 247-286.
8. P.-L. Lions, Mathematical Topics in Fluid Mecanics, vol.1, IncompressibleModels, Claredon Press, 1996.
9. P.-L. Lions, Mathematical Topics in Fluid Mecanics, vol.2, Compressible Models, Cleredon Press, 1998.
10. Y. Amirat, K. Hadamche, A. Ziani, Mathematical Analysis for Compressible Miscible Displacement Models in Porous Media, Math. Models and Methods in Appl. Sc., 6 (1996), 729-747.
11. R.J. DiPerna, Convergence of the Viscosity Method for Isentropic Gas Dynamics, Comm. Math. Phys., 91 (1983), 1-30.
12. R.J. DiPerna, P.-L. Lions, On the Cauchy Problem for Bolzmann Equations: Global Existence and Weak Stability, Ann. of Math., 130 (1989), 321-366.
13. J. Malek, J. Necas, M. Rokyta, M. Ruzicka, Weak and Measure-valued Solutions to Evolutionary PDE's, Chapman-Hall, 1996.
14. D. Hoff, Global Well-posedness of the Cauchy Problem for Nonisentropic Gas Dynamics with Discontinuous Initial Data, J.Diff.Eq's., 5 (1992), 33-73.
15. E. Feireisl, A. Novotny, H. Petzeltova, On the Existence of Globally Defined Weak Solutions to the Navier-Stokes Equations of Compressible Isentropic Fluids, J.Math.Fluid Mech.,3 (2001), 358-392.
16. A.V. Kazhikhov, Recent Developments in the Global Theory of Two-dimensional Compressible Navier-Stokes Equations, Seminar on Math., Keio University, 25 (1998).
17. V.V. Shelukhin, Bingham Viscoplastic Limit of Non-Newtonian Fluids, J. Math. Fluid Mech., 4 (2002), 109-127.
18. M.A. Krasnosel'skij, Ya. B. Rutitskij, Convex functions and Orlicz spaces, Noordhoff, (1961)
19. E.B. Bykhovsky, N.V. Smirnov, On orthogonal decomposition of the spaceof vector-functions, square integrable on given domain, and operators of vector analysis, Proceedings of Steklov Math. Institute, 59 (1960),5-36.(in Russian)
20. Ladyzhenskaya O.A. The Mathematical Theory of Viscous Incompressible Flow, Gordon and Breach, 1969.
21. G.P. Galdi, An Introduction to the Mathematical Theory of the Navier- Stokes Equations, v.1, Springer Tracts in Natural Philosophy, Springer-Verlag, 1994.
22. N.I. Akhiezer, I.M. Glazman, Theory of Linear Operators in Hilbert Space, Nauka, 1966.(in Russian).
23. R.J. DiPerna, P.-L. Lions, Ordinary Differential Equations, Transport Theory and Sobolev Spaces, Invent.Math., 98 (1989), 511-547.

24. A. Kufner, S. Fucik, O. John, Function Spaces, Prague, Academia, 1977.
25. A.V. Kazhikhov, A.E. Mamontov, On a certain class of convex functions and the exact well-posedness of the Cauchy problem for the transport equations in Orlicz spaces, Siberian Math. Journ., 39, N.4 (1998), 716-734.
26. P. Hartman, Ordinary differential equations, John Wiley and Sons, 1964.
27. A. Dolcini, A div-curl result in Orlicz spaces, Rend.Accad.Sci. Fis. Mat. Napoli (4), v.60, (1993), 113-120.
28. C. Sbordone, New Estimates for Div-Curl Products and Very Weak Solutions of PDEs, Ann.Scuola Norm.Sup. Pisa Cl.Sci.(4), v.XXV, (1997), 739-756.
29. I.B. Simonenko, The boundedness of singular integrals in Orlicz spaces, Dokl.Akad.Nauk USSR, v.130, N.6.(1960), 984-987.
30. I.B. Simonenko The interpolation and extrapolation of linear operators in Orlicz spaces, Mat.Sbornik, v.63(105),N.4,(1964), 536-553.
31. Y. Amirat, A. Ziani, Asymptotic Behavior of the Solutions of an Elliptic-Parabolic System Arising in Flow in Porous Media, Zeitschrift für Analysis und ihre Anwendungen, Journal for Analysis and its Applications, v.23 (2004), N 2, 1-7.

Oscillating Patterns in Some Nonlinear Evolution Equations

Yves Meyer

CMLA, Ecole normale supérieure de Cachan
61, avenue du Président-Wilson, F-94235 CACHAN Cedex, France
ymeyer cmla.ens-cachan.fr

1 Introduction

My interest in Navier-Stokes equations arose from the wavelet revolution. I was puzzled by (1) a series of talks and preprints by Marie Farge and (2) an intriguing paper by Paul Federbush entitled *'Navier and Stokes meet the wavelets'*. Both Marie Farge and Paul Federbush were convinced that wavelets could play an important role in fluid dynamics.

Let me first support Marie Farge and Paul Federbush's views with two remarks (**A** and **B** below).

A. Turbulent flows are active over a full range of scales. Understanding or modelling the nonlinear interactions between different scales or investigating the *energy transfers across scales* are serious scientific issues. For the sake of concision, I will only examine one model. This model was proposed by P. Frick and V. Zimin in [43]. The velocity field v is written as an expansion $v = \sum_{-\infty}^{+\infty} v_j$ where v_j corresponds to scales around 2^{-j} and to frequencies (or wavelengths) around 2^j. Then the Navier-Stokes equations become a sequence E_j, $j \in \mathbf{Z}$, of equations where E_j governs the evolution of the component v_j of the flow. Finally the nonlinear interactions between distinct scales are modelled by an adequate coupling between these equations E_j.

B. In the thirties J. E. Littlewood and R. Paley elaborated an algorithm which permits to travel across scales while keeping an eye on the frequency content and the model proposed by P. Frick and V. Zimin can be viewed as a nonlinear version of Littlewood-Paley analysis. This precisely defines the paradifferential calculus. Other good news are coming. Indeed paradifferential operators [22], [30] have been successfully applied to Euler or Navier-Stokes equations by Jean-Yves Chemin and his students. Gustavo Ponce and Tosio Kato [66] used Calderón's commutator estimates in their work on Navier-Stokes equations. Calderón's bilinear operators are not *stricto sensu* paradifferential operators but are similar in spirit. Finally wavelet analysis is a fast and flexible alternative to the paradifferential calculus and times were ripe for using wavelets in fluid dynamics.

We now better understand why Marie Farge and Paul Federbush thought that wavelets are relevant in turbulence. Moreover they believed that *wavelet based Galerkin schemes* could overcome pseudospectral algorithms which are acknowledged to be the best solvers for Navier-Stokes equations.

It is surprising that this program did not become a success story. As often in science, something else was found. Marco Cannone was working on Paul Federbush's paper while he was preparing his Ph.D. He noticed that Federbush should have used Littlewood-Paley expansions instead of wavelet expansions. The proofs in [40] would have gained in concision and clarity. Then M. Cannone observed that a strategy due to Tosio Kato [62] but also used by Frederic Weissler in [114], was even more powerful. These authors did not use microlocal analysis or paradifferential operators. Size estimates on the velocity field only mattered in their proofs and oscillations were ignored. Wavelets seemed to be useless.

But M. Cannone reversed this trend when he discovered that *oscillations matter*. Here is the story of this achievement. As it was already mentioned, T. Kato was studying solutions $v(x,t) \in \mathcal{C}([0,\infty); L^3(\mathbf{R}^3))$ to the Navier-Stokes equations. In 1984, he proved global existence (in time) of such solutions when the L^3 norm of the initial condition is small enough. M. Cannone replaced T. Kato's sufficient condition $\|v_0\|_3 < \epsilon$ by a much weaker one which is satisfied whenever v_0 *is oscillating*. These oscillations are defined by the smallness of a Besov norm in a suitable Besov space B_p, the exact bound η_p depending on $p \in (3,\infty)$. An equivalent condition is given by simple size estimates on the wavelet coefficients of v_0 (Theorem 19.3). These methods did not yield the limiting case $p = \infty$. The best result in this direction was finally obtained by Herbert Koch and Daniel Tataru [67]. The Koch and Tataru criterion is given by simple size estimates on the wavelet coefficients of v_0 (Theorem 19.2).

My interest in nonlinear evolution equations oriented my research in another direction. In a joint work with Patrick Gérard and Frédéric Oru, I discovered new Gagliardo-Nirenberg inequalities which, in a certain sense, are related to the blow-up of solutions of the nonlinear heat equation. Here also, the lifetime of a solution is infinite (no blow-up) when the initial condition is oscillating and our new Gagliardo-Nirenberg inequalities are explaining the role played by these oscillations. Our approach does not yield new results here but places classical facts in a new perspective. The limiting case in these sharp Gagliardo-Nirenberg inequalities concerned functions of bounded variation and could not be reached by my approach. At that time Albert Cohen was investigating the properties of the wavelet coefficients of functions of bounded variation. In a joint work with Wolfgang Dahmen, Ingrid Daubechies, and Ron DeVore, he proved that these coefficients belong to weak-l^1 in two dimensions and obtained similar results in any dimension. This was exactly what I needed for completing my program. Then Michel Ledoux looked for a direct proof which would not use wavelet analysis. He succeeded and his remarkable approach will be detailed in Section 17.

It is time to return to the main problem raised by Marie Farge and Paul Federbush on wavelet based Galerkin schemes. Instead of asking for an efficient wavelet based solver, I asked myself another question. Let us assume that the initial velocity v_0 has a sparse wavelet expansion. Is this property preserved by the nonlinear evolution? Lorenzo Brandolese solved this problem in his Ph.D. dissertation. L. Brandolese proved that a property like $v_0 \in \mathcal{S}(\mathbf{R}^3)$ is immediately lost by the evolution unless v_0 satisfies an infinite number of algebraic identities. A sharp version of this result will be proved in Section 11.

Let us observe that L. Brandolese's work did not ruin Marie Farge's claims. Indeed she stressed that one should focus on the vorticity instead of the velocity. We will back her claims by proving that 3-D vortex patches exist in the following sense: if the initial vorticity ω_0 belongs to BV (the space of functions with bounded variation), then this property persists along the nonlinear evolution. We will also prove that a property like $\omega_0 \in \mathcal{S}(\mathbf{R}^3)$ is preserved by the evolution.

These notes are organized as follows. In Section 2, some classical facts about the nonlinear heat equation are presented and the connection with the sharp Gagliardo-Nirenberg inequalities is explained. Sections 3 to 12 are devoted to the Navier-Stokes equations. Some new results are given here with their proofs. The main results of L. Brandolese's Ph.D. are enunciated without proofs in Section 12. Finally Sections 13 to 18 should be viewed as an appendix. The functional analysis tools (wavelets, Littlewood-Paley analysis and BV) which are being used in the first part of these notes are given there. Two proofs of the sharp Gagliardo-Nirenberg inequalities are detailed.

These notes are dedicated to Lorenzo Brandolese, Marco Cannone, Giulia Furioli and Elide Terraneo.

2 A Model Case the Nonlinear Heat Equation

Two nonlinear heat equations, two nonlinear Schrödinger equations, and the Navier-Stokes equations are sharing the same translation and dilation invariance. The methods and tools which will be used on the Navier-Stokes equations will be tested on these four evolution equations. Here is the first nonlinear heat equation:

$$\begin{cases} \frac{\partial u}{\partial t} = \Delta u + u^3, & (x,t) \in \mathbf{R}^3 \times (0,\infty) \\ u(x,0) = u_0(x) \end{cases} \tag{2.1}$$

where $u = u(x,t)$ is a real valued function of (x,t), $x \in \mathbf{R}^3$ and $t \geq 0$.

In the second one $+u^3$ is replaced by $-u^3$ and in the last two ones, this nonlinearity becomes $\pm iu|u|^2$.

If $u(x,t)$ is a solution of (2.1), so are $\lambda u(\lambda x, \lambda^2 t)$ for $\lambda > 0$ and $u(x - x_0, t + \tau)$ for $x_0 \in \mathbf{R}^3, \tau \geq 0$. The initial condition is modified accordingly.

We now focus on (2.1). In the following calculation, u is assumed to be a classical solution to (2.1) with enough regularity and decay at infinity. It implies that L^p-norms will be finite by assumption and integrations by parts will be legitimate.

Multiplying (2.1) by u and integrating over $\mathbf{R}^3$ yields $\frac{d}{dt}\|u\|_2^2 = -2\|\nabla u\|_2^2 + 2\|u\|_4^4$ which suggests that the evolution depends on the competition between $\|u\|_4^4$ and $\|\nabla u\|_2^2$.

The following theorem by J. Ball, H. A. Levine and L. Payne [4] says more:

Theorem 2.1. *If u_0 is a smooth compactly supported function which does not vanish identically and if*

$$\|\nabla u_0\|_2 \leq \frac{1}{\sqrt{2}}\|u_0\|_4^2 \tag{2.2}$$

then the corresponding solution of (2.1) *blows up in finite time: there exists a finite T_0 such that $\|u(.,t)\|_2$ is unbounded as t approaches T_0.*

If $u_0(x) = \phi(\epsilon x)$ where ϕ is any testing function, then (2.2) holds when ϵ is small enough. Then the corresponding solution of (2.1) blows up in finite time. On the other hand, if $u_0(x) = \cos(\omega \cdot x)\phi(x)$ and $|\omega|$ tends to infinity, then the left-hand side of (2.2) tends to infinity while the right-hand side is a constant. The following conjecture complements Theorem 2.1 and says that such oscillations in the initial condition prevent blow-up.

Conjecture 2.1. There exists a Banach space $\mathcal{Z}$ whose norm is denoted by $\|.\|_{\mathcal{Z}}$ such that

$(\alpha)\ \|f\|_4^2 \leq C\|\nabla f\|_2\|f\|_{\mathcal{Z}}$
(β) If the initial condition u_0 is smooth, compactly supported and satisfies $\|u_0\|_{\mathcal{Z}} \leq \eta$, then the lifetime of the corresponding solution of (2.1) is infinite (here η is a positive constant).
$(\gamma)\ \|f_\lambda\|_{\mathcal{Z}} = \|f\|_{\mathcal{Z}}$ if $f_\lambda(x) = \lambda f(\lambda x), \quad \lambda > 0$.
$(\delta)\ \|f\|_{\mathcal{Z}}$ is small if f is oscillating.

If we accept this conjecture, then (2.2) implies that $\|u_0\|_{\mathcal{Z}}$ is large. Property (α) is a special case of some new Gagliardo-Nirenberg estimates which will be unveiled in these notes (Section 13 and 15). If the norm in $\mathcal{Z}$ is translation invariant, then we have $\dot{B}_1^{2,1} \subset \mathcal{Z} \subset \dot{B}_\infty^{-1,\infty}$. These two Besov spaces are the minimal and the maximal ones for which (γ) holds. We do not know whether or not this maximal Besov space fulfills (β) while (α) will be proved in Section 13.

Property (γ) and the invariance of (2.1) by suited scalings are obviously related. Let us be more precise. A solution $u(x,t)$ to (2.1) generates a family $u_\lambda, \lambda > 0$, of solutions by $u_\lambda(x,t) = \lambda u(\lambda x, \lambda^2 t)$. If E is a Banach space and if the norm in E obeys a scaling law where the exponent is not the one in (γ), then a condition like $\|u_0\|_E \leq \eta$ for some small η is meaningless. Indeed it suffices to replace the initial condition u_0 by $\lambda u_0(\lambda \cdot)$ and to choose the right value of λ to meet this smallness requirement. We then would reach a contradiction. Indeed if u is a global solution to (2.1), so is u_λ. This remark explains (γ). Property (γ) paves the road to the construction of self-similar solutions and two instances of this program will be studied in the case of Navier-Stokes equations (Theorem 6.2 and 10.1).

The simplest norm fulfilling (γ) is the L^3 norm for which Frederic Weissler [113] proved the following.

Theorem 2.2. *There exists a positive number η, such that $\|u_0\|_3 < \eta$ implies the existence of a global solution $u(x,t) \in \mathcal{C}([0,\infty); L^3(\mathbf{R}^3))$ to (2.1).*

Uniqueness was proved by F. Weissler under the condition that u belongs to a Banach space Y slightly smaller than the 'natural space' which is $\mathcal{C}([0,\infty); L^3(\mathbf{R}^3))$. This smaller space is defined by imposing a second condition on $u(x,t)$. This condition reads

$$t^{1/4}\|u(.,t)\|_6 \leq \eta', \quad t > 0 \tag{2.3}$$

where η' is a 'small' positive number.

Let us denote the heat semigroup $\exp(t\Delta)$, $t \geq 0$, by $S(t)$. We have $S(t)[f] = f \star \Phi(t)$ where $\Phi(t) = t^{-3/2}\Phi(x/\sqrt{t})$ and $\Phi(x)$ is the usual Gaussian function. Imposing (2.3) to the linear evolution $S(t)u_0$ is then equivalent to saying that u_0 belongs to the Besov space $\dot{B}_6^{-1/2,\infty}$ (a definition will be given in this section). This is not a restriction since this Besov space contains L^3.

In her thesis, Elide Terraneo constructed a striking counter-example showing that uniqueness of solutions $u(x,t)$ to (2.1) in $\mathcal{C}([0,\infty); L^3(\mathbf{R}^3))$ could not be expected in general [111]. This explains the role of (2.3).

We now follow F. Weissler and prove Theorem 2.2. The proof depends on the following remarks. In the first place we consider the linear space $\mathcal{H}$ of all solutions $u(x,t), x \in \mathbf{R}^3, t \in (0,\infty)$, of the linear heat equation $\frac{\partial u}{\partial t} = \Delta u$. Then many functional norms are equivalent on $\mathcal{H}$. For instance the three norms $\|u(\cdot,0)\|_3$, $\sup\{\|u(\cdot,t)\|_3, \ t \geq 0\}$ and $\|\sup_{t\geq 0}|u(\cdot,t)|\|_3$ are equivalent on $\mathcal{H}$. Secondly F. Weissler observed that this property is somehow preserved by the Nonlinearity in (2.1). This second observation will be specified in a moment. Let us mention that Weissler's proof preludes T. Kato's approach to the Navier-Stokes equations.

Using a technique which will be detailed below in Lemma 10, F. Weissler transformed the nonlinear heat equation $\frac{\partial u}{\partial t} = \Delta u + u^3$, with $u(x, 0) = u_0$ into

$$u(t) = S(t)u_0 + \int_0^t S(t - \tau)u^3(\tau)\, d\tau \qquad (2.4)$$

where $S(t) = \exp(t\Delta)$ is the heat semigroup and $u(x, t)$ is viewed as a vector-valued function of the time variable t.

Before moving further, let us stop and say a few words on these vector-valued functions of the time variable. The general methodology which will be used through these notes consists in looking for solutions $u(\cdot, t)$ belonging to the Banach space $X = \mathcal{C}([0, \infty); E)$, where E is a suitable Banach space of functions of x. The continuity in t will be given a precise definition. The norm of $u(\cdot, t)$ in $X = \mathcal{C}([0, \infty); E)$ is denoted by $\|u\|_X$ and defined as

$$\|u\|_X = \sup\{\|u(\cdot, t)\|_E;\ t \geq 0\} \qquad (2.5)$$

This norm will be called the *natural norm* in what follows. A solution $u \in X$ of (2.4) is called a mild solution of (2.1).

The collection of Banach spaces E which are being used splits into two classes. The first class consists of separable Banach spaces and the condition $u \in \mathcal{C}([0, \infty); E)$ means that u, as a function of t, is continuous when E is equipped with the topology defined by its norm.

The second class consists of Banach spaces E which are not separable but are dual spaces of some separable Banach space F. Then the continuity requirement will be more involved. The $\sigma(E, F)$ topology would have been the most natural choice. However this weak continuity will not suffice since it is not compatible with the nonlinearities of the problem. The following definition applies to both cases:

Definition 2.1. *If E is a separable Banach space, $u \in \mathcal{C}([0, \infty); E)$ simply means (a) the ordinary continuity with respect to the norm and (b) the boundedness of $\|u(t)\|_E$, $t \geq 0$. With these notations, the norm of $u \in \mathcal{C}([0, \infty); E)$ is $\sup\{\|u(t)\|_E,\ t > 0\}$.*

If the Banach space $E = F^\star$ is a nonseparable dual space, then the property $u \in \mathcal{C}([0, \infty); E)$ is defined by (c) and (d):

(c) at every $t_0 > 0, u(t)$ is continuous with respect to the topology defined by the norm

(d) at 0, we only impose a weak-star convergence: $u(t) \rightharpoonup u(0)$, $t \to 0$.

The norm is defined as in the first case.

F. Weissler used Picard's fixed-point theorem to solve (2.4). We postpone the proof of Theorem 2.2 and focus on Picard's theorem. A sharp version of this fixed-point theorem is given by the following lemma

Lemma 2.1. *Let X be a Banach space, let $\|\cdot\|$ denote the norm in X and let $B : X \times X \times X \to X$ be a trilinear mapping such that*

$$\|B(x,y,z)\| \leq C_0 \|x\| \, \|y\| \, \|z\| \tag{2.6}$$

If $\|x_0\| < \frac{2}{3\sqrt{3}} C_0^{-1/2}$, then the equation

$$x = x_0 + B(x,x,x), \ x \in X \tag{2.7}$$

has a unique solution satisfying $\|x\| < \frac{1}{\sqrt{3}} C_0^{-1/2}$ and this solution is the limit of the sequence x_n, $n \in \mathbf{N}$, defined by

$$x_{n+1} = x_0 + B(x_n, x_n, x_n) \tag{2.8}$$

Finally the mapping Φ defined by $\Phi(x_0) = \lim_{n \to \infty} x_n$ is analytic in the ball $\|x\| < \frac{2}{3\sqrt{3}} C_0^{-1/2}$.

The proof is straightforward. It consists in estimating $\|x_{n+1} - x_n\|$ by the numerical sequence r_n defined by

$$\begin{cases} r_{n+1} = C_0(3s_n^2 r_n + 3s_n r_n^2 + r_n^3) \\ s_{n+1} = s_n + r_n \\ s_0 = \|x_0\| \end{cases} \tag{2.9}$$

Then an obvious induction yields $\|x_{n+1} - x_n\| \leq r_n$, $n \in \mathbf{N}$. On the other hand the sequence s_n converges to the smallest positive solution of $s = \|x_0\| + C_0 s^3$ if and only if $\|x_0\| < \frac{2}{3\sqrt{3}} C_0^{-1/2}$. These two remarks imply Lemma 1. This argument will be repeated with more details in Section 5, Lemmata 5.1 and 5.2.

For applying Lemma 2.1 , one has to prove the estimate

$$\|\Gamma(u_1, u_2, u_3)\|_X \leq C \|u_1\|_X \|u_2\|_X \|u_3\|_X \tag{2.10}$$

for the trilinear operator defined by

$$\Gamma(u_1, u_2, u_3) = \int_0^t S(t-\tau) u_1(\tau) u_2(\tau) u_3(\tau) \, d\tau \tag{2.11}$$

We wrote $u(t) = u(\cdot, t)$ for keeping the notations as simple as possible.

If (2.10) holds and if $\|u_0\|_E < \frac{2}{3\sqrt{3}} C^{-1/2}$, Picard's fixed-point theorem applies and yields a mild solution. The constant C is defined by (2.10).

It is time to return to Weissler's proof of Theorem 2.2. We are looking for solutions $u(x,t) \in \mathcal{C}([0,\infty); L^3(\mathbf{R}^3))$ of (2.4). We therefore apply Lemma 2.1 to $E = L^3(\mathbf{R}^3)$.

Bad news is coming. *If Lemma 2.1 is applied to $E = L^3(\mathbf{R}^3)$, the fundamental trilinear estimate (2.10) is definitely incorrect.*

F. Weissler proposed three remedies for fixing this problem. In the **first remedy**, X is replaced by the Banach space $Y \subset X$ consisting of all functions such that

$$\begin{cases} u(\cdot, t) \in \mathcal{C}([0, \infty); L^3(\mathbf{R}^3)) \\ t^{1/4} u(\cdot, t) \in \mathcal{C}([0, \infty); L^6(\mathbf{R}^3)) \\ \lim_{t \to 0} t^{1/4} \|u(\cdot, t)\|_6 = 0 \\ \lim_{t \to \infty} t^{1/4} \|u(\cdot, t)\|_6 = 0 \end{cases} \tag{2.12}$$

The second and the third remedies will be discussed later on. *Three distinct norms* will be used. As above, the *natural norm* is

$$\|u\|_X = \sup\{\|u(\cdot, t)\|_3, \, t > 0\}$$

The second norm $\|u\|_Y$ is named the *strong norm* and is defined by

$$\|u\|_Y = \|u\|_X + \sup\{t^{1/4} \|u(\cdot, t)\|_6; \, t \geq 0\} \tag{2.13}$$

The third norm is the *weak norm* and is defined by

$$\|u\|_* = \sup\{t^{1/4} \|u(\cdot, t)\|_6, \, t \geq 0\} \tag{2.14}$$

The success of F. Weissler's approach relies on the following two facts:

Proposition 2.1. *Keeping the preceding notations, we have*

(a) (2.10) holds when $\|u\|_X$ is replaced by $\|u\|_Y$
(b) there exists a constant C such that $\|u\|_Y \leq C\|u\|_X$ when u is a solution of the linear heat equation.

Since $\|u\|_X \leq \|u\|_Y$, the strong norm and the natural norm are equivalent ones on $\mathcal{H}$.

The proof of Proposition 2.1 is simple. By Hölder's inequality $\|u_1 u_2 u_3(\cdot, t)\|_2 \leq t^{-3/4} \|u_1\|_Y \|u_2\|_Y \|u_3\|_Y$. Then the operator $S(t - \tau)$ pulls back L^2 to L^3 with an operator norm which is $C(t-\tau)^{-1/4}$ and also pulls back L^2 to L^6 with an operator norm which is $C(t - \tau)^{-1/2}$. The proof ends with $\int_0^t (t - \tau)^{-1/4} \tau^{-3/4} \, d\tau = C$ and $\int_0^t (t - \tau)^{-1/2} \tau^{-3/4} \, d\tau = C t^{-1/4}$. The proof of (b) reduces to Young inequalities if one writes $u = S(t)u_0$. Proposition 2.1 is proved and Picard's fixed point theorem implies Theorem 2.2.

In the **second remedy**, the L^6 norm is replaced by the L^∞ norm and (2.12) is replaced by $\|u(\cdot, t)\|_\infty \leq C t^{-1/2}$. The proofs are similar.

The **third remedy** is more original since the natural space and the corresponding norm are forgotten. Instead we consider the Banach space $\mathcal{M}$ of all measurable functions $f(x, t)$ on $\mathbf{R}^3 \times (0, \infty)$ such that $\sup\{t^{1/4} \|f(\cdot, t)\|_6; \, t \geq 0\}$ is finite. This norm is denoted by $\|f\|_*$. If $f(\cdot, t) = S(t)g$ where $g(x)$ is a function defined on $\mathbf{R}^3$, then $\|f\|_*$ is equivalent to the norm of g in the Besov space $\dot{B}_6^{-1/2, \infty}$. As above, one obtains

$$\|\Gamma(u_1, u_2, u_3)\|_Y \leq C\|u_1\|_*\|u_2\|_*\|u_3\|_* \tag{2.15}$$

and

$$\|S(t)u_0\|_Y \leq C\|u_0\|_3 \tag{2.16}$$

Moreover we also have

$$\|\Gamma(u_1, u_2, u_3)\|_{\dot{H}^{1/2}} \leq C\|u_1\|_*\|u_2\|_*\|u_3\|_* \tag{2.17}$$

Here $\dot{H}^{1/2} \subset L^3$ denotes the homogeneous Sobolev space.

The proof of (2.17) begins with Hölder's inequality as above. Then we need to compute the operator norm of $S(t) : L^2 \mapsto \dot{H}^{1/2}$. A simple calculation shows that this norm is $Ct^{-1/4}$ and the exponent here does not depend on the dimension. The proof ends with $\int_0^t (t - \tau)^{-1/4} \tau^{-3/4} d\tau = C$.

Finally Picard's fixed point theorem can be applied to (2.4) and to the Banach space $\mathcal{M}$. This third remedy will be used below in the proof of Theorem 2.5.

Some variants and improvements on Theorem 2.2 will be given now. One motivation is the construction of self-similar solutions of (2.1). A first group of self-similar solutions of (2.1) were discovered by A. Haraux and F. Weissler. They proved the following theorem:

Theorem 2.3. *There exists a (nontrivial) function $w(x)$ in the Schwartz class $\mathcal{S}(\mathbf{R}^3)$ such that*

$$u(x, t) = \frac{1}{\sqrt{t}} w\left(\frac{x}{\sqrt{t}}\right), \ t > 0, \ x \in \mathbf{R}^3, \tag{2.18}$$

is a global solution of (2.1).

The corresponding initial value $u_0(x) = 0$ identically. We observe that this solution belongs to $L^\infty((0, \infty); L^3)$. Therefore uniqueness does not hold in this setting. But the stronger condition $u \in \mathcal{C}([0, \infty); L^3)$ does not imply uniqueness either, as it was proved by Elide Terraneo. She constructed two distinct solutions $u(x, t)$ and $v(x, t)$ of (2.1) such that $u, v \in \mathcal{C}([0, \infty); L^3)$ and $u(\cdot, 0) = v(\cdot, 0)$. In her counter-example, u is the solution given by Theorem 2.2 while $\|v(\cdot, t)\|_6 = +\infty$ for $t > 0$. [111]

This situation sharply contrasts with what happens for Navier-Stokes equations. As it will be told in Section 5, T. Kato proved the analogue of Theorem 2.2. T. Kato's proof is close to F. Weissler's approach and (2.3) is playing a very important role in the construction. For quite a long time, uniqueness of Kato's solutions $v(x, t) \in C([0, \infty), L^3(\mathbf{R}^3))$ was an open problem. Finally uniqueness was proved by Giulia Furioli, Pierre-Gilles Lemarié-Rieusset and Elide Terraneo without assuming (2.3). The interested reader is referred to [47], [78] or to Section 7.

We now return to the problems raised by Conjecture 1. Theorem 2.2 says that the space $\mathcal{Z} = L^3(\mathbf{R}^3)$ and the norm $\|.\|_3$ fulfill (β). Moreover one has $\|f\|_4^2 \leq C\|\nabla f\|_2\|f\|_3$ (see Section 13). In other words, the space $\mathcal{Z} = L^3$ is meeting the requirements $(\alpha), (\beta), (\gamma)$. However (δ) is not satisfied. But the L^3 norm can be replaced by a much weaker one for which $(\alpha), (\beta), (\gamma)$ and (δ) hold. This weaker norm is a Besov norm. The relevance of Besov norms in (α) is explained by the following Gagliardo-Nirenberg inequality which will be proved in Section 13.

Lemma 2.2. *For any function f belonging to the homogeneous Sobolev space $\dot{H}^1$, we have*

$$\|f\|_4^2 \le C\|\nabla f\|_2 \|f\|_B \qquad (2.19)$$

where B is the homogeneous Besov space $\dot{B}_\infty^{-1,\infty}$.

Lemma 2.2 suggests that the norm $\|f\|_B$ in the Besov space $\dot{B}_\infty^{-1,\infty}$ might be the one we are looking for. This Besov norm is the weakest one since $\dot{B}_\infty^{-1,\infty}$ is the largest function space whose norm is translation invariant and fulfills (γ). But we do not know if (β) holds for this maximal Besov space. Nevertheless the following theorem gives an example of a norm fulfilling conditions (α) to (δ) which is weaker than the L^3 norm but unfortunately stronger than the norm in the Besov space $\dot{B}_\infty^{-1,\infty}$. In the following theorem, $\dot{W}^{1,3/2}$ denotes the homogeneous Sobolev space of functions whose gradient belongs to $L^{3/2}$. This homogeneous Sobolev space $\dot{W}^{1,3/2}$ is contained in L^3. The homogeneous Besov space $\dot{B}_6^{-1/2,\infty}$ is also needed. We start with $\varphi(x) = (2\pi)^{-3/2}\exp(-|x|^2/2)$ and let S_j be the convolution operator with $2^{3j}\varphi(2^j x), j \in \mathbf{Z}$. Then we have (see Section 18)

Definition 2.2. *A tempered distribution f belongs to $\dot{B}_6^{-1/2,\infty}$ if and only if a constant C exists such that $\|S_j(f)\|_6 \le C2^{j/2}, \ j \in \mathbf{Z}$. The optimal C being the norm of f in $\dot{B}_6^{-1/2,\infty}$.*

With these notations and definitions we have

Theorem 2.4. *Let $\|.\|_{\mathcal{Z}}$ denote the norm in the homogeneous Besov space $\mathcal{Z} = \dot{B}_6^{-1/2,\infty}$. There exists a positive number η such that, if the initial condition u_0 satisfies $u_0 \in L^3$ and $\|u_0\|_{\mathcal{Z}} \le \eta$, then the corresponding solution of the nonlinear heat equation is global in time and belongs to $\mathcal{C}([0,\infty); L^3(\mathbf{R}^3))$. Moreover $u(t) = S(t)u_0 + w(t)$ where $w(\cdot, t) \in \mathcal{C}([0,\infty); \dot{W}^{1,3/2})$,*

$$\|w(\cdot,t)\|_{\dot{W}^{1,3/2}} \le C\|u_0\|_3 \qquad (t \ge 0) \qquad (2.20)$$

and $\|w(\cdot,t)\|_{\dot{W}^{1,3/2}} \to 0 \ (t \to 0)$.

Theorem 2.4 implies $u(\cdot,t) \in \mathcal{C}([0,\infty); \dot{W}^{1,3/2})$ iff $u_0 \in \dot{W}^{1,3/2}$. In M. Cannone's Ph.D. [17], $S(t)u_0$ is called the *trend* and $u(t) - S(t)u_0$ is the *fluctuation*. Here the fluctuation is more regular than the trend.

Theorem 2.4 is not optimal but it improves on Theorem 2.2. On the one hand,

$$L^3 \subset \dot{B}_6^{-1/2,\infty} \subset \dot{B}_\infty^{-1,\infty} \qquad (2.21)$$

and these embeddings are provided by Bernstein's inequalities. On the other hand, Theorem 2.4 is consistent with the guess that the oscillating character of the initial condition implies the global (in time) existence of the corresponding solution. Indeed one can easily check that

$$\limsup_{\omega \to \infty} |\omega|^{1/2} \| \cos(\omega x)\varphi(x)\|_{\mathcal{Z}} \leq C\|\varphi\|_6 \qquad (2.22)$$

Here ϕ has the same meaning as in Definition 2.2. The smallness requirement is met when $|\omega|$ is large enough, and the corresponding solution is global in time. A last observation concerns scale invariance. The norm in $B_6^{-1/2,\infty}$ has the same invariance as the L^3 norm does and this invariance is consistent with the one we found in the nonlinear heat equation.

The proof of Theorem 2.4 will permit to obtain more. Let us denote by $\dot{H}^{1/2}$ the usual homogeneous Sobolev space defined by the norm $(\int |\widehat{f}(\xi)|^2 |\xi| d\xi)^{1/2}$ where $\widehat{f}$ is the Fourier transform of f. This Sobolev space is contained in L^3. Then we have

Theorem 2.5. *Let us replace the assumption $u_0 \in L^3$ in Theorem 2.4 by $u_0 \in \mathcal{Z} = \dot{B}_6^{-1/2,\infty}$ and $\|u_0\|_{\mathcal{Z}} \leq \eta$. Then the algorithm used in the proof of Theorem 2.4 yields a solution $u \in \mathcal{C}([0,\infty); \mathcal{Z})$ with the following property: $u(\cdot,t) = S(t)u_0 + w(\cdot,t)$, $w(\cdot,t) \rightharpoonup 0$, $(t \to 0)$ and*

$$\|w(\cdot,t)\|_{\dot{H}^{1/2}} \leq C\|u_0\|_{\dot{B}_6^{-1/2,\infty}} \quad (t \geq 0) \qquad (2.23)$$

Here the continuity refers to the weak* topology of $\mathcal{Z}$. As a simple counterexample will show one cannot expect $w \in \mathcal{C}([0,\infty); \dot{H}^{1/2})$. We have instead $w \in L^\infty([0,\infty); \dot{H}^{1/2})$.

Let us sketch the proofs of Theorem 2.4 and of Theorem 2.5. It paves the way to the proof of Theorem 6.2. For proving Theorem 2.4, we forget $\mathcal{C}([0,\infty); L^3(\mathbf{R}^3))$ for a while. Instead we consider the Banach space $\mathcal{M}$ of all measurable functions $f(x,t)$ on $\mathbf{R}^3 \times (0,\infty)$ such that $\sup\{t^{1/4}\|f(\cdot,t)\|_6; t \geq 0\}$ is finite. We already noticed that the trilinear operator Γ satisfies the following two estimates

$$\|\Gamma(u_1,u_2,u_3)\|_* \leq C\|u_1\|_*\|u_2\|_*\|u_3\|_* \qquad (2.24)$$

and

$$\|\Gamma(u_1,u_2,u_3)(\cdot,t)\|_3 \leq C\|u_1\|_*\|u_2\|_*\|u_3\|_* \qquad (2.25)$$

Let us denote by $\mathcal{M}_0$ the closure in $\mathcal{M}$ of the space of compactly supported continuous functions. Then the information given by (2.24) and (2.25) can be completed with $\Gamma(u_1,u_2,u_3)(\cdot,t) \in \mathcal{C}([0,\infty); L^3(\mathbf{R}^3))$ if u_1,u_2,u_3 belong to $\mathcal{M}_0$. Therefore the 'artificial space' Y is not needed and Lemma 2.1 can be applied to $\mathcal{M}_0$.

The proof of (2.20) is slightly more involved. The estimates $\|u(\cdot,t)\|_3 \leq C$, $\|u(\cdot,t)\|_\infty \leq C't^{-1/2}$ yield $\|u^3(\cdot,t)\|_{3/2} \leq Ct^{-1/2}$. We plug this information into $\nabla \int_0^t S(t - \tau)u^3(\tau)\,d\tau$ and we glue together $\nabla S(t - \tau)$. The resulting operator is a convolution with a function in L^1. This L^1 norm does not exceed $C(t - \tau)^{-1/2}$. The proof ends with $\int_0^t (t - \tau)^{-1/2}\tau^{-1/2}\,d\tau = \pi$.

The proof of Theorem 2.5 opens a new and important debate. The question is: *What is the meaning of u^3 when u is a distribution?* We all know that distributions or generalized functions can be solutions of a linear PDE. Is this possible with nonlinear PDE? Jean-Michel Bony created the theory of paradifferential operator for solving this problem. Here we use another strategy. The Littlewood-Paley theory tells us that the oscillating behavior of a function or a distribution $f(x)$ is equivalent to size estimates on the linear evolution $S(t)f$. That is why we forget once more the natural space $\mathcal{C}([0, \infty); \mathcal{Z})$ and use the space $\mathcal{M}$ instead. The subspace $\mathcal{M}_0$ does not suffice here. Then the meaning of u^3 is clear and the proof of Theorem 2.5 is similar to the preceding one. We play with L^6 norms and we have $\|u^3(\cdot, t)\|_2 \le Ct^{-3/4}$. Then the operator $S(t - \tau)$ maps L^2 into $\dot{H}^{1/2}$ with an operator norm not exceeding $C(t - \tau)^{-1/4}$. The proof ends with $\int_0^t (t - \tau)^{-1/4}\tau^{-3/4}\,d\tau = C$.

Theorem 2.4 and Theorem 2.5 being proved, we now treat an example. Let us now assume that the initial condition u_0 is a self-similar function: $\lambda u_0(\lambda x) = u_0(x)$, $\lambda > 0$. In fact any function which is homogeneous of degree -1 and smooth on $\mathbf{R}^3 \setminus \{0\}$ belongs to $\mathcal{Z}$. An optimal result can be found in [17]. Then the solution of (2.1) associated to this self-similar initial condition is also self-similar. Using this remark, one easily builds a self-similar solution which is radial and positive. We start with $u_0(x) = \eta|x|^{-1}$ (η being small enough) which is the simplest example of a function in $\dot{B}_6^{-1/2, \infty}(\mathbf{R}^3)$ and we apply the iterative scheme which is described by Lemma 2.1. Each iteration provides us with a function which is positive and radial and so is the limit. From this example we clearly see that oscillating patterns are not needed for building self-similar solutions. We also observe that these self-similar solutions are far away from the ones discovered by Haraux and F. Weissler. Indeed our self-similar solutions are given by $u(x, t) = S(t)[u_0](x) + \frac{1}{\sqrt{t}}W\left(\frac{x}{\sqrt{t}}\right)$ where $W(x)$ belongs to $\dot{H}^{1/2}$. We see that $w(x, t) = \frac{1}{\sqrt{t}}W\left(\frac{x}{\sqrt{t}}\right)$ tends to 0 in the weak sense but does not tend to 0 in norm. This is the counter-example which was announced in Theorem 2.5. Finally u_0 is the weak* limit of $u(x, t)$ as t tends to 0. This contrasts with the Haraux and F. Weissler self-similar solution which weakly tends to 0 as t tends to 0.

Before moving to the Navier-Stokes equations, let us say a few words on the nonlinear heat equation with the *wrong sign*. It reads

$$\begin{cases} \frac{\partial u}{\partial t} = \Delta u - u^3, & (x, t) \in \mathbf{R}^3 \times (0, \infty) \\ u(x, 0) = u_0(x) \end{cases} \tag{2.26}$$

Then $\|u(\cdot, t)\|_2$ is a decreasing function of the time variable and this excludes blow-up in finite time or self-similar solutions given by $u(x, t) = \frac{1}{\sqrt{t}}U\left(\frac{x}{\sqrt{t}}\right)$ where U belongs to L^2. Indeed, for such a solution $\|u(\cdot, t)\|_2 = C\sqrt{t}$ which is increasing. On the other hand, the theorems which are based on perturbation arguments are still valid in this setting.

We now consider the nonlinear Schrödinger equation which obeys the same scaling laws as the two preceding nonlinear PDE's. There are indeed two such equations depending on a $\pm$ sign. The Schrödinger equations with critical nonlinearity are the following evolution equations:

$$\begin{cases} i\frac{\partial u}{\partial t} + \Delta u = \epsilon|u|^2 u \\ u(x,0) = u_0, \quad x \in \mathbf{R}^3, \, t \in [0,\infty) \end{cases} \tag{2.27}$$

where ϵ is either -1 or 1 and $u = u(x,t)$ is a complex valued function defined on $\mathbf{R}^3 \times (0,\infty)$. If λ is any positive scale factor, then for every solution $u(x,t)$ of (2.27), $\lambda u(\lambda x, \lambda^2 t)$ is also a solution of (2.27) for which the initial condition is $\lambda u_0(\lambda x)$. Therefore it is not unnatural to expect some similarities with both the nonlinear heat equation and the Navier-Stokes equations.

More precisely we follow T. Kato and H. Fujita and expect (2.27) to be well posed for the critical Sobolev space $\dot{H}^{1/2}(\mathbf{R}^3)$. Cazenave and F. Weissler [20] proved that it was the case under a smallness condition on the norm of the initial condition in $\dot{H}^{1/2}(\mathbf{R}^3)$. Fabrice Planchon [92] extended this theorem and replaced the smallness condition $\|u_0\|_{\dot{H}^{1/2}} \leq \eta$ by a weaker requirement which reads $\|u_0\|_* \leq \eta$ where the norm $\| \cdot \|_*$ is the homogeneous Besov norm in $\dot{B}_2^{1/2,\infty}$.

Theorem 2.6. *With the preceding notations, there exists a positive constant η such that for every initial condition u_0 in $\dot{H}^{1/2}(\mathbf{R}^3)$ satisfying $\|u_0\|_* \leq \eta$, there exists a solution $u(\cdot,t)$ to the Schrödinger equation (2.27) which belongs to $\mathcal{C}([0,\infty); \dot{H}^{1/2}(\mathbf{R}^3))$.*

This theorem can be compared to Theorem 2.4. However the Besov space which is now used has a positive regularity index. Theorem 2.6 implies the existence of many self-similar solutions to the nonlinear Schrödinger equation. Such solutions were previously proved to exist by Cazenave and F. Weissler [21] under much more restrictive regularity assumptions.

The experience we gained on these four nonlinear equations will now be used to attack the much more difficult Navier-Stokes equations.

3 Navier-Stokes Equations

We now consider the Navier-Stokes equations decribing the motion of some incompressible fluid. The fluid is assumed to be filling the space and there are no exterior forces. Then the Navier-Stokes equations can be written as follows

$$\begin{cases} (a) & \frac{\partial v}{\partial t} = \Delta v - (v.\nabla)v - \nabla p \\ (b) & \nabla.v = 0 \\ (c) & v(x,0) = v_0 \end{cases} \tag{3.1}$$

Here $v(x,t) = (v_1(x,t), v_2(x,t), v_3(x,t))$ is the velocity at time $t \geq 0$ and position $x \in \mathbf{R}^3$. The pressure $p = p(x,t)$ is a scalar and the Navier-Stokes equations are a system of four equations with four unknown functions v_1, v_2, v_3 and p.

The vector ∇p is the gradient of the pressure p. We have $v \cdot \nabla = v_1\partial_1 + v_2\partial_2 + v_3\partial_3$. It yields $(v \cdot \nabla)v = v_1\partial_1 v + v_2\partial_2 v + v_3\partial_3 v$ which is a vector. The divergence of v is $\operatorname{div} v = \nabla \cdot v = \partial_1 v_1 + \partial_2 v_2 + \partial_3 v_3$.

If the velocity $v(x,t)$ is not a smooth function of x, then multiplying some components of v with derivatives of some other components might be impossible. That is why $(v \cdot \nabla)v$ should be rewritten as $\partial_1(v_1 v) + \partial_2(v_2 v) + \partial_3(v_3 v)$ which makes sense whenever v is locally square integrable.

The initial condition makes sense when $v(x,t)$ is a continuous function of the time variable. For a Leray solution, it is the case since $v \in \mathcal{C}([0,T], L_w^2(\mathbf{R}^3))$ where L_w^2 means L^2 equipped with the weak topology. Then $v(\cdot, 0)$ makes sense.

Navier-Stokes equations have some remarkable scale invariance properties. First they commute with translations in x and $t \geq 0$. Moreover if the pair $(v(x,t), p(x,t))$ is a solution of (3.1) and if for every $\lambda > 0$ we dilate this solution into

$$\begin{cases} v_\lambda(x,t) = \lambda u(\lambda x, \lambda^2 t) \\ p_\lambda(x,t) = \lambda^2 p(\lambda x, \lambda^2 t) \end{cases} \tag{3.2}$$

then $(v_\lambda(x,t), p_\lambda(x,t))$ is also a solution to the Navier-Stokes equations. The initial condition is replaced by

$$v_\lambda(x,0) = \lambda v_0(\lambda x) \tag{3.3}$$

We observe that this scale invariance is exactly the same as the one we met in the nonlinear heat equation.

The Navier-Stokes system consists of 4 equations with 4 unknown functions. If one gets rid of the pressure, the dimensionality is reduced to 3. There exist three distinct algorithms for doing it. A first approach consists in applying the curl operator to (3.1). A second one uses the Leray-Hopf operator and will be treated next. The last one consists in applying the divergence operator to (3.1).

Let us analyze the first algorithm. The curl of the velocity is named the *vorticity* and will be relevant in Section 8. This first algorithm is using the following definitions and remarks.

Definition 3.1. *If $u(x) = (u_1(x), u_2(x), u_3(x)), x \in \mathbf{R}^3$, is a vector field, then* $\operatorname{curl}(u) = \omega$ *is defined by* $\omega_1 = \partial_2 u_3 - \partial_3 u_2$, $\omega_2 = \partial_3 u_1 - \partial_1 u_3$, $\omega_3 = \partial_1 u_2 - \partial_2 u_1$.

For inverting the curl operator, it suffices to use the following lemma:

Lemma 3.1. *If u is a divergence free vector field, we then have*

$$\text{curl}(\text{curl}\,u) = -\Delta u \tag{3.4}$$

Let us now assume that u vanishes at infinity in some weak sense. Then $\text{curl}(u) = \omega$ can be inverted into

$$u = -(\Delta)^{-1}\text{curl}(\omega) \tag{3.5}$$

The operator $(\Delta)^{-1}\text{curl}$ is an integral operator whose kernel is explicit and (3.5) reads

$$-4\pi u(x) = \int_{\mathbf{R}^3} |x - y|^{-3}(x - y) \times \omega(y)dy \tag{3.6}$$

This identity is the Biot-Savart law.

For instance, the operator $\mathcal{C} : u \mapsto \text{curl}(u)$ is an isomorphism between the Hilbert space of divergence free vector fields belonging to the Sobolev space $\dot{H}^1$ and the Hilbert space of divergence free vector fields belonging to L^2.

This isomorphism and the following lemma will yield our first algorithm. We plan to apply $\mathcal{C}$ to both sides of (a) in (3.1). The calculation relies on the following:

Lemma 3.2. *If $\text{div}\,u = 0$, then*

$$\text{curl}[(u \cdot \nabla)u] = (u \cdot \nabla)\omega - (\omega \cdot \nabla)u \tag{3.7}$$

The proof is a simple calculation. Using (3.7) and the isomorphism $\mathcal{C}$, we obtain a new formulation fo the Navier-Stokes equations. It reads

Lemma 3.3. *The evolution of the velocity v is governed by the Navier-Stokes equation if and only if the corresponding vorticity ω satisfies the following:*

$$\begin{cases} \frac{\partial \omega}{\partial t} = \Delta \omega + \partial_j(\omega_j v - v_j \omega) \\ \nabla \cdot \omega_0 = 0 \\ \omega(x, 0) = \omega_0(x) \\ v = -(\Delta)^{-1}\text{curl}(\omega) \end{cases} \tag{3.8}$$

We got rid of the pressure, as announced.

The second algorithm uses the *Leray-Hopf projection*.

Definition 3.2. *The Leray-Hopf operator $\mathbb{P}$ is defined as the orthogonal projector from $(L^2(\mathbf{R}^3))^3$ onto the closed subspace of divergence free vector fields.*

We then have

Lemma 3.4. *The Leray-Hopf operator can be computed through the following identity:*

$$\mathbb{P} = (-\Delta)^{-1}\,\text{curl}\,\text{curl} \tag{3.9}$$

The proof is simpler on the Fourier transform side. Let us denote by ξ the frequency vector, by $\hat{v}(\xi) = \int_{\mathbf{R}^3} \exp(-ix \cdot \xi) v(x)\, dx$ the Fourier transform of v and by Π_ξ the plane passing through 0 which is orthogonal to ξ. Then the Fourier transform of curl(v) is the exterior product $i\xi \times \hat{v}$ while $\mathbb{P}$ is the orthogonal projection upon Π_ξ. These remarks being made, (3.9) is obvious.

The Leray-Hopf operator is a pseudodifferential operator of order 0. It belongs to the algebra generated by the Riesz transformations $R_j, j = 1, 2, 3$. The Riesz transformation are defined by $R_j = \partial_j(-\Delta)^{-1/2}$ where Δ is the ordinary Laplace operator. Let us be more precise.

Lemma 3.5. *Let $f(x) = (f_1(x), f_2(x), f_3(x))$ be a vector field in $L^2(\mathbf{R}^3)$. We then have*

$$\mathbb{P}(f_1, f_2, f_3) = (\sigma - R_1\sigma, \ \sigma - R_2\sigma, \ \sigma - R_3\sigma) \tag{3.10}$$

where $\sigma = \sigma(f) = R_1 f_1 + R_2 f_2 + R_3 f_3$. It implies that $\mathbb{P}$ acts boundedly on all spaces which are preserved by the Riesz transforms R_1, R_2 and R_3.

Let $M(\xi)$ be the 3×3 matrix of the orthogonal projection P_ξ on the plane Π_ξ. We then have

Lemma 3.6. *The Leray Hopf projector $\mathbb{P}$ is the pseudodifferential operator of order 0, acting on $(L^2)^3$, and defined by one of the equivalent conditions:*

(a) the operator $\mathbb{P}$ is the orthogonal projection from $(L^2)^3$ onto the closed subspace consisting of divergence free vector fields
(b) the symbol of $\mathbb{P}$ is $M(\xi)$

The relation between the translation invariant operator $\mathbb{P}$ and its symbol M is given by $\mathcal{F}T(v)(\xi) = M(\xi)\mathcal{F}v(\xi)$. Here $\mathcal{F}$ denotes the Fourier transformation.

Finally we obtain

Lemma 3.7. *The Navier-Stokes equations can be written as follows*

$$\frac{\partial v}{\partial t} = \Delta v - \mathbb{P}[\partial_1(v_1 v) + \partial_2(v_2 v) + \partial_3(v_3 v)] \tag{3.11}$$

The third algorithm consists in applying the divergence operator to (3.1). We obtain

$$\Delta p = -\partial_j \partial_k (u_j u_k) \tag{3.12}$$

with the usual summation convention on repeated indices.

Assuming that the pressure p tends to 0 at infinity, (3.12) can be inverted into $p = -R_j R_k(u_j u_k)$ and it suffices to plug p into (3.1) to obtain (3.11). In other words, this third algorithm is, in a sense, identical to the second one.

This section will end with the semigroup formulation of Navier-Stokes equations. It begins with the following lemma were H^{-3} denotes the dual space of the Sobolev space H^3. The heat semigroup is $S(t) = \exp(t\Delta)$.

Lemma 3.8. *Let $p \in L^\infty([0,T]; H^{-2})$, $g = (g_1, g_2, g_3) \in L^\infty([0,T]; H^{-3})$, $v_0 \in H^{-3}$ be given. Then there exists a unique solution $v(x,t) \in \mathcal{C}([0,T]; H^{-3})$ of the linear heat equation*

$$\begin{cases} \frac{\partial v}{\partial t} = \Delta v + g - \nabla p \\ \nabla \cdot v = 0 \\ v(x,0) = v_0 \end{cases} \tag{3.13}$$

and this solution is

$$v(x,t) = S(t)v_0 + \mathbb{P} \int_0^t S(t - \tau)g(\cdot, \tau)\, d\tau \tag{3.14}$$

The Sobolev space H^{-3} is here used for convenience. In the application we have in mind, any H^s, $s < -5/2$ suffices. Before proving Lemma 3.8, let us explain the role played by the Sobolev space H^{-3}. We plan to apply Lemma 3.8 to the Navier-Stokes equations. We will first focus on Leray solutions which belong to $\mathcal{C}([0,T]; L_w^2)$ where the index means that L^2 is equipped with its weak topology. Then g will be defined as $g = \partial_j(v_j v)$. It implies that the Fourier transform $G(\xi, t)$ of $g(x,t)$ can be written as $G(\xi, t) = \xi H(\xi, t)$ where $H(\xi, t)$ belongs to the Wiener algebra (as a Fourier transform of products $u_j u_k$ of two functions in L^2). Therefore $H(\xi, t)$ is bounded. Its product with ξ is $O(|\xi|)$ at infinity and $\xi H(\xi, t)$ is square integrable against $(1 + |\xi|)^{-6}$ as announced. Similar remarks apply to v_0.

Lemma 3.8 is also relevant in the following situation. We now consider solutions $v \in \mathcal{C}([0,T]; L^3)$ of Navier-Stokes equations. Then all products $v_j v_k$ belong to $L^{3/2}$ uniformly in time. Their Fourier transforms belong to L^3 (Hausdorff-Young theorem) and this suffices for obtaining $g \in L^\infty([0,T]; H^{-3})$.

The proof of Lemma 3.8 is routine but we will sketch it for the convenience of the reader. We perform a Fourier transformation with respect to x. The partial Fourier transform of $v(x,t)$ is denoted by $V(\xi, t)$, $V_0(\xi)$ is the Fourier transform of $v_0(x)$ and $\lambda(\xi, t)$ denotes the partial Fourier transform of the pressure p. We then obtain an ordinary differential equation which reads

$$\frac{\partial V}{\partial t} = -|\xi|^2 V + G(\xi, t) + i\xi\lambda \tag{3.15}$$

Let $M(\xi)$ be the 3×3 matrix of the orthogonal projection P_ξ on the plane defined by $\xi \cdot y = 0$. This projector is given by $P_\xi(y) = y - |\xi|^{-1}(\xi \cdot y)$. The entries of $M(\xi)$ are obviously bounded by 1. We know that $V(\xi, t) \cdot \xi = 0$. Applying $M(\xi)$ or P_ξ to (3.15), one gets rid of $i\xi\lambda$ and obtains

$$\frac{\partial V}{\partial t} = -|\xi|^2 V + M(\xi)G(\xi, t), \quad V(\xi, 0) = V_0(\xi) \tag{3.16}$$

This is an ordinary differential equation where V and G belong to $L^2(\mathbf{R}^3, (1 + |\xi|)^{-6})$ uniformly in the time variable t. Therefore (3.15) implies

$$V(\xi, t) = \exp(-t|\xi|^2)V_0(\xi) + M(\xi)\int_0^t \exp[-(t-\tau)|\xi|^2]G(\xi, \tau)\, d\tau \qquad (3.17)$$

An inverse Fourier transform gives (3.14).

Returning to solutions of the Navier-Stokes equations, we have

Lemma 3.9. *If* $v \in \mathcal{C}([0, T], L_w^2(\mathbf{R}^3))$ *or* $v \in \mathcal{C}([0, T]; L^3)$, *then* $v(x, t)$ *is a solution of Navier-Stokes equations if and only if*

$$v(x, t) = S(t)v_0 - \int_0^t S(t-\tau)\mathbb{P}\partial_j[v_j v](\cdot, \tau)\, d\tau \qquad (3.18)$$

This is an obvious corollary of Lemma 3.8.

The bilinear version of the quadratic operator $\int_0^t S(t-\tau)\mathbb{P}\partial_j[v_j v](\cdot, \tau)\, d\tau$ will be denoted by $\mathcal{B}(v^{(1)}, v^{(2)})$ and (3.18) will be rewritten as

$$v = \tilde{v}_0 - \mathcal{B}(v, v) \qquad (3.19)$$

where $\tilde{v}_0 = S(t)v_0$. This bilinear operator $\mathcal{B}(v^{(1)}, v^{(2)})$ is playing a key role in our understanding of the Navier-Stokes equations.

4 The L^2-theory is Unstable

One of the most fascinating problem concerning Navier-Stokes equations is the uniqueness of Leray's solutions. We will shortly recall what a Leray solution is. Then we will prove that the L^2 theory is somehow unstable. It means that one cannot find a $\mathcal{C}^2$ mapping $\Phi : L^2(\mathbf{R}^3) \mapsto \mathcal{C}([0, T], L_w^2(\mathbf{R}^3))$ which maps the initial u_0 data to a Leray solution $u(x, t)$. *We will prove that the only stable theory is based on L^3 or on spaces with the same scaling.* This remark preludes the following sections.

Among other requirements, a Leray solution to the Navier-Stokes equations satisfies the following properties,

$$\begin{cases} v \in L^2([0, T]; H^1(\mathbf{R}^3)) \\ v \in \mathcal{C}([0, T], L_w^2(\mathbf{R}^3)) \end{cases} \qquad (4.1)$$

Again L_w^2 means L^2 equipped with its weak topology.

The lack of stability of Leray's solutions has several meanings. We could not succeed in proving the following:

Conjecture 4.1. For $t_0 > 0$, there exists a Leray solution $u(x, t)$ with $u(\cdot, 0) = u_0 \in L^2$, and a positive η such that, for each positive ϵ, one can find a second Leray solution $\tilde{u}(x, t)$ with $\tilde{u}(\cdot, 0) = \tilde{u}_0$ satisfying the following two properties:

$$\|\tilde{u}_0 - u_0\|_2 \leq \epsilon \qquad (4.2)$$

while

$$\|\tilde{u}(\cdot, t_0) - u(\cdot, t_0)\|_2 \geq \eta \tag{4.3}$$

This conjecture is obviously related to the problem of the uniqueness of Leray's solutions: a bifurcation obviously implies (4.3)

Unable to reach this goal, we will prove a weaker theorem where the lack of continuity is replaced by the lack of differentiability.

We first need to define the regularity of a mapping between two Banach spaces E and F.

Definition 4.1. *A mapping $\Phi : E \mapsto F$ is C^2 at 0 if one can find a bounded linear operator $T : E \mapsto F$ and a bicontinuous bilinear operator $B : E \times E \mapsto F$ such that*

$$\begin{cases} \Phi(x) = \Phi(0) + T(x) + B(x,x) + \rho(x) \\ \|\rho(x)\|_F = o(\|x\|_E^2) \end{cases} \tag{4.4}$$

when x belongs to some neighborhood of 0 in E.

The lack of stability of the L^2 theory is detailed in the following theorem:

Theorem 4.1. *Let us consider the Banach spaces $E = L^2(\mathbf{R}^3)$ and $F = \mathcal{C}([0,T], L_w^2(\mathbf{R}^3))$. If a mapping $\Phi : E \mapsto F$ has the property that, for each $u_0(x) \in L^2(\mathbf{R}^3), u(x,t) = \Phi(u_0)$ is a Leray solution of Navier-Stokes equations with initial condition u_0, then Φ cannot be $C^2(E; F)$ at 0.*

Since we do not know whether a Leray solution is uniquely defined by its initial condition, several such mappings Φ might exist.

The proof relies on the theory of bilinear operators $B : E \times E \mapsto F$, where E and F are two functional Banach spaces. Let us assume that translations act continously on E and F. Let us also assume that B commutes with simultaneous translations. It means the following property. We denote by $\quad x_0 : E \mapsto E$ (or $F \mapsto F$) the translation by $x_0 \in \mathbf{R}^3$ defined by $\quad x_0 f(x) = f(x - x_0)$ and we impose the following property: $B(\quad x_0 f, \quad x_0 g) = \quad x_0 B(f,g), \; f, g \in E$. Then such a bilinear operator is uniquely defined by its bilinear symbol. The bilinear symbol of B is a function $m(\xi, \eta), \xi \in \mathbf{R}^3, \eta \in \mathbf{R}^3$, which is defined by the following property:

Lemma 4.1. *With the preceding notations and definition, the relation between the bilinear operator and its symbol is given by*

$$B(e_\xi, e_\eta) = m(\xi, \eta)e_{\xi+\eta}, \; e_\xi(x) = \exp(i\xi \cdot x) \tag{4.5}$$

or in a more elaborate way by

$$B(f,g) = (2\pi)^{-6} \int \int \exp(ix \cdot (\xi + \eta)) m(\xi, \eta) \hat{f}(\xi) \hat{g}(\eta) \, d\xi \, d\eta \tag{4.6}$$

where $\hat{f}$ denotes the Fourier transform of f.

Getting a bound on $\|B(f,g)\|_F$ when $\|f\|_E \leq 1, \|g\|_E \leq 1$, is not as easy as one might guess.

Returning to Leray solutions, Lemma 3.8 implies

$$v(x,t) = S(t)v_0 - \int_0^t S(t-\tau)\mathbb{P}\partial_j[v_j v](\cdot,\tau)\,d\tau \qquad (4.7)$$

and a scalar and bilinear version of this quadratic operator is

$$B(f,g) = \int_0^t S(t-\tau)\Lambda(fg)\,d\tau \qquad (4.8)$$

where

(a) Λ is either the Calderón operator $(-\Delta)^{1/2}$ or $R_j\Lambda$
(b) R_j is a Riesz transformation
(c) f and g are real valued functions.

Let us write $B(f,g) = \Lambda\Gamma(f,g)$. The following lemma will help us in estimating some error terms

Lemma 4.2. *The bilinear operator* $\Gamma(f,g) : L^\infty([0,T],L^2) \times L^\infty([0,T],L^2) \mapsto \mathcal{C}([0,T],L^2)$ *is continuous.*

The proof is almost trivial. The product fg belongs to $L^\infty([0,T],L^1)$ and the convolution operator $S(t-\tau)$ pulls back L^1 to L^2. The operator norm is the L^2 norm of the gaussian kernel scaled to the scale $(t-\tau)^{1/2}$. This L^2 norm is $(t-\tau)^{-3/4}$ which leads to a convergent integral.

We now return to the proof of Theorem 4.1. We denote by λ a small parameter and form the Taylor expansion of $\Phi(\lambda v_0) = v_\lambda$ at 0. On the one hand Φ is $\mathcal{C}^2$ at 0 and we have

$$v_\lambda = \lambda v^{(1)} + \lambda^2 v^{(2)} + o(\lambda^2) \qquad (4.9)$$

where the error term is measured in $L^\infty([0,T];L^2)$.

On the other hand, we can use (3.9) and write $v_\lambda = \lambda S(t)v_0 - B(v_\lambda,v_\lambda) = \lambda S(t)v_0 - \lambda^2 B(v^{(1)},v^{(1)}) + O(\lambda^3)$ where the error term is measured in $\mathcal{C}([0,T];H^{-1})$ and is estimated by Lemma 4.2. The uniqueness of Taylor expansion yields

$$\begin{cases} v^{(1))} = S(t)v_0 \\ v^{(2)} = -B(v^{(1)},v^{(1)}) \end{cases} \qquad (4.10)$$

Since $\Phi \in \mathcal{C}^2(E,X)$, where $E = L^2, X = \mathcal{C}([0,T],L^2_w(\mathbf{R}^3))$, we have $v^{(2)} \in \mathcal{C}([0,T],L^2_w(\mathbf{R}^3))$.

We now simplify the notations and treat a model case where functions are real valued and the Leray-Hopf projector is forgotten. The proof will adapt to Navier-Stokes equations as well, since it is based on some inconsistency in the scalings. Let us be more precise

Lemma 4.3. *Let us freeze the time variable t and consider the bilinear operator $\Gamma_t : L^2 \times L^2 \mapsto L^2$ defined by*

$$h = \Gamma_t(f, g) = \int_0^t \partial_j [S(t - \tau)[S(\tau)(f)(S(\tau)(g)] \, d\tau \qquad (4.11)$$

where $f, g, h \in L^2$. Then the operator norm of this bilinear operator is $t^{-1/4}$ which blows up as $t \to 0$.

The proof of Lemma 4.3 is almost obvious. One computes the bilinear symbol of $\Gamma_t(f, g)$ which is

$$m_t(\xi, \eta) = \int_0^t \exp[-(t - \tau)|\xi + \eta|^2 - \tau(|\xi|^2 + |\eta|^2)](\eta_j + \xi_j) \, d\tau \qquad (4.12)$$

We then obtain $m_t(\xi, \eta) = t^{1/2} \omega(t^{1/2}\xi, t^{1/2}\eta)$. We have

$$\omega(\xi, \eta) = (\xi_j + \eta_j) \frac{\exp[-(|\xi|^2 + |\eta|^2)] - \exp[-|\xi + \eta|^2]}{|\xi + \eta|^2 - |\xi|^2 - |\eta|^2}$$

This exact computation of $\omega(\xi, \eta)$ does not play any role. We observe that $\omega(\xi, \eta) \in L^p(\mathbf{R}^3 \times \mathbf{R}^3)$, $p > 3/2$ and an obvious rescaling proves Lemma 4.3.

We can even be more precise. We return to (4.11) and compute $\Gamma_t(\phi_t, \phi_t)$ when $\phi_t = t^{-3/4}\phi(xt^{-1/2})$, ϕ being any testing function. On the Fourier transform side we obtain

$$t^2 \int \int \omega(t^{1/2}\xi, t^{1/2}\eta)\hat{\phi}(t^{1/2}\xi)\hat{\phi}(t^{1/2}\eta) \exp(ix \cdot (\xi + \eta)) \, d\xi \, d\eta = t^{-1}\psi(t^{-1/2}x)$$

$$(4.13)$$

The L^2 norm of this function is $t^{-1/4}$, up to a multiplicative constant, as announced. We are concerned by the localization of the functions which are used in the counter example. The small scales are responsible for the lack of stability.

The following observation might be interesting

Lemma 4.4. *If $m(\xi, \eta) \in L^4(\mathbf{R}^3 \times \mathbf{R}^3)$, then the bilinear operator defined by $B(f, g) = \int \int \exp(ix \cdot (\xi + \eta))m(\xi, \eta)\hat{f}(\xi)\hat{g}(\eta) d\xi d\eta$ is continuous from $L^2 \times L^2$ to L^2.*

The proof of Lemma 4.4 is a simple exercise on Young inequalities and is left to the reader.

The proof of Theorem 4.1 would work as well if L^2 is replaced by L^p. When $2 \le p < 3$, the same argument proves that the bilinear operator is unbounded when $t \to 0$. When $p > 3$, the norm of the bilinear operator is unbounded as $t \to \infty$.

We now consider a class of Banach spaces E for which one would like to prove stability. A functional Banach space E is, by definition, included inside the space of tempered distributions and contains the space of Schwartz testing functions. Both inclusions are assumed to be continuous.

The proof of Theorem 4.1 leads to the following definition:

Definition 4.2. *We say that E is a regular Banach space if the norm in E is translation invariant and if, for each positive λ and $f \in E$, we have*

$$\lambda \| f(\lambda \cdot) \| = \| f(\cdot) \| \tag{4.14}$$

Theorem 4.1 tells us that we cannot expect stability with some other scaling exponent in (4.14). Let us provide the reader with a string of such spaces. The homogeneous Besov space $\dot{B}_1^{2,1}$ is the smallest one. We then have

$$\dot{B}_1^{2,1} \subset \dot{L}^{3/2,1} \subset \dot{H}^{1/2} \subset L^3 \subset KT \subset \dot{B}_\infty^{-1,\infty} \tag{4.15}$$

where the definition of the Koch&Tataru space will be provided in Section 10. Another interesting example of a Banach space satisfying (4.14) is provided by the Morrey-Campanato spaces.

Definition 4.3. *If $1 < p < \infty$, $\alpha \geq 0$, a measurable function $f(x), x \in \mathbf{R}^3$ belongs to the Morrey-Campanato space $M^{p,\alpha}$ if a constant C exists such that for any ball B centered at x_0 with radius R we have*

$$\left(R^{-3} \int_B |f(x)|^p dx \right)^{1/p} \leq C \, R^{-\alpha} \tag{4.16}$$

The norm in the Morrey-Campanato space satisfies (4.14) if $\alpha = 1$. When $p = 1$, the definition is slightly modified since we now allow f to be a Borel measure μ and $\int_B |f| dx$ is replaced by $|\mu|(B)$. An example is the surface measure on a Lipschitz graph. The spaces $M^{p,\alpha}$ are increasing if p is decreasing. P. Federbush used the space $M^{2,1}$ in his work on Navier-Stokes equations [41]. If $1 \leq p < 3$, the space $M^{p,1}$ is containing the subsapce of functions which belong locally to L^p and are homogeneous of degree -1.

We can now complete the string (4.15) with

$$\dot{B}_1^{2,1} \subset \dot{L}^{3/2,1} \subset \dot{H}^{1/2} \subset L^3 \subset M^{2,1} \subset M^{1,1} \subset \dot{B}_\infty^{-1,\infty} \tag{4.17}$$

The relevance of these Morrey-Campanato spaces is illustrated by the following remark answering a problem raised by Peter Constantin.

Proposition 4.1. *Let us assume that $f \in \dot{B}_\infty^{-1,\infty}$ has the following property: for each function $m(x)$ in L^∞, the pointwise product $m(x)f(x)$ still belongs to $\dot{B}_\infty^{-1,\infty}$. Then f belongs to $M^{1,1}$.*

Indeed the closed graph theorem implies $\| mf \|_{\dot{B}_\infty^{-1,\infty}} \leq C \| m \|_\infty$ and the duality between $\dot{B}_\infty^{-1,\infty}$ and $\dot{B}_1^{1,1}$ yields $| \int \phi_{a,b}(x) m(x) f(x) \, dx | \leq C \| m \|_\infty$ when $\phi_{a,b}(x) = a^{-2} \phi(\frac{x-b}{a})$, $a > 0, b \in \mathbf{R}^3$. This implies $\int |\phi_{a,b}(x) f(x)| \, dx \leq C$ which is the definition of $M^{1,1}$.

From now on we will concentrate on these *regular spaces* E and we will investigate the behavior of solutions $v(\cdot, t) \in \mathcal{C}([0, \infty), E)$ to the Navier-Stokes equations. This behavior will be quantified using several functional norms.

Some norms are describing energy estimates, other norms provide us with quantitative information about the regularity of solutions, and finally some norms are related to the oscillating properties of solutions. Are these norms somehow preserved under the evolution or should we expect some blow-up? Do all norms blow up simultaneously? We will show that a global solution exists whenever the weak norm of the initial condition is small. Then we will end with the following property which is our main goal:

If the initial condition is (everywhere) oscillating, then the corresponding solution to the Navier-Stokes equations is global in time. Moreover this global solution will keep forever the additional regularity of the initial condition.

For instance an initial condition which is C^∞ and is sufficiently oscillating will lead to a global solution which will also be C^∞ in the time-space variables. Such a résult seems inconsistent. Indeed a function cannot at the same time be smooth and oscillating. However this objection disappears at a quantitative level since the (strong) norm which measures the smoothness can be arbitrarily large while the norm describing the oscillations is as small as we like. We will impose $\|v_0\|_* < \eta$ where $\|f\|_*$ is a norm which is small whenever f is oscillating. In other words, we will assume that our initial condition v_0 is a function which has a small norm in a function space containing generalized functions. The norm of a function f in such a function space takes advantage of the oscillating character of f. At the same time our initial condition v_0 may be extremely large in function spaces like the Hölder or Sobolev spaces.

We will denote by B a Banach space of smooth functions. For instance B can be the Sobolev space H^m or the usual C^m, $m \geq 1$. Then our discussion can be summarized as follows:

Conjecture 4.2. There exists a Banach space $\mathcal{Z} \subset \mathcal{S}'(\mathbf{R}^3)$ and a (small) positive number η such that

(a) if v_0 is smooth and compactly supported and if $\|v_0\|_{\mathcal{Z}} \leq \eta$, then there exists a global (in time) smooth solution $v(x,t)$ to (3.1) (no blow-up)

(b) $\|f_\lambda\|_{\mathcal{Z}} = \|f\|_{\mathcal{Z}}$ if $f_\lambda(x) = \lambda f(\lambda x)$

(c) $\|f\|_{\mathcal{Z}}$ is small if f is oscillating.

Once more we are not requiring that $\|v_0\|_B$ be small. This conjecture will be one of our guide line in these notes.

As Roger Temam pointed out, the first result along these lines has been proved by H. Fujita and T. Kato in 1964 [46]. These authors proved (a) and (b) when $\mathcal{Z} = \dot{H}^{1/2}(\mathbf{R}^3)$. However (c) is not satisfied. Let us be more specific

Theorem 4.2. Let $\dot{H}^{1/2}(\mathbf{R}^3)$ *denote the usual homogeneous Sobolev space. Then there exists a positive number η such that if v_0 belongs to $H^1(\mathbf{R}^3)$ and fulfills*

$$\nabla . v_0 = 0 \qquad (4.18)$$

$$\|v_0\|_{\dot{H}^{1/2}(\mathbf{R}^3)} \leq \eta \qquad (4.19)$$

there exists a unique global solution $v \in C([0, \infty), H^1(\mathbf{R}^3))$ to the Navier-Stokes equations.

Roger Temam told me that this theorem is an interesting example illustrating our conjecture. Indeed this theorem is especially attractive if $\|v_0\|_{\dot{H}^{1/2}}$ is much smaller than $\|v_0\|_{H^1}$. This is likely to happen since the first Sobolev norm is less demanding than the second one.

The following lemma tells us when $\|f\|_{\dot{H}^{1/2}}$ is small while $\|f\|_{H^1}$ is large.

Lemma 4.5. *Let B be the homogeneous Besov space $\dot{B}_2^{-1/2,\infty}$. Then there exists a constant C such that*

$$\|f\|_{\dot{H}^{1/2}} \leq C\|f\|_{\dot{H}^1}^{2/3}\|f\|_B^{1/3} \tag{4.20}$$

This lemma will be proved in Section 13. The weak norm is much smaller than the square root of the strong norm when our Besov norm is small. Does this Gagliardo-Nirenberg estimate mean that $\dot{B}_2^{-1/2,\infty}$ is the space which needs to be used in our heuristic approach? It cannot be so since this Besov space does not enjoy the right scaling property. Indeed $f(x)$ and $f_\lambda(x) = \lambda f(\lambda x)$, $\lambda > 0$, do not have the same norm in B.

A second option consists in using the homogeneous space $\dot{H}^{1/2}$ which enjoys the right scale invariance. But an oscillating initial condition has a large norm in this Sobolev space and the Fujita-Kato theorem does not meet our expectations.

Let us observe that $\dot{H}^{1/2}(\mathbf{R}^3) \subset L^3(\mathbf{R}^3)$. It is therefore natural to replace (4.19) by the corresponding condition $\|u_0\|_3 < \eta$ and to expect a global solution. It is what T. Kato did in 1984. T. Kato's achievement is so important that it deserves a full section.

5 T. Kato's Theorem

In 1984, T. Kato discovered a remarkable existence theorem. Uniqueness was proved much later by Giulia Furioli, Pierre-Gilles Lemarié-Rieusset and Elide Terraneo. We then have:

Theorem 5.1. *There exists a positive number η with the following property: if the initial velocity v_0 belongs to $L^3(\mathbf{R}^3)$, fulfills $\|v_0\|_3 < \eta$ and $\nabla.v_0 = 0$, there exists a unique global solution $v \in \mathcal{C}([0, \infty); L^3(\mathbf{R}^3))$ to the Navier-Stokes equations such that $v(\cdot, 0) = v_0$. Moreover the mapping $\Phi : v_0 \mapsto v \in \mathcal{C}([0, \infty); L^3(\mathbf{R}^3))$ is analytic in the open ball $\|v_0\|_3 < \eta$.*

The proof of Theorem 5.1 mimics F. Weissler's approach to Theorem 2.2. One uses Picard's fixed point theorem and solves Navier-Stokes equations as written in (3.18). The versions of Picard's fixed point theorem we are using in the proofs of Theorem 5.1 are given by the following lemmata which are similar in spirit to Lemma 2.1.

Lemma 5.1. *Let $(X, \|\cdot\|)$ be a Banach space and $B : X \times X \to X$ a bilinear mapping such that*

$$\|B(x,y)\| \leq C_0 \|x\| \, \|y\| \tag{5.1}$$

If $\|x_0\| < \frac{1}{4C_0}$, then the equation

$$x = x_0 + B(x,x) \tag{5.2}$$

has a unique solution satisfying $\|x\| < \frac{1}{2C_0}$ and this solution is the limit of the sequence x_n, $n \in \mathbf{N}$, defined by

$$x_{n+1} = x_0 + B(x_n, x_n) \tag{5.3}$$

The proof of this lemma is straightforward but will be detailed for the reader's convenience. We write $y_{n+1} = x_{n+1} - x_n$ and (5.3) reads

$$y_{n+1} = B(x_{n-1}, y_n) + B(y_n, x_{n-1}) + B(y_n, y_n) \tag{5.4}$$

Then (5.1) yields

$$\|y_{n+1}\| \leq 2C_0 \|y_n\| \left(\|x_0\| + \sum_1^{n-1} \|y_k\| \right) + C_0 \|y_n\|^2 \tag{5.5}$$

An obvious induction implies that $\|y_n\| \leq \epsilon_n$ where ϵ_n is the solution of the corresponding induction

$$\epsilon_{n+1} = 2C_0 \epsilon_n (\|x_0\| + \epsilon_1 + \ldots + \epsilon_{n-1}) + C_0 \epsilon_n^2 \tag{5.6}$$

Moving backwards in the above computations, we obtain $\epsilon_n = \eta_n - \eta_{n-1}$ where

$$\eta_{n+1} = \|x_0\| + C\eta_n^2 \tag{5.7}$$

If $\|x_0\| < \frac{1}{4C_0}$, then the sequence η_n is increasing and converges to the smallest solution of $u = \|x_0\| + C_0 u^2$. This implies $\sum_0^\infty \|y_n\| < \infty$ as announced.

This lemma suffices for proving T. Kato's theorem. However an improvement which will be used in the next section reads as follows

Lemma 5.2. *Keeping the notations of Lemma 5.1, we assume the following properties:*

(1) we are given a 'strong norm' $\|\cdot\|_1 \geq \|\cdot\|$ on a Banach space $V \subset X$
(2) the bilinear operator B satisfies the assumptions of Lemma 5.1
(3) the bilinear operator $B(x,y)$ is symmetric and there exists a constant C_1 such that

$$\|B(x,y)\|_1 \leq C_1 \|x\| \, \|y\|_1, \quad x \in X, \; y \in V \tag{5.8}$$

(4) x_0 belongs to V with $\|x_0\| < \frac{1}{4C_0}$.

Then the sequence x_n defined in Lemma 5.1 converges for the strong norm.

It should be stressed that the smallness condition $\|x_0\| < \frac{1}{4C_0}$ only concerns the 'weak norm'. In the applications with have in mind, the weak norm will be a norm in a Besov space with a negative regularity.

The proof of Lemma 5.2 is not difficult. We write $\|y_n\|_1 = \theta_n$ and (5.8) yields $\|y_{n+1}\|_1 \leq 2C_1\|x_{n-1}\|_1\|y_n\| + C_1\|y_n\|_1\|y_n\| \leq 2C_1\epsilon_n(\|x_0\|_1 + \theta_1 + \ldots + \theta_n)$. This implies

$$\|x_0\|_1 + \theta_1 + \ldots + \theta_n \leq \Pi_1^\infty(1 + 2C_1\epsilon_n)\|x_0\|_1 \tag{5.9}$$

The right-hand side of (5.9) is a constant C' and we have $\theta_{n+1} \leq C'\epsilon_n$ which ends the proof.

In some instances, the bilinear operator $B(x, y)$ satisfies

$$\|B(x, y)\|_1 \leq C_1\|x\|\,\|y\|, \quad x, y \in X \tag{5.10}$$

which is more than needed in Lemma 5.2.

Since we want to construct solutions $u \in \mathcal{C}([0, \infty); L^3)$, one is tempted to apply Lemma 5.1 to the 'natural space' $X = \mathcal{C}([0, \infty); L^3)$. Unfortunately the bilinear mapping $\mathcal{B}$ is not bicontinuous from $X \times X \mapsto X$. To overcome this issue, F. Weissler and T. Kato introduced an 'artificial norm' $\|\cdot\|_{\mathcal{X}}$ which is the sum $\|u\|_X + \sup\{t^{1/2}\|u(\cdot, t)\|_\infty; t \geq 0\}$. The subspace of X defined by the finiteness of this 'artificial norm' is the 'artificial space' and is denoted by $\mathcal{X}$. Then the proof of Theorem 5.1 reduces to proving (5.1) and to checking that the natural norm and the artificial norm are equivalent ones when these two norms are restricted to the closed subspace of X consisting of all $S(t)v_0$, $v_0 \in L^3$, .

For proving the first assertion, one uses the following:

Lemma 5.3. *For $j = 1, 2$ and 3, the operator $S(t)\partial_j\mathbb{P}$ is a convolution operator with the function $(t - \tau)^{-2}w_j(\frac{x-y}{(t-\tau)^{1/2}})$ where w_j is a continuous function which is $O(|x|^{-4})$ at infinity. The same remark applies to $S(t)\partial_j$.*

The proof of Lemma 5.3 is obvious if calculations are made on the Fourier transform side.

We now prove (5.1).

Proposition 5.1. *The bilinear operator defined by*

$$w(\cdot, t) = \int_0^t S(t - \tau)\mathbb{P}\partial_j[v_j^{(1)}v^{(2)}](\cdot, \tau)\,d\tau \tag{5.11}$$

satisfies, uniformly in t,

$$\|w(\cdot, t)\|_3 \leq C\|v^{(1)}\|_{\mathcal{X}}\|v^{(2)}\|_{\mathcal{X}} \tag{5.12}$$

and

$$\|w(\cdot, t)\|_\infty \leq Ct^{-1/2}\|v^{(1)}\|_{\mathcal{X}}\|v^{(2)}\|_{\mathcal{X}} \tag{5.13}$$

Moreover, if $v^{(1)}, v^{(2)} \in \mathcal{C}([0, \infty); L^3)$, then $w \in \mathcal{C}([0, \infty); L^3)$ and

$$\|w(\cdot, t)\|_3 \to 0 \quad (t \to 0) \tag{5.14}$$

The proof of Proposition 5.1 is easy. Combining the information given by the norm in $\mathcal{X}$, we have $\|(v_j^{(1)}v^{(2)}(\cdot,\tau)\|_3 \leq C\tau^{-1/2}$ while the operator $S(t)\partial_j\mathbb{P}$ maps L^3 to L^3. Indeed this operator is a convolution with a function in L^1 (lemma 19). The operator norm is bounded by this L^1 norm which is $(t-\tau)^{-1/2}$. The proof of the first assertion ends with $\int_0^t (t-\tau)^{-1/2}\tau^{-1/2}\, d\tau = \pi$.

We now compute L^∞-norms. We begin as above and estimate the L^6-norm of $v_j^{(1)}v^{(2)}$. The L^6-norm of $v^{(1)}$ is $O(t^{-1/4})$ since it can be estimated by the geometrical mean between the L^3 and the L^∞-norm. Then the L^6-norms of the products $v_j^{(1)}v_k^{(2)}$, $1 \leq j,k \leq 3$, are $O(\tau^{-3/4})$. Finally the operator $S(t)\partial_j\mathbb{P}$ pulls back L^6 to L^∞. The operator norm is given by Young inequalities and we obtain $C(t-\tau)^{-3/4}$. The proof ends with $\int_0^t (t-\tau)^{-3/4}\tau^{-3/4}\, d\tau = Ct^{-1/2}$.

The proof of the continuity at 0 relies on the following observation: if $f \in L^3$, then $\lim_{t\to 0} t^{1/2}\|S(t)f\|_\infty = 0$. Therefore if $v(\cdot,t) \in \mathcal{C}([0,\infty); L^3)$, and if one writes $v(\cdot,t) = S(t)v_0 + w(\cdot,t)$, then $\|w(\cdot,t)\|_\mathcal{X}$ tends to 0 with t. Finally the computation of the right-hand side of (5.11) yields four terms and the proof is easily concluded with the preceding remarks and (5.12). The proof of the continuity at $t > 0$ is similar and will be omitted. We will return to this proof in Section 7.

Finally the $\mathcal{X}$ norm of $S(t)u_0$ and $\|u_0\|_3$ are equivalent norms and the proof of Proposition 5.1 is complete. It implies Theorem 5.1.

Existence and uniqueness were obtained inside the 'artificial space' $\mathcal{X}$. Therefore uniqueness requires the condition

$$t^{1/2}\|u\|_\infty \leq C_0 \tag{5.15}$$

For quite a long time, uniqueness of Kato's solutions $v(x,t)$ belonging to the natural space $\mathcal{C}([0,\infty), L^3(\mathbf{R}^3))$ was an open problem. Finally uniqueness was proved by Giulia Furioli, Pierre-Gilles Lemarié-Rieusset and Elide Terraneo without assuming (5.15). The interested reader is referred to [47], [78], or to Section 7.

6 The Kato Theorem Revisited by Marco Cannone

Some important improvements on Theorem 4.2 and 5.1 were obtained by M. Cannone and F. Planchon. They proved that the norm in the Sobolev space $\dot{H}^{1/2}$ which was used by H. Fujita and T. Kato can be replaced by a weaker norm. This weaker norm is the norm in the homogeneous Besov space $\dot{B}_q^{-(1-3/q),\infty}$ when $3 \leq q < \infty$. For simplifying the notations, this Besov space will be denoted B_q. Theorem 6.1 can be interpreted the following way. We know from Kato's theorem that the map $\Phi : v_0 \mapsto v(\cdot,t) \in \mathcal{C}([0,\infty); L^3(\mathbf{R}^3))$ is analytic on a neighborhood V of 0. We plan to construct the analytic continuation of Φ and to replace V by a much larger open set which will be defined by a weaker norm. The weakest one will be unveiled in Section 10.

We will focus on Theorem 5.1. The discussion on Theorem 4.2 is fully similar and will be treated again in Theorem 10.2.

M. Cannone and F. Planchon proved the following theorem [17]

Theorem 6.1. *For $q \in (3, \infty)$, there exists a positive constant η_q such that the three conditions*

$$\nabla.v_0 = 0, \quad v_0 \in L^3(\mathbf{R}^3) \quad and \quad \|v_0\|_{B_q} < \eta_q \tag{6.1}$$

imply the existence of a global solution $v(x,t)$ to the Navier-Stokes equations, belonging to $\mathcal{C}([0, \infty); L^3(\mathbf{R}^3))$ and such that $v(x,0) = v_0$. The same is true if L^3 is replaced by the Sobolev space H^1.

The homogeneous Besov space B_q is defined exactly the same way as in the special case $q = 6$ (see Theorem 2.4). We let φ be a function in the Schwartz class $\mathcal{S}(\mathbf{R}^3)$ such that $\int_{\mathbf{R}^3} \varphi(x)\, dx = 1$. Then we write $\varphi_j(x) = 2^{3j}\varphi(2^j x)$ and denote by S_j convolution operator with $\varphi_j(x)$. Finally if $\alpha > 0$, a function or a distribution f belongs to the homogeneous Besov space $\dot{B}_q^{-\alpha,\infty}$ if and only if $\|S_j(f)\|_q \le C2^{j\alpha}, j \in \mathbf{Z}$.

As it was said before, $B_q = \dot{B}_q^{-(1-3/q),\infty}$

The Banach spaces B_q increase with q in such a way that the conditions (6.1) seem to be less demanding as q grows. However the positive constant η_q which appears in (6.1) tends to 0 as q tends to infinity. Therefore comparing these distinct conditions is a delicate matter, as F. Planchon noticed: if q increases, the norm becomes less demanding but the diameter of the ball $\|\cdot\|_{B_q} < \eta_q$ is decreasing.

Theorem 6.1 is, in some sense, a corollary of the following theorem:

Theorem 6.2. *For $q \in (3, \infty)$, there exists a positive number η_q with the following property: if the initial condition v_0 belongs to the Besov space $B_q = \dot{B}_q^{-(1-3/q),\infty}$ and satisfies $\|v_0\|_{B_q} \le \eta_q$, there exists a unique global solution $v(\cdot, t)$, $t \in [0, \infty)$, of the Navier-Stokes equations enjoying the following six properties for $t \ge 0$*

$$v \in \mathcal{C}([0, \infty); B_q) \tag{6.2.a}$$
$$v(x,0) = v_0 \tag{6.2.b}$$
$$\|v(\cdot, t)\|_{B_q} \le C_q \|v_0\|_{B_q} \tag{6.2.c}$$
$$\|v(\cdot, t) - S(t)v_0\|_3 \le C_q \|v_0\|_{B_q} \tag{6.2.d}$$
$$\|v(\cdot, t)\|_\infty \le C_q\, t^{-1/2} \tag{6.2.e}$$
$$If\ v_0 \in \dot{B}_3^{0,\infty},\ then\ \|\nabla v(\cdot, t)\|_3 \le C\, t^{-1/2}\|v_0\|_{B_3} \tag{6.2.f}$$

The Besov space B_q is a dual space and is equipped with its weak-star topology. The meaning of the continuity of $v \in \mathcal{C}([0, \infty); B_q)$ is given by Definition 2.1

As it was said when Theorem 2.5 was discussed, Theorem 6.2 is raising a fundamental problem. Can distributions be solutions of some nonlinear PDEs?

The answer is clearly no and the main obstacle concerns the pointwise multiplication between two distributions. Here this problem is avoided by imposing some other estimates to $v(x,t)$. These estimates are consistent with the properties of the linear evolution. In other words, (6.2.a) and (6.2.c) will be ignored during the proof. We instead focus on (6.2.d) which is a stronger statement.

Theorem 6.2 implies the existence of self-similar solutions to Navier-Stokes equations. The proof of this remark relies on the two following facts.

(i) If we apply Theorem 6.2 to $\lambda v_0(\lambda x)$, $\lambda > 0$, the invariance of the Navier-Stokes equations implies that $\lambda v(\lambda x, \lambda^2 t)$ is a solution.

(ii) Any function which is homogeneous of degree -1 and smooth away from 0 belongs to B_q (See [17] for an optimal regularity assumption).

The proof of the existence of self-similar solutions is now immediate. We assume that v_0 is homogeneous of degree -1 and is smooth away from the origin. We multiply v_0 by a small positive number which yields $\|v_0\|_{B_q} \leq \eta_q$. We use the uniqueness of the solution described in Theorem 6.2 and conclude that $\lambda v(\lambda x, \lambda^2 t) = v(x, t)$. This implies the existence of a profile U such that $u(x,t) = \frac{1}{\sqrt{t}} U(\frac{x}{\sqrt{t}})$. This elegant argument is due to M. Cannone.

For proving Theorems 10 and 11, we assume $p = 6$ (the general case is left to the reader). In Theorem 6.2, we will focus on statements (6.2.a) to (6.2.d). The last two facts (6.2.e) and (6.2.f) need an other proof. We will use the semigroup formulation of the Navier-Stokes equations which is given by (3.18) or (3.19).

Lemma 5.1 will now be applied to the bilinear operator $\mathcal{B}(v^{(1)}, v^{(2)})$ defined by

$$w(\cdot, t) = \mathcal{B}(v, v)(\cdot, t) = \int_0^t \mathbb{P}S(t - \tau)[\partial_j(v_j v)](\tau)d\tau \qquad (6.3)$$

As we did in the proof of Theorem 2.4, we consider the Banach space $\mathcal{M}$ of all measurable functions $f(x,t)$ on $\mathbf{R}^3 \times (0, \infty)$ such that

$$\|f\|_{\mathcal{M}} = \sup\{t^{1/4}\|f(\cdot, t)\|_6; \ t \geq 0\} \qquad (6.4)$$

is finite. The space $\mathcal{C}([0, T]; B_q)$ would have been inadequate since the nonlinear terms in the Navier-Stokes equations could not be given a meaning. As in the proof of Theorem 2.4, $\mathcal{M}_0$ will denote the closure in $\mathcal{M}$ of the subspace of continuous functions with compact support. We have

Proposition 6.1. *The bilinear operator defined by (6.3) is bicontinuous from* $\mathcal{M} \times \mathcal{M}$ *into* $\mathcal{M} \cap L^\infty((0, \infty); L^3)$. *Moreover* $w(\cdot, t) \to 0$ *as* $t \to 0$. *The bilinear operator* $\mathcal{B}$ *is also bicontinuous from* $\mathcal{M}_0 \times \mathcal{M}_0$ *into* $\mathcal{M}_0 \cap \mathcal{C}([0, \infty); L^3)$.

Let us accept Proposition 6.1 and prove Theorem 6.1, If $v_0 \in L^3$, then $S(t)v_0(\cdot, t) \in \mathcal{M}_0$. This trivial observation follows from Young inequalities and does not even need a proof. Moreover if $v_0 \in B_q$, we have $S(t)v_0(\cdot, t) \in \mathcal{M}$. Indeed this is the definition of B_q. This being said, Theorem 6.1 immediately follows from Proposition 6.1 and Lemma 5.1. The same remark applies to Theorem 6.2, but (6.2.e) and (6.2.f) need another proof.

Now we return to the proof of Proposition 6.1. We first prove $\|w(\cdot, t)\|_3 \leq C$. The weak continuity in t at 0 will be postponed until the end of the proof. We know from the definition of the norm that $\|[v_j v](\cdot, \tau)\|_3 \leq C\tau^{-1/2}$. The operator $S(t)\partial_j \mathbb{P}$ pulls back L^3 to L^6. The operator norm is given by Young inequalities and is $C(t-\tau)^{-3/4}$. The proof of the first assertion in Proposition 6.1 ends with the following obvious calculation: $\int_0^t (t-\tau)^{-3/4}\tau^{-1/2}\, d\tau = ct^{-1/4}$.

We now observe that the uniform L^3 estimate on $\mathcal{B}(v, v)$ is given for free. Indeed the computation starts as above and ends with $\int_0^t (t-\tau)^{-1/2}\tau^{-1/2}\, d\tau = \pi$. Finally the X norm of $S(t)u_0$ coincides with the norm of u_0 in the Besov space $\dot{B}_6^{-1/2,\infty}$.

The weak continuity of $v(\cdot, t)$ at 0 will follow from

$$w(\cdot, t) \rightharpoonup 0, \quad t \to 0 \tag{6.5}$$

For proving (6.5) one writes $w(\cdot, t) = \partial_j w_j(\cdot, t)$ where $w_j(\cdot, t) = \int_0^t \mathbb{P}S(t - \tau)(v_j v)(\tau)d\tau$. We then argue as in the proof of the uniform estimates and obtain $\|w_j(\cdot, t)\|_3 \leq Ct^{1/2}$. Therefore $w_j(\cdot, t)$ tends to 0 in the distributional sense when t tends to 0 and so does $w(\cdot, t)$. The last assertion in Proposition 6.1 is a variant on Proposition 5.1 and left to the scrupulous reader. The argument will be detailed in Section 7. The proof of Proposition 6.1 is now complete and Theorem 6.1 will follow when we will show that $S(t)v_0 \in \mathcal{M}_0$. This is given by the following lemma

Lemma 6.1. *If $f \in B_q$, and if $S(t)$ is the heat semigroup, then the mapping defined by $t \in (0, \infty) \mapsto S(t)f \in L^q(\mathbf{R}^3)$ is continuous. If $f \in L^3$, then $t^{1/4}\|S(t)f\|_6 \to 0$, $(t \to 0)$. The same holds as $t \to \infty$.*

This simple observation will be proved for the reader's convenience. Indeed $\frac{\partial}{\partial t}S(t)f = t^{-1}t\Delta S(t)f = t^{-1}g_t$ Since $f \in B_q$, we have $\|g_t\|_q \leq Ct^{-(1-3/q)/2}$ and it yields $\|t\frac{\partial}{\partial t}S(t)f\|_q \leq Ct^{-(1-3/q)/2}$ which proves the continuity on $(0, \infty)$.

We now prove (6.2.e) and (6.2.f) in Theorem 6.2. Getting the L^∞ estimate is a harder task. The computation starts as above but ends with the divergent integral $\int_0^t (t - \tau)^{-1}\tau^{-1/2}\, d\tau$. Instead we use Lemma 5.2 where the 'strong norm' will be the sum between the 'weak norm' defined by (6.4) and $\sup\{t^{1/2}\|v(\cdot, t)\|_\infty, t \geq 0\}$. The strategy of the proof is now identical to the preceding one. The products $v_j v$ are viewed as products between a function in L^∞ and a function in L^6. We finally use the duality between L^6 and $L^{6/5}$. It suffices to use Lemma 5.3 and to compute the $L^{6/5}$ norm of $(t - \tau)^{-2}w_j(\frac{x-y}{(t-\tau)^{1/2}})$. We obtain $C(t - \tau)^{-3/4}$ and the proof ends with $\int_0^t (t - \tau)^{-3/4}\tau^{-3/4}\, d\tau = ct^{-1/2}$.

We now turn to (6.2.f) in Theorem 6.2. Once more Lemma 5.2 will be used and the 'strong norm' is $\|v\|_{\mathcal{M}} + \sup\{t^{1/2}\|\nabla v\|_3, t \geq 0\}$. The argument is identical to the preceding one and we analyze the bilinear operator $\partial_k \mathcal{B}(v,v)$. When the function $v_j \in L^6$ is multiplied by the function $\partial_j v \in L^3$, we obtain a function in L^2 whose norm is $O(\tau^{-3/4})$. The operator $\partial_k \mathbb{P} S(t-\tau)$ is a convolution with the integrable function $(t-\tau)^{-2} w_{j,k}(\frac{x-y}{(t-\tau)^{1/2}})$. Since $w_{j,k}$ belongs to $L^{6/5}$, this convolution pulls back L^2 to L^3 with an operator norm which is $C(t-\tau)^{-3/4}$. As above we have $\int_0^t (t-\tau)^{-3/4} \tau^{-3/4} \, d\tau = c \, t^{-1/2}$. Lemma 5.2 applies if the linear evolution belongs to the Banach space defined by the strong norm. That is why we need $v_0 \in \dot{B}_3^{0,\infty}$. This ends the proof of Theorem 6.2.

The proof of Theorem 6.1 can be seen, in a sense, as a corollary of the preceding proof. Indeed, in both theorems, the bilinear operator maps $\mathcal{M} \times \mathcal{M}$ into L^3 when the $\mathcal{M}$ norm is defined by (6.4). The linear evolution is responsible for the differences between the two theorems. There exists another difference. In Theorem 6.2, the 'fluctuation' $w(\cdot, t)$ belongs to $L^\infty((0,\infty), L^3)$ but cannot belong to $\mathcal{C}([0,\infty); L^3)$. A simple counter-example is given by self-similar solutions. Let us assume in Theorem 6.2 that the initial condition v_0 is smooth away from the origin and is homogeneous of degree -1. Then v_0 belongs to all the homogeneous Besov spaces B_q. We apply Theorem 6.2 to ηv_0 where η is a small positive number. Then the scheme defined in Lemma 5.1 generates a self-similar solution to Navier-Stokes equations. It is given by

$$v(x,t) = S(t)[v_0](x) + \frac{1}{\sqrt{t}} W\left(\frac{x}{\sqrt{t}}\right) \tag{6.6}$$

where $W(x)$ belongs to L^3. It is clear that the L^3 norm of the 'fluctuation' $w(\cdot, t)$ is a constant. However $w(\cdot, t)$ weakly tends to 0 with t.

M. Cannone and F. Planchon proved the following theorem:

Theorem 6.3. *Let $v(x,t)$ be a solution of (3.1) in $\mathcal{C}([0,\infty); L^3)$, with initial data $v_0 \in L^3$ and denote by w the function $w = v - S(t)v_0$. Then*

$$w \in \mathcal{C}([0,\infty); \dot{W}^{1,3/2}) \tag{6.7}$$

Let $\omega(\cdot, t)$ denote the vorticity with $\omega(\cdot, 0) = \omega_0$. Then $\omega(\cdot, t) - S(t)\omega_0$ is continuous in time with values in $L^{3/2}$.

Here also, what M. Cannone and F. Planchon call the 'fluctuation' is more regular than the trend, as given by $S(t)u_0$. This theorem by M. Cannone and F. Planchon paves the way to Theorem 9.1 of Section 9 where we will prove that the vorticity ω is continuous in time with values in BV when the initial vorticity belongs to BV. Let us observe that BV is contained in $L^{3/2}$. If ω_0 belongs to $L^{3/2}$, then Theorem 6.3 says that $\omega(\cdot, t) \in L^{3/2}$, uniformly in time. We will prove the same result with $L^{3/2}$ replaced by BV. Let us observe that $\dot{W}^{1,3/2} \subset \dot{H}^{1/2}$ and Theorem 6.3 should be compared to Theorem 2.5.

For proving Theorem 6.3, we denote by X the Banach space consisting of all functions $f(x,t)$ defined on $\mathbf{R}^3 \times (0,\infty)$ belonging to $\mathcal{C}([0,\infty); L^3(\mathbf{R}^3))$, such that

$$\|f(\cdot,t)\|_6 \le Ct^{-1/4}, \quad \|\nabla f(\cdot,t)\|_3 \le Ct^{-1/2} \tag{6.8}$$

The norm in X is the sum between the natural norm in $\mathcal{C}([0,\infty); L^3(\mathbf{R}^3))$ and the optimal constants in (6.8). Then we have

Lemma 6.2. *The bilinear operator $\mathcal{B}$ defined by (3.18) maps $X \times X$ to the homogeneous Sobolev space $\dot{W}^{1,3/2}$.*

The proof is straightforward. We return to (3.18) and write $\partial_j(u_j u) = u_j \partial_j u = w$. Then $\|w(\cdot,\tau)\|_{3/2} \le \|u\|_X^2/\sqrt{\tau}$. We plug this estimate inside (3.18) and estimate the $L^{3/2}$ norm of $\partial_k \mathcal{B}(v^{(1)}, v^{(2)})$. We are led to estimating

$$\|\partial_k \int_0^t S(t-\tau)\mathbb{P}\partial_j[v_j v](\cdot,\tau)\,d\tau\|_{3/2} \tag{6.9}$$

Then we glue together $\partial_k S(t-\tau)\mathbb{P}$ and the resulting operator is a convolution with a function in L^1. The L^1 norm is $C(t-\tau)^{-1/2}$ and the proof ends with $\int_0^t (t-\tau)^{-1/2}\tau^{-1/2}\,d\tau = \pi$.

Once Lemma 6.2 is proved, Theorem 6.3 follows from the proof of Lemma 18. Indeed the 'strong norm' here is the sum between the 'weak norm' and $\sup\{\|f(\cdot,t)\|_{\dot{W}^{1,3/2}};\ t \ge 0\}$ and the space X is defined accordingly. Here we do not have $x_0 \in X$. However the proof of Lemma 5.2 applies and yields Theorem 6.3.

7 The Kato Theory with Lorentz Spaces

Let us remind the reader with the definition of the Lorentz space $L^{(p,1)}(\mathbf{R}^n)$ and $L^{(q,\infty)}(\mathbf{R}^n)$.

Definition 7.1. *A function f belongs to $L^{(p,1)}, 1 \le p \le \infty$ if and only if*

$$f = \sum_0^\infty \alpha_k a_k(x) \tag{7.1}$$

where the coefficients α_k and the normalized atoms $a_k(x)$ satisfy the following properties:

(a) $a_k(x)$ is supported by a ball B_k with volume v_k
(b) $|a_k(x)| \le v_k^{-1/p}$
(c) $\sum|\alpha_k| \le C$

The norm of f in $L^{(p,1)}$ is the lower bound of the sums $\sum|\alpha_k|$ computed on all possible expansions (7.1) of f.

The dual space of $L^{(p,1)}$ is $L^{(q,\infty)}$ if $1 \le p < \infty$ and $\frac{1}{p} + \frac{1}{q} = 1$. The Lorentz space $L^{(q,\infty)}$ can be defined by the usual condition: there exists a constant C such that for every positive λ, the measure of the set of points x where $|f(x)| >$

λ does not exceed λ^{-q}. The unit ball of the Lorentz space $L^{(q,\infty)}$, $1 < q \le \infty$ is convex. It is not the case if $q = 1$.

We already noticed that the fundamental bilinear operator $\mathcal{B}(\cdot,\cdot)$ defined by (3.18) is not continuous from $X \times X \mapsto X$ when $X = \mathcal{C}([0,\infty);L^3)$. That is why F. Weissler and T. Kato gave up the *natural norm* in $\mathcal{C}([0,\infty);L^3)$ and used instead the *artificial norm* where a growth condition on $\|v(\cdot,t)\|_\infty$ is imposed. Therefore it was difficult to prove uniqueness of T. Kato's solutions in the natural space $\mathcal{C}([0,\infty);L^3)$.

We here use another approach and replace the natural space $\mathcal{C}([0,\infty);L^3)$ by the natural space $\mathcal{C}([0,\infty);L^{(3,\infty)})$. As it was said above, the topology which is used in $\mathcal{C}([0,\infty);L^{(3,\infty)})$ is a combination between weak and strong continuity. The strong continuity is imposed when $t \in (0,\infty)$ while the continuity at 0 takes the form $u(\cdot,t) \rightharpoonup u(\cdot,0)$ as $\to 0$. We will study the bilinear operator $\int_0^t \mathbb{P}S(t-\tau)[\partial_j(v_jv)](\tau)d\tau$.

Theorem 7.1. *Let X be the space $\mathcal{C}([0,\infty);L^{(3,\infty)})$ and let $\mathcal{B}$ be the bilinear operator defined by*

$$w(\cdot,t) = \mathcal{B}(v^{(1)},v^{(2)}) = \int_0^t \mathbb{P}S(t-\tau)[\partial_j(v_j^{(1)}v^{(2)})](\tau)d\tau, \quad v^{(1)},v^{(2)} \in X \quad (7.2)$$

Then $\mathcal{B} : X \times X \mapsto X$ is bicontinuous and $w(\cdot,0) = 0$.

The proof relies on the theory of linear operators acting on Lorentz spaces. Indeed if $v \in \mathcal{C}([0,\infty);L^{(3,\infty)})$, then all the products u_ju_k will belong to $L^\infty([0,\infty),L^{(3/2,\infty)}) \cap \mathcal{C}((0,\infty),L^{(3/2,\infty)})$. Therefore it will suffice to study the following linear operator

$$\mathcal{L}(f) = \int_0^\infty P_sf(\cdot,s)\,ds, \quad P_sf(x) = \int_{\mathbf{R}^3} K_s(x,y)f(y)\,dy \qquad (7.3)$$

where $K_s(x,y)$ is a kernel fulfilling the following estimation

$$|K_s(x,y)| \le Cs^{-3}(1 + |x-y|/s)^{-4} \qquad (7.4)$$

Let us prove this remark. In our situation, $P_\tau = \sqrt{\tau}S(\tau)\mathbb{P}\partial_j$ which is a convolution operator with $\tau^{-3}w_j(x/\tau)$ as we already observed many times. The functions $w_j, j = 1,2,3$, are continuous and decay as $|x|^{-4}$ at infinity. With these new notations, the scalar version $\tilde{B}(f,g)$ of the bilinear operator $\mathcal{B}$ can be written as $\tilde{B}(f,g) = \int_0^t P_{t-\tau}(fg)\frac{d\tau}{\sqrt{t-\tau}}$. We now perform the change of variable $t - \tau = s^2$ and define $h(\cdot,s) = fg(\cdot,t-s^2)$ if $0 \le s \le \sqrt{t}$ and by 0 elsewhere. Finally the time variable t is frozen and everything reduces to the linear operator $\int_0^\infty P_sh(\cdot,s)\,ds$. For easing notations the variable s will be rewritten t and we now sum up our discussion as follows. For proving Theorem 7.1, we consider a slightly more general class of linear operators $\mathcal{L}_\alpha(f) = \int_0^\infty P_tf(\cdot,t)\,t^{\alpha-1}dt$ and everything reduces to the following theorem:

Theorem 7.2. *If $\alpha \in (0,3)$ and $p \in (1, 3/\alpha)$, then*

$$\|\mathcal{L}_\alpha f\|_{(q,\infty)} \leq C \sup_{t \geq 0} \|f(\cdot, t)\|_{(p,\infty)} \tag{7.5}$$

where $q = \frac{3p}{3-\alpha p}$.

The proof of Theorem 7.2 is straightforward. Let $\lambda > 0$ be a threshold and E_λ denote the set of points x where $|g(x)| > \lambda$ when $g = \mathcal{L}_\alpha f$. We then split $\int_0^\infty P_t f \, t^{\alpha-1} \, dt$ into $\int_0^\tau () + \int_\tau^\infty () = u + v$. The value of τ will be unveiled in the proof. We first estimate $\|v\|_\infty$. Indeed, as a function of y, $K_t(x,y)$ belongs to $L^{p',1}$ where $1/p + 1/p' = 1$. Moreover $\|K_t(x, \cdot)\|_{(p',1)} \leq Ct^{-3/p}$. This implies

$$\|P_t f(\cdot, t)\|_\infty \leq Ct^{-3/p} \sup_{s \geq 0} \|f(\cdot, s)\|_{(p,\infty)} \tag{7.6}$$

Therefore

$$\|v\|_\infty \leq C \int_\tau^\infty t^{(\alpha-1-3/p)} \, dt = C'\tau^{\alpha-3/p} \tag{7.7}$$

We now define τ by $C'\tau^{\alpha-3/p} = \lambda/2$ where C' is defined by (7.7). It implies $\|v\|_\infty \leq \lambda/2$. Then $|g(x)| > \lambda$ implies $|u(x)| > \lambda/2$. We now turn to u and use the fact that the norm in $L^{(p,\infty)}$ is convex when $p > 1$. We then obtain

$$\|u\|_{(p,\infty)} \leq C \int_0^\tau \|f(\cdot, t)\|_{(p,\infty)} t^{\alpha-1} dt = C''\tau^\alpha \tag{7.8}$$

Here we used the following lemma

Lemma 7.1. *We have $|P_t(f)| \leq C|f| \star \phi_t$ where $\phi(x) = (1+|x|)^{-4}$, $\phi_t(x) = t^{-3}\phi(x/t)$ which implies the uniform boundedness of P_t on all the Lorentz spaces $L^{(p,\infty)}$, $1 < p < \infty$.*

Returning to our proof, we observed that $|g(x)| > \lambda$ implies $|u(x)| > \lambda/2$ and finally the measure of this set does not exceed $C''(\frac{\lambda}{\tau^\alpha})^{-p} = C_0\lambda^{-q}$ as announced.

We now treat the continuity of the operator $B(f,g)$. The main issue comes form the following. If $v(\cdot, t) \rightharpoonup v_0$ as $t \to 0$, then there are no reasons why $v(\cdot, t)v_j(\cdot, t) \rightharpoonup v(\cdot, 0)v_j(\cdot, 0)$ as t tends to 0. However this issue happens to be completely irrelevant in what follows. The following lemma will suffice to our needs.

Lemma 7.2. *Let $\mathcal{B}$ be defined by (3.18) and $h(\cdot, t) = \mathcal{B}(f,g)$, $(f,g) \in X$. Then we have $h(\cdot, t) \to 0$ as $t \to 0$.*

The proof consists in proving a stronger estimate which reads $\|h(\cdot, t)\|_{(3/2,\infty)} \to 0$ as $t \to 0$. The proof is almost trivial. We know that $\|fg\|_{(3/2,\infty)} \leq C$. But the convexity of the norm implies that

$$\|h\|_{(3/2,\infty)} \leq C \int_0^t (t-\tau)^{-1/2} \|fg\|_{(3/2,\infty)} \, d\tau \leq Ct^{1/2} \tag{7.9}$$

We used the following observation: the operator $\mathbb{P}S(t)\partial_j$ is a convolution with a function whose L^1 norm is $Ct^{-1/2}$.

The proof of the continuity on $(0,\infty)$ for the strong norm is much easier since the conflict between nonlinearities and weak convergence does not appear. We can directly go to the linear operator $\mathcal{L}_\alpha$, denote by f one of the components of $v_j v$ and prove the strong required continuity. For proving the continuity at t_0, we suppose $t, t' \in (t_0/2, 2t_0)$. Then we split f into a sum $f_1 + f_2$ where $f_1(\cdot, t) = 0$ for $t \geq t_0/3$, $f_2(\cdot, t) = 0$ for $t \leq t_0/4$ and where $\|f_1\|_{(3/2,\infty)} \leq C, \|f_2\|_{(3/2,\infty)} \leq C$. Then we write $g = \mathcal{L}f = \int_0^t S(t-\tau)\Lambda f\, d\tau$ where Λ stands for $\mathbb{P}\partial_j$ and is a pseudodifferential operator of order 1. Let us first treat $u(\cdot, t) = \int_0^t S(t-\tau)\Lambda f_1(\cdot, \tau).\,d\tau$ We have

$$\|\frac{\partial}{\partial t}u(\cdot, t)\|_{(3,\infty)} \leq \int_0^\infty \|\Delta^2 S(t-\tau)\Lambda^{-1}f_1\|_{(3,\infty)}d\tau \qquad (7.10)$$

We now use the boundedness of $\Lambda^{-1} : L^{(3/2,\infty)} \mapsto L^{(3,\infty)}$ and the computation ends with $\int_0^{t_0/3}(t-\tau)^{-2}\,d\tau \leq C/t_0$. Concerning v, we write $v(\cdot, t') - v(\cdot, t) = \int_0^{t'} S(\tau)\Lambda f_2(\cdot, t'-\tau)\,d\tau - \int_0^t S(\tau)\Lambda f_2(\cdot, t-\tau)\,d\tau = U + V$. Concerning U we factor out the Calderón's operator. We obtain

$$U = \int_t^{t'} [S(\tau)\Delta]\Lambda^{-1}f_2(\cdot, t'-\tau)\,d\tau$$

which implies

$$\|U\|_{(3,\infty)} \leq \int_t^{t'} \|Q(\tau)\Lambda^{-1}f_2(\cdot, t'-\tau)\|_{(3,\infty)}\frac{d\tau}{\tau} \qquad (7.11)$$

The operators $Q(t) = t\Delta\exp(t\Delta), t \geq 0$, are uniformly bounded on $L^{(3,\infty)}$. We finally obtain

$$\|U\|_{(3,\infty)} \leq C\log(\frac{t'}{t}) \qquad (7.12)$$

Concerning V, it suffices to use Theorem 7.2 once more together with the continuity of f_2 in t. We then have $\|f_2(\cdot, t'-\tau) - f_2(\cdot, t-\tau)\|_{(3/2,\infty)} \leq \epsilon$ when $|t'-t| \leq \eta$ and $0 \leq \tau \leq t \leq t' \leq 2t_0$. It implies $\|V\|_{(3,\infty)} \leq C\epsilon$ as announced.

We now turn to the main theorem by Furioli, Lemarié-Rieusset, Terraneo [47]. It reads

Theorem 7.3. *Let* $u(x,t), v(x,t) \in \mathcal{C}([0,T]; L^3(\mathbf{R}^3))$ *be two solutions of the Navier-Stokes equations. If* $u(x,0) = v(x,0)$, *then* $u(x,t) = v(x,t)$ *on* $[0,T]$.

Theorem 7.3 easily follows from a seamingly weaker property: there exists a positive ϵ such that

$$u(t) = v(t) \quad t \in [0,\epsilon]. \qquad (7.13)$$

For proving (7.12) we return to (3.18) and write

$$u = S(t)u_0 + \mathcal{B}(u, u), \quad v = S(t)u_0 + \mathcal{B}(v, v) \tag{7.14}$$

Let $v = u + w$. Then

$$w = \mathcal{B}(u + w) - \mathcal{B}(u, u) = \mathcal{B}(v, w) + \mathcal{B}(w, u) \tag{7.15}$$

We define $\eta(t) = \sup\{\|w(\cdot, s)\|_{(3,\infty)}; 0 \leq s \leq t\}$ and we shall prove the existence of a positive ϵ such that

$$\eta(t) \leq \frac{1}{2}\eta(t), \quad 0 \leq t \leq \epsilon \tag{7.16}$$

This obviously imply (7.13).

For estimating $\eta(t)$ we write

$$w = \mathcal{B}(S(t)u_0, w) + \mathcal{B}(v - S(t)u_0, w) + \mathcal{B}(w, u - S(t)u_0) + \mathcal{B}(w, S(t)u_0) \tag{7.17}$$

For estimating $\|\mathcal{B}(w, S(t)u_0)\|_{(3,\infty)}$, we return to the definition of the bilinear operator $\mathcal{B}$ and glue together $\mathbb{P}\partial_j S(t)$. As was already observed, this product is a convolution operator with a kernel whose L^1 norm is $Ct^{-1/2}$. The operator bound on $L^{(3,\infty)}$ of $\mathbb{P}\partial_j S(t-\tau)$ does not exceed $C(t-\tau)^{-1/2}$. Finally the L^∞ norm of $S(\tau)u_0$ will be written $\tau^{-1/2}\|\tau^{1/2}S(\tau)u_0\|_\infty$. Altogether we obtain

$$\|B(w, S(t)u_0\|_{(3,\infty)} \leq C\eta(t)\sup\{\|\tau^{1/2}S(\tau)u_0\|_\infty\} \tag{7.18}$$

A similar treatment applies to the first term. Concerning the second and the third terms in (7.16), we simply apply Theorem 7.1 and observe that $\|u(t) - S(t)u_0\|_{(3,\infty)} \leq \|u(t) - S(t)u_0\|_3 = \alpha(t)$ where $\alpha(t)$ tends to 0 with t. Altogether it yields

$$\|w(t)\|_{(3,\infty)} \leq \gamma(t)\eta(t) \tag{7.19}$$

where $\gamma(t)$ tends to 0 with t. One should observe that $\|t^{1/2}S(t)u_0\|_\infty$ tends to 0 with t when $u_0 \in L^3$. Since (7.19) implies (7.16), this end the proof of Theorem 7.3.

8 Vortex Filaments and a Theorem by Y. Giga and T. Miyakawa

We know from Lemma 3.3 of Section 3 that the evolution of the velocity v is governed by the Navier-Stokes equation if and only if the corresponding vorticity ω satisfies the following:

Lemma 8.1.

$$\begin{cases} \frac{\partial\omega}{\partial t} = \partial_j(\omega_j v - v_j\omega) + \Delta\omega \\ \nabla \cdot \omega_0 = 0 \\ \omega(x, 0) = \omega_0(x) \\ v = -(\Delta)^{-1}\mathrm{curl}(\omega) \end{cases} \tag{8.1}$$

The right-hand side is meant as a sum over j (usual convention on repeated indices). We have

$$\operatorname{div}(\partial_j(\omega_j v - v_j \omega)) = 0 \tag{8.2}$$

and the evolution of ω is keeping the divergence free property.

We now turn to an important theorem by Y. Giga and T. Miyakawa. The motivation of Giga and Miyakawa was twofold. They wanted to model vorticity filaments in order to understand the nonlinear evolution of such filaments. These vorticity filaments appear in numerical simulations of Navier-Stokes equations. At the same time Giga and Miyakawa wanted to construct some self-similar solutions to the Navier-Stokes equations. In order to achieve these goals, they modeled these vorticity filaments with measures μ satisfying the following *Guy David condition*

Definition 8.1. *A Borel measure μ satisfies the Guy David condition if and only if a constant C exists such that, for every ball B with radius R, we have*

$$|\mu|(B) \le CR \tag{8.3}$$

We then write $\mu \in M$ and the norm in M is defined as the lower bound of all constants C in (8.3).

Let us provide the reader with an example of a vorticity field $\omega(x)$ belonging to M. We start with the definition of a Guy David curve. It is a rectifiable open oriented curve $\Gamma \subset \mathbf{R}^3$ satisfying the following condition where s is the arclength on Γ.

Definition 8.2. *An oriented rectifiable curve Γ is a Guy David curve if constant C exists such that for each ball $B(x_0, r)$ centered at x_0 with radius r, the total length of $\Gamma \cap B(x_0, r)$ does not exceed Cr.*

We then define a vorticity filament by the following.

Definition 8.3. *Let Γ be a Guy David curve. The corresponding vorticity filament is defined by $\omega = \mathbf{t}\, ds$ where $\mathbf{t}$ denotes the oriented unit tangent vector to Γ.*

Let us check that $\operatorname{div} \omega = 0$. Indeed if $u(x)$ is a testing function, it suffices to prove that $\int_\Gamma \nabla u \cdot \mathbf{t}\, ds = 0$. But the integrand is $\frac{\partial u}{\partial s}$ and then the proof reduces to $\int_\Gamma \frac{\partial u}{\partial s}\, ds = 0$ which is obvious.

Returning to the general case, the space of all measures satisfying the Guy David condition will be denoted by M and the corresponding norm is $\|\mu\|_M$. The space SMS is a dual space and will always be equipped with its weak* topology and $v \in C([0, \infty), M)$ is defined by Definition 2.1. One observes that M is a particular Morrey-Campanato space. Indeed $M = M_1^{3/2}$.

Giga and Miyakawa proved the following theorem [54]

Theorem 8.1. *There exists a (small) positive number η such that whenever the initial condition $\omega_0(x)$ satisfies*

$$\begin{cases} \operatorname{div}(\omega_0) = 0 \\ \|\omega_0\|_M < \eta \end{cases} \tag{8.4}$$

then there exists a solution $\omega(x,t) \in \mathcal{C}([0,\infty), M)$ to the Navier-Stokes equation which agrees with this initial condition. Moreover there exists a constant C_1 such that the corresponding velocity v satisfies

$$t^{1/2}\|v(\cdot,t)\|_\infty \le C_1\|\omega_0\|_M \tag{8.5}$$

For proving this theorem, we first rewrite (8.1) as an integral equation. As in section 4, this is achieved by solving the linear heat equation and we obtain

$$\begin{cases} \omega(t) = S(t)\omega_0 + \int_0^t S(t-\tau)[\partial_j(\omega_j v - v_j\omega)](\tau)d\tau \\ \omega(x,0) = \omega_0 \\ v = -(\Delta)^{-1}\operatorname{curl}(\omega) \end{cases} \tag{8.6}$$

As it was previously said, there is another way of getting rid of the pressure. One directly applies the Leray Hopf operator to (3.1) and one solves the corresponding heat equation. It yields

$$\begin{cases} u(t) = S(t)u_0 + \int_0^t \mathbb{P}S(t-\tau)[\partial_j(v_j v)](\tau)d\tau \\ u(x,0) = u_0 \end{cases} \tag{8.7}$$

Lemmata 3.1 and 3.4 (section 3) imply the following

Lemma 8.2. *We have*

$$(-\Delta)^{-1}\operatorname{curl}[\partial_j(\omega_j v - v_j\omega)] = \mathbb{P}[\partial_j(v_j v)] \tag{8.8}$$

In other terms, (8.7) can be seen as a pullback of (8.6) if the Biot-Savart law is being used.

In order to use Lemma 5.1 and Lemma 5.2, (8.6) will be rewritten in a more concise form as

$$\omega = h + \tilde{B}(\omega,v), \ v = I(\omega) \tag{8.9}$$

or equivalently as

$$\omega = h + B(\omega,\omega) \tag{8.10}$$

where ω is now a vector inside some function space X and h is a given vector in X. The difficult part of the proof is the construction of this Banach space X to which Picard's fixed point theorem will be applied.

In our proof, X is defined by imposing two unrelated conditions. The first one says that we are looking for solutions which are continuous with respect to the time variable with values in the space of Guy David measures. The second is more artificial since it concerns the L^∞ norm of the velocity field.

Definition 8.4. *The Banach space X consists of all divergence free vector fields $\omega(x,t) \in C([0,\infty), M)$ such that the corresponding velocity v satisfies*

$$\|v(\cdot, t)\|_\infty \le Ct^{-1/2} \tag{8.11}$$

This velocity is calculated by the Biot-Savart law. The norm in X is defined as $\sup\{\|\omega(\cdot, t)\|_M, t \ge 0\} + \sup\{t^{1/2}\|v(\cdot, t)\|_\infty, t \ge 0\}$.

For proving Theorem 8.1, it suffices to apply Picard's fixed point theorem to (8.6) or equivalently to (8.10). The quadratic nonlinearity in (8.10) is

$$B(\omega, \omega) = \int_0^t S(t-\tau)[\partial_j(\omega_j v - v_j \omega)](\tau)d\tau \tag{8.12}$$

and the corresponding bilinear operator $B(f,g)$ is given by $2B(f,g) = B(f + g, f + g) - B(f,f) - B(g,g)$.

Two facts are required. In the first place, we need to prove that the linear evolution h in (8.10) belongs to X whenever ω_0 is a Guy David measure. This is an easy consequence of Lemma 9.1. This lemma will play another role and will be postponed until the heart of the proof is reached.

In the second place we treat the nonlinear evolution. We need to prove that the bilinear operator B defined by (8.12) maps $X \times X$ to X. This concerns the Guy David norm of $B(\omega, \omega)$ and the L^∞ norm of the corresponding velocity.

For treating the first norm, we use two facts. On the first hand we observe that a function $v_j \in L^\infty$ is a pointwise multiplier of a Guy David measure (in our case ω). On the second hand, if we also glue together the operators $S(t-\tau)$ and ∂_j, the product $S(t-\tau)\partial_j$ is a convolution with a kernel whose L^1 norm is $C(t-\tau)^{-1/2}$. But the convolution between a Guy David measure and a function in L^1 is still a Guy David measure. Combining this remark with the bound on the L^∞ norm of $v(\cdot, t)$ leads to $\|B(\omega, \omega)\|_{BV} \le C \int_0^t \frac{1}{\sqrt{\tau(t-\tau)}} d\tau = C\pi$.

The L^∞ bound on the velocity is slightly more involved. As it was said in Definition 8.4, the velocity is computed by the Biot-Savart law. Then our problem concerns the L^∞ norm of $w(\cdot, t) = (-\Delta)^{-1}\mathrm{curl}\, B(\omega, \omega)$. We use (8.12) again and obtain

$$w(\cdot, t) = \int_0^t (-\Delta)^{-1}\mathrm{curl}\, S(t-\tau)[\partial_j(\omega_j v - v_j \omega)](\tau)d\tau \tag{8.13}$$

For easing notations, the integrand in (8.13) will be denoted by $W_t(\cdot, \tau)$.

We first take advantage on our previous estimate on the Guy David norm of $\omega_j v - v_j \omega$ and we glue together $(-\Delta)^{-1}\mathrm{curl}\,\partial_j S(t-\tau)$ into an operator named $K_j(t-\tau)$. The kernel of $K_j(t-\tau)$ is $(t-\tau)^{-3/2}g(\frac{x-y}{(t-\tau)^{1/2}})$ where g is continuous and decays as $|x|^{-3}$ at infinity. We do not have $g_j \in L^1$ but we do not care. Instead we use the following lemma

Lemma 8.3. *Let $w(x)$ be a nonnegative continuous function defined on $\mathbf{R}^3$ such that $w(x) \leq C(1+|x|)^{-\alpha}$ where $\alpha > 1$. Then for any Guy David measure μ there exists a constant C' such that for any $x_0 \in \mathbf{R}^3$ and $t > 0$ we have*

$$\int w(\frac{x - x_0}{t}) d\mu \leq C' t \tag{8.14}$$

For proving this remark it suffices to decompose the domain of integration into dyadic rings of size $2^m t$, $m \in \mathbf{N}$, centered at x_0, and to apply to each such ring the definition of a Guy David measure.

Let us postpone the proof of Theorem 8.1 and comment on Lemma 8.4. This lemma implies the following important observation. For a nonnegative Borel measure, the norm in M as defined by (8.3) is equivalent to its norm in the homogeneous Besov space $\dot{B}_\infty^{-2,\infty}$. The Biot-Savart law implies that the velocity belongs to $\dot{B}_\infty^{-1,\infty}$ whenever the vorticity belongs to M. But we have a better estimate, since an easy calculation shows that the velocity admits a representation as $v = \partial_1 g_1 + \partial_2 g_2 + \partial_3 g_3$ where g_1, g_2, g_3 belong to BMO.

We now return to the proof of Theorem 8.1. Lemma 8.3 applies and the operator norm of $K_j(t - \tau)$ acting from Guy-David measures to L^∞ is exactly $(t - \tau)^{-1}$. Altogether it yields

$$\|W_t(\cdot, \tau)\|_\infty = O((t - \tau)^{-1} \tau^{-1/2}) \tag{8.15}$$

This estimate is useless since it would lead to a divergent integral in (8.13). But we also know that the Biot-Savart pullback of $\partial_j(\omega_j v - v_j \omega)$ is $\mathbb{P}(v_j v)$. We then forget Guy David measures and focus on L^∞ estimates. In a first place we have $\|v_j v(\cdot, \tau)\|_\infty \leq C/\tau$. Then we glue together $\mathbb{P}S(t - \tau)\partial_j$. This new operator is again a convolution with a kernel of the form $(t - \tau)^{-2} w_j(\frac{x-y}{(t-\tau)^{1/2}})$ where w_j is a continuous function which is $O(|x|^{-4})$ at infinity. Therefore the operator norm of $\mathbb{P}S(t - \tau)\partial_j$ acting on L^∞ is $O((t - \tau)^{-1/2})$. We then obtain

$$\|W_t(\cdot, \tau)\|_\infty = O((t - \tau)^{-1/2} \tau^{-1}) \tag{8.16}$$

The geometric mean between (8.15) and (8.16) is $O((t - \tau)^{-3/4}(\tau)^{-3/4})$ which integrated over $[0, t]$ yields the required $t^{-1/2}$. This ends the proof of Theorem 8.1.

The Biot-Savart law enables us to lift Theorem 8.1 from vorticities to velocities, but we are not going to be more precise about the Banach space Γ describing these velocities. This space is a dual space, its norm is compatible with the scaling properties of the Navier-Stokes equations and it contains functions which are homogeneous of degree -1, which permitted Giga and Miyakawa to build self similar solutions to the Navier Stokes equations. Seven years later, M. Cannone and F. Planchon proposed another construction of self-similar solutions, based on Theorem 6.2.

Theorem 6.2 and Theorem 8.1 complement each other. Indeed the space Γ is not contained inside any space B_q, $3 \leq q < \infty$ and conversely $u \in B_q$ does not imply $u \in \Gamma$.

Before moving to the special case of vortex patches, we would like to incorporate Theorem 8.1 in a more general framework. The Banach space Γ used by Giga and Miyakawa is a subspace of the functional space used by H. Koch and D. Tataru. Indeed H. Koch and D. Tataru are assuming that the initial velocity v_0 is the divergence of a vector field (g_1, g_2, g_3) where these three components belong to BMO. Let us be more precise   .

Definition 8.5. *We denote by* $\mathcal{K} = \mathcal{K}_\infty$ *the Banach space consisting of all generalized functions* f *which can be written as* $f = \partial_1 g_1 + \partial_2 g_2 + \partial_3 g_3$ *where* g_j, $j = 1, 2$ *and* 3, *belong to BMO.*
The norm in $\mathcal{K} = \mathcal{K}_\infty$ *is denoted by* $\|f\|_\mathcal{K}$ *and is defined as the infimum of the sums of the three BMO norms.*

The definition of the John and Nirenberg space BMO runs as follows

Definition 8.6. *A function* f *belongs to BMO if and only if* f *is locally square integrable and if there exists a uniform bound for all the variances of* f *over all the balls* $B \subset \mathbf{R}^n$. *Each such ball* B *is viewed as a probability space, the probability measure being* $dP = |B|^{-1} dx$.

We will return to the theorem by H. Koch and D. Tataru in Section 10. For the time being, this theorem says the following: there exists a positive constant η such that the following two conditions (a) $\|v_0\|_\mathcal{K} \leq \eta$, and (b) $\nabla.v_0 = 0$ imply the existence of a global solution $v \in \mathcal{C}([0, \infty); \mathcal{K})$ of the Navier-Stokes equations. The space $\mathcal{K}$ is equipped with its weak-star topology.

It remains to say why this theorem generalizes Theorem 8.1. Indeed we have

Lemma 8.4. *If* μ *is a Guy-David measure, then* $\mu \star |x|^{-1}$ *belongs to BMO.*

The proof of Lemma 9.1 is straightforward. We take an arbitrary ball B on which the BMO condition needs to be checked. Without loosing generality we can assume that B is centered at 0. The double ball is denoted by $\tilde{B}$. Then we decompose μ into a sum between μ_1 which is carried by $\tilde{B}$ and μ_2 which vanishes on this double ball. Now $h_1(x) = \mu_1 \star |x|^{-1} = \mu_1 \star g$ on B where $g(x) = |x|^{-1}$ on the triple ball $3B$ and $g(x) = 0$ elsewhere. The L^2 norm of g is $Cr^{1/2}$ where r is the radius of B. On the other hand the total mass of the measure μ_1 does not exceed Cr (this is the Guy David condition). Therefore the L^2 norm of h_1 does not exceed $C'r^{3/2}$. It remains to treat $h_2 = \mu_2 \star |x|^{-1}$. This is achieved by estimating $|h_2(x) - h_2(0)|$ on B. It suffices to bound $||x - y|^{-1} - |y|^{-1}|$ by $r|y|^{-2}$ whenever $|x| \leq r$ and to apply once more the definition of a Guy David measure (or Lemma 8.1).

Lemma 9.1 is proved and we return to Theorem 8.1 and to the H. Koch & D. Tataru theorem. If the vorticity ω is a Guy David measure, then the Biot-Savart law and Lemma 9.1 show that the corresponding velocity belongs to the H. Koch and D. Tataru space $\mathcal{B}$.

In the next sections, we will focus on a special case of Theorem 8.1 where the initial vorticity ω_0 belongs to the space BV of functions with bounded variation. This space BV is indeed contained in M and our next theorem says that if the initial vorticity is a function of bounded variation (a vortex patch), this property persists over time. The few properties of functions with bounded variations which will be used are listed in Section 14.

9 Vortex Patches

The properties of the space $BV(\mathbf{R}^3)$ of functions with bounded variation will be freely used in this section. The reader who is not familiar with this space may consult [2] or go to Section 14 to be given some more information.

The following theorem describes the evolution of the vorticity of a solution $v(\cdot, t) \in \mathcal{C}([0, \infty); L^3(\mathbf{R}^3))$ given by Theorem 5.1. Let us assume, for instance, that the initial vorticity ω_0 is supported by a domain and has a jump discontinuity across the boundary. We would like to know whether or not this geometrical information is preserved under the nonlinear evolution. Do 'vortex patches' exist as it is the case for Euler equations? We now give up the specific example which opened this discussion and treat a slightly more general situation where ω_0 belongs to the space BV.

Theorem 9.1. *Let η be defined as in Theorem 5.1 and let us assume that the initial velocity satisfies $\|v_0\|_3 < \eta$. If the initial vorticity ω_0 belongs to BV, then the unique solution $\omega(x, t)$ of the Navier-Stokes equations given by Theorem 5.1 satisfies the following properties*

$$\begin{cases} \omega(x, t) \in \mathcal{C}([0, \infty); BV) \\ \omega(x, 0) = \omega_0(x) \\ v(x, t) \in \mathcal{C}([0, \infty); L^3(\mathbf{R}^3)) \\ \|v(\cdot, t)\|_\infty \le Ct^{-1/2} \end{cases} \tag{9.1}$$

where $v(x, t) = (-\Delta)^{-1} \operatorname{curl} \omega$.

In Theorem 9.1, BV is given its weak-star topology. Theorem 9.1 complements T. Kato's theorem. But Theorem 9.1 also complements Theorem 8.1. The proofs of these observations depend on the following remarks.

Lemma 9.1. *The space BV is contained in the space M of Guy David measures.*

The two continuous embeddings $BV \subset L^{3/2}$ and $L^{3/2} \subset M$ suffice to conclude.

Lemma 9.2. *If ω belongs to the Lebesgue space $L^{3/2}$, then u belongs to L^3.*

This is an obvious consequence of the Biot-Savart law and of the Sobolev embedding theorem.

We now turn to the proof of Theorem 9.1.The proof of the Cannone-Planchon theorem will be adapted to our situation. In other words, we play with two norms. The first one $\|v(\cdot, t)\| = \sup\{t^{1/4}\|v(\cdot, t)\|_6 \ t \geq 0\}$ is the norm defined by (6.4) which was used in the proof of Theorem 6.2 and this norm be the weak one in our discussion. The strong norm $\|v(\cdot, t)\|_1$ will be defined by

$$\|v(\cdot, t)\|_1 = \sup\{t^{1/4}\|v(\cdot, t)\|_6 \ t \geq 0\} + \sup\{\|\omega(\cdot, t)\|_{BV} \ t \geq 0\} \qquad (9.2)$$

We are given a solution $v(x, t) \in \mathcal{C}([0, \infty); L^3(\mathbf{R}^3))$ such that ω_0 belongs to BV and we aim at proving that this property persists under the evolution. We apply lemma 18 and it suffices to prove (5.10). We now concentrate on that proof. In other words, we shall show that the bilinear operator defined by the right-hand side of (8.6) satisfies the mixed continuity property of Lemma 5.2.

We then need to estimate the BV norm of the integrand in (8.6). On the first hand, in terms like $v_j\omega$, v can be viewed as a pointwise multiplier of BV. But Theorem 14.2 of Section 14 takes care of that problem and tells us that the corresponding operator norm can be estimated by $\|\nabla v\|_3$. We now turn to (6.2.f) which yields $\|\nabla v\|_3 \leq C\tau^{-1/2}$. Finally we have $\|v_j\omega(\cdot, \tau)\|_{BV} \leq C\tau^{-1/2}$.

On the second hand, we glue together $S(t - \tau)$ and ∂_j. The resulting operator is a convolution with an integrable function. The L^1 norm of this function is $C(t - \tau)^{-1/2}$. The operator norm on BV will be estimated by this trivial bound. Once again the proof of (5.10) ends with $\int_0^t (t-\tau)^{-1/2}\tau^{-1/2}\,d\tau = \pi$.

Here are some examples of vortex patches. We start with a (bounded and connected) domain $\Omega \subset \mathbf{R}^3$ with a smooth boundary $\partial\Omega$ and we let $g(x) = (g_1(x), g_2(x), g_3(x))$ be a divergence free vector field belonging to $\mathcal{C}^1(\overline{\Omega})$. Let us denote by $\mathbf{n}$ the exterior normal vector to $\partial\Omega$ and assume that $\mathbf{n} \cdot g = 0$ on $\partial\Omega$. We let $\omega_0(x)$ be the ordinary product between $g(x)$ and the indicator function of Ω. This $\omega_0(x)$ provides us with an example of an initial vorticity with bounded variation. Moreover $\operatorname{div}\omega_0 = 0$ in the distributional sense. For constructiong this vector field g, one can assume that $\partial\Omega$ is the level set of a smooth real valued function θ. Then $g_1 = \partial_2\theta - \partial_3\theta, g_2 = \partial_3\theta - \partial_1\theta, g_3 = \partial_1\theta - \partial_2\theta$ fulfills the requirements we imposed on the pair (ω, g). Many other examples can be constructed in a similar way.

A last remark concerns the role of the smallness condition $\|\omega_0\|_{BV} < \alpha$ in Theorem 9.1. We will follow M. Cannone and F. Planchon and replace it by the weaker condition $\|v_0\|_{B_q} < \eta_q$ for some $q \in (3, \infty)$. Lemma 5.2 will be used again. The weak norm is the sum between $\sup\{t^{1/2}\|u(\cdot, t)\|_\infty, t \geq 0\} + \sup\{t^{1/2}\|\nabla u(\cdot, t)\|_3, t \geq 0\}$. The calculations are identical to the preceding one and the weaker condition suffices, as in Theorem 6.2.

We now consider wavelet expansions and assume that the initial vorticity belongs to $\dot{B}_1^{1,1}(\mathbf{R}^3)$ with a small norm. Let us observe that this Besov space is contained in BV. What can be said about the evolution under the Navier-Stokes equations?

Theorem 9.2. *There exists a positive constant α such that if $div\,\omega_0 = 0$ and $\|\omega_0\|_{\dot{B}_1^{1,1}} < \alpha$, then the vorticity $\omega(\cdot, t)$ belongs to $\mathcal{C}([0, \infty), \dot{B}_1^{1,1})$.*

Here the situation is much simpler than the preceding one. Indeed classical pseudodifferential operators are bounded on Besov spaces. It implies that the initial velocity belongs to $\dot{B}_1^{2,1}$ with a small norm. It then suffices to study the evolution of the velocity and to pull back the corresponding estimates to the vorticites.

We then need the following simple observation:

Lemma 9.3. *The Banach space $\dot{B}_1^{3,1}$ is an algebra of continuous functions in $\mathbf{R}^3$ and each such continous functions is a pointwise multiplier for $\dot{B}_1^{2,1}$.*

The proof which is easy relies on paraproduct techniques. This lemma allows us to mimic the proof of T. Kato's theorem and details are left to ther reader.

10 The H. Koch & D. Tataru Theorem

The H. Koch & D. Tataru theorem is, in a sense the limiting case, as $q \to \infty$, of the M. Cannone-F. Planchon theorem (Theorem 6.2). Indeed the Besov spaces B_q which are used to define oscillatory patterns are increasing with q and are now replaced by their limit. This limit will not be the space $\dot{B}_\infty^{-1,\infty}$ as one would expect. The limit space introduced by H. Koch and D. Tataru is slightly smaller and one does not know if their theorem can be extended to $B_\infty^{-1,\infty}$.

Definition 10.1. *We denote by $\mathcal{K} = \mathcal{K}_\infty$ the Banach space consisting of all generalized functions f which can be written as $f = \partial_1 g_1 + \partial_2 g_2 + \partial_3 g_3$ where g_j, $j = 1, 2$ and 3, belong to BMO.*
The norm in $\mathcal{K}_\infty$ is denoted by $\|f\|_B$ and is defined as the infimum of the sums of the three BMO norms.

We now arrive to the main theorem

Theorem 10.1. *There exists a positive constant η such that the following two conditions (a) $\|v_0\|_{\mathcal{K}} \leq \eta$, (b) and (b) $\nabla.v_0 = 0$ imply the existence of a global solution v of the Navier-Stokes equations belonging to $\mathcal{C}([0,\infty);\mathcal{K})$ with $\omega(\cdot,0) = \omega_0$. Here the continuity refers to the weak* topology of $\mathcal{K}$.*

As it was already stressed, the H. Koch and D. Tataru space contains all the previous Besov spaces $\dot{B}_q$ which were used in Theorem 6.2. Moreover the H. Koch and D. Tataru theorem implies the Giga-Miyakawa result. We have already observed (Lemma 8.4) that $\Lambda^{-1}(\mu)$ belongs to BMO whenever μ satisfies the Guy David condition. Here $\Lambda = (-\Delta)^{1/2}$.

A combination between the H. Koch & D. Tataru theorem [67] and an elegant remark by Pierre-Gilles Lemarié-Rieusset and his students [48] yields the following theorem

Theorem 10.2. *There exists a positive number η such that, for every positive integer m, the following three conditions (a) $\|v_0\|_{\mathcal{K}} \leq \eta$, (b) $v_0 \in H^m(\mathbf{R}^3)$ and (c) $\nabla.v_0 = 0$ imply the existence of a global solution v of the Navier-Stokes equations. This global solution belongs to $\mathcal{C}([0,\infty);H^m(\mathbf{R}^3))$ and is unique.*

Before ending this section, we would like to say a few more words about the proof of the H. Koch and D. Tataru theorem. This proof follows the general organization which was pioneered by T. Kato and F. Weissler. The Navier-Stokes equations are rewritten as an integral equation. This is achieved by solving the linear heat equation. Then we obtain, as above,

$$v(t) = S(t)v_0 + \mathbb{P} \int_0^t S(t-\tau) \sum_1^3 \partial_j(v_j v)(\tau)d\tau \qquad (10.1)$$

Here $\mathbb{P}$ denotes the Leray-Hopf projector on divergence-free vector fields. Two points should be made. First the pressure $p(x,t)$ has disappeared from the Navier-Stokes equations and next the initial condition has been incorporated inside (10.1). Indeed the kernel of the Leray-Hopf operator is precisely the collection ∇p of curl-free vector fields.

We then rewrite (10.1) in a more condensed way as

$$v = h + \mathcal{B}(v,v) \qquad (10.2)$$

where v is now a vector inside some function space X and h is a given vector in X. In the application we have in mind $X = \mathcal{C}([0,\infty);E)$ is a Banach space of functions $f(x,t)$ which are defined on $\mathbf{R}^3 \times (0,\infty)$ and are viewed as E valued continuous functions of the time variable.

The difficult part of the proof is the construction of this Banach space X to which Picard's fixed point theorem will be applied.

H. Koch and D. Tataru used a function space X describing the expected behavior of a solution $u(x,t)$ when the initial data belongs to $\mathcal{K}_\infty$. The definition of X is driven by the behavior of the linear evolution. In the H. Koch and D. Tataru theorem, the Banach space X in which the iteration scheme is applied is defined as follows.

Definition 10.2. *The Banach space X consists of all functions $f(x,t)$ which are locally square integrable on $\mathbf{R}^3 \times (0,\infty)$ and which satisfy the following condition*

$$\|f\|_X = \sup \|t^{1/2}f(.,t)\|_\infty + \sup \left(|B(x,R)|^{-1} \int_{Q(x,R)} |f|^2 dydt\right)^{1/2} < \infty$$

As usual $B(x,R)$ denotes the ball centered at x with radius R while $Q(x,R)$ is the 'Carleson box' $B(x,R) \times [0,R^2]$. The supremum is computed over all such Carleson boxes and the right-hand side is the norm of f in X.

Two facts need to be proved. First the function $h = S(t)v_0$ in (10.2) should belong to X. Next the bilinear operator $B(v,v)$ should act boundedly from $X \times X$ to X.

The first fact is an easy consequence of the characterization of BMO by Carleson measures. This characterization can also be interpreted as a characterization of BMO by size conditions on wavelet coefficients. The second part of the proof is much deeper and the reader is referred to the beautiful paper by H. Koch and D. Tataru [67].

11 Localized Velocity Fields

Theorem 6.2 shows that oscillating patterns play some role in our understanding of Navier-Stokes equations. I hoped more and believed that the following conjecture might be true:

Conjecture 11.1. Let $\mathcal{S}_0(\mathbf{R}^3)$ be the subspace of $\mathcal{S}(\mathbf{R}^3)$ consisting of all functions f in the Schwartz class such that

$$\int f(x)x^\alpha \, dx = 0, \qquad \alpha \in \mathbf{N}^n \tag{11.1}$$

Let us assume that the initial condition $v_0(x)$ belongs to $\mathcal{S}_0(\mathbf{R}^3)$. Then there exists a positive T such that the corresponding solution $v(x,t)$ of the Navier-Stokes equations still belongs to $\mathcal{S}(\mathbf{R}^3)$ uniformly in t when $0 \leq t \leq T$.

If the Navier-Stokes equations are replaced by the linear heat equation, this is obviously true and moreover any norm on $v(x,t)$ has a rapid decay in t as t tends to ∞. If the heat equation were driving the Navier-Stokes equations, then our conjecture would obviously be correct and this would pave the way for finding wavelet based solvers for Navier-Stokes equations. Let us say a few more words about such bases [69].

Lemma 11.1. *There exists an orthonormal wavelet basis for divergence free vectors in $L^2(\mathbf{R}^3)$. These wavelets have the ordinary structure $2^{3j/2}\psi(2^j x - k)$, $j \in \mathbf{Z}$, $k \in \mathbf{Z}^3$, $\psi \in F$. The finite set F contains 14 vector-valued mother wavelets $\psi(x)$ belonging to $\mathcal{S}_0(\mathbf{R}^3)$.*

Expanding the initial condition v_0 into this basis seems to be adapted to the Navier-Stokes equations since such an expansion agrees with the translation and scale invariance of the Navier-Stokes equations. If v_0 belongs to the Schwartz class and satisfies the vanishing moments conditions, then the expansion of v_0 in our wavelet basis is sparse. Saying that there exists an effective wavelet based solver of Navier-Stokes equations implies that this sparsity should be preserved under the evolution. In other words there is no hope finding some effective wavelet solver unless our conjecture is true.

A second motivation comes from the Faedo-Galerkin schemes. In such a scheme, we are looking for a solution $v \in E$ of some equation $F(v) = g$ where the nonlinear mapping F and the right-hand side g are given. Then the functional space E containing the true solution v is replaced by an approximation space V. This approximation space V comes equipped with a basis v_m, $1 \leq m \leq M$. Then the equation $F(v) = g$ is replaced by

$$< F(\bar{v}), v_m >=< g, v_m >, \quad \bar{v} = \sum_1^M c_m v_m, \quad 1 \leq m \leq M \qquad (11.2)$$

Paul Federbush [40] has been trying to apply this scheme to Navier-Stokes equations when V is the space V_j coming from a multiresolution analysis. Here also everything depends on the sparsity of some matrices.

Bad news is coming. S.Y. Dobrokhotov and A.I. Shafarevich [34], [35], proved that the conjecture is wrong. Indeed the properties we require for the solution imply infinitely many algebraic conditions. Among these conditions, the simplest ones read

$$\int_{\mathbf{R}^3} v_{0,m}(x)v_{0,n}(x)\, dx = c\, \delta_{m,n} \qquad (11.3)$$

where $v_{0,m}$, $m = 1, 2, 3$, are the components of the velocity field and $\delta_{m,n}$ is the Kronecker symbol.

Using another approach, L. Brandolese [14] proved a deeper result.

Theorem 11.1. *Let us assume that the initial velocity $v_0(x)$ belongs to $L^2(\mathbf{R}^3, (1 + |x|)dx)$ and also to $L^1(\mathbf{R}^3, (1 + |x|)dx)$ and that $\mathrm{div}(v_0) = 0$. If there exists a positive T, constant C and solution $v(x, t) \in \mathcal{C}([0, T], L_w^2)$ of the Navier-Stokes equations which satisfies $v(x, 0) = v_0(x)$ and the uniform estimates*

$$\int_{\mathbf{R}^3} |v(x, t)|(1 + |x|)dx \leq C, \quad \int_{\mathbf{R}^3} |v(x, t)|^2(1 + |x|)dx \leq C, \ 0 \leq t \leq T$$

$$(11.4)$$

then (11.3) holds.

The following property will provide a better understanding of Theorem 11.1.

Proposition 11.1. *Let us assume that the initial velocity $v_0(x)$ satisfies the estimate $|v_0(x)| \leq C(1+|x|)^{-4}$. Then there exists a positive T such that the Kato's solution $v(\cdot, t)$ of the Navier-Stokes equation still satisfies $|v(x,t)| \leq C(1+|x|)^{-4}$, $0 \leq t \leq T$.*

In other words an $O(|x|^{-4})$ decay is preserved by the nonlinear evolution. But this exponent is sharp and cannot be replaced by $\gamma > 4$.

In other words relating coherent structures in fluid dynamics to localized and oscillating patterns of the velocity field is hopeless, simply because localization is not preserved by the nonlinear evolution.

Brandolese's theorem can be slightly improved by introducing a new functional space E containing $L^1(\mathbf{R}^3, (1+|x|)dx)$.

Let E denote the vector space consisting of all functions in $L^1(\mathbf{R}^3)$ such that the two following conditions hold

$$\int_{|x| \geq r} |f(x)|dx \leq C/r \tag{11.5}$$

for every $r \geq 1$, together with

$$r \int_{|x| \geq r} |f(x)|dx \to 0, \; r \to \infty \tag{11.6}$$

The norm in E is defined as the sum between $\int_{|x| \leq 1} |f(x)|dx$ and the constant C in (11.5). Condition (11.6) implies that testing functions are dense in E.

We then have

Proposition 11.2. *If there exists a positive T, and a solution $v(x,t)$ of the Navier-Stokes equations belonging to $\mathcal{C}([0,T], L_w^2)$ which satisfies $v(x,0) = v_0(x)$ and the two conditions*

$$v(x,t) \in L^\infty([0,T]; E) \tag{11.7}$$

$$|v(x,t)|^2 \in L^\infty([0,T]; E) \tag{11.8}$$

then (11.3) *holds.*

Here L_w^2 denotes the Hilbert space $L^2(\mathbf{R}^3)$ equipped with its weak-star topology.

This implies that a $o(|x|^{-4})$ decay at infinity is not preserved by the nonlinear evolution.

Let us sketch the proof of Proposition 11.2. Here are the main tools. We first observe that the Banach space E is a convolution algebra. The second observation concerns the definition of E. For proving $f \in E$, it suffices to assume

$$r \int_{r \leq |x| \leq 2r} |f(x)|dx \to 0, \; r \to \infty \tag{11.9}$$

Let Y denote the Banach space consisting of locally integrable functions f such that

$$\int_{r \le |x| \le 2r} |f(x)| dx \to 0, \; r \to \infty \tag{11.10}$$

Then the proof of Proposition 11.2 relies on the following two lemmata:

Lemma 11.2.

$$\nabla f \in E \quad f - c \in Y \tag{11.11}$$

for some constant c.

This fact does not depend on the dimension n as long as $n \ge 2$.
The second lemma reads as follows:

Lemma 11.3. *If*

$$\Delta f = -\sum_{j,k} \partial_j \partial_k (g_{j,k}) \tag{11.12}$$

where f is assumed to belong to Y and $g_{j,k}$ belong to $L^1(\mathbf{R}^3)$, then

$$\int g_{j,k}(x) dx = c\delta_{j,k} \tag{11.13}$$

where $\delta_{j,k}$ is the Kronecher symbol.

The proofs of theese lemata will be omitted but can be found in [14].

Then the proof of Proposition 18.1 runs as follows. We use the semigroup formulation of the Navier-Stokes equations which reads

$$v(t) = S(t)v_0 + \int_0^t S(t-\tau) \sum_1^3 \partial_j(v_j v)(\tau)d\tau + \int_0^t S(t-\tau)\nabla p(\tau)d\tau. \tag{11.14}$$

All terms in (11.14) but the last belong to $L^\infty([0, T]; E)$. Then (11.11) implies that, up to some additive constant, the pressure belongs to $L^\infty([0, T]; Y)$. We then apply the divergence operator to the Navier-Stokes equations. It yields $\Delta p = -\sum_{j,k} \partial_j \partial_k (v_j v_k)$. Here we observe that all products between two components of the velocity belong to $L^1(\mathbf{R}^3)$. This together with (11.13) ends the proof.

We now turn to other vanishing moment conditions. The integral $I(t) = \int v(x, t)dx$ is preserved during the evolution. If $v(x, t)$ is $O(|x|^{-4})$ uniformly on $0 \le t \le T$, then we obtain $I(t) = I(0) = 0$. But other vanishing moments conditions would raise the same localization issue we already discussed. Indeed the convergence of the integral $\int v(x, t)x_m dx$, $m = 1, 2, 3$, requires the localization condition $v(x, t) \in L^1(\mathbf{R}^3, (1+|x|)dx)$ which is precisely forbidden by Brandolese's theorem.

We would like to conclude with some remarkable examples by L. Brandolese of localized velocity fields satisfying the Navier-Stokes equations. We begin with the following definition.

Definition 11.1. *Let $u = (u_1, u_2, u_3)$ be a vector field on $\mathbf{R}^3$. We say that u is symmetric vector field if the following two properties are satisfied*

i) for $j = 1, 2$ or 3, $u_j(x_1, x_2, x_3)$ is an even function in each coordinate x_k, $k \neq j$, and is an odd function in x_j
ii) $u_1(x_1, x_2, x_3) = u_2(x_3, x_1, x_2) = u_3(x_2, x_3, x_1)$.

If $u \in L^2(\mathbf{R}^3)$, these conditions obviously imply (11.3).

L. Brandolese proved that this symmetry is preserved by the nonlinear evolution of Leray solutions [13]. We then have

Theorem 11.2. *Let us assume $\gamma \in (4, 5)$. Let v_0 be a divergence free symmetric vector field. Let us assume that*

$$|v_0(x)| \leq C_0(1 + |x|)^{-\gamma}. \tag{11.15}$$

If C_0 is sufficiently small, there exists a solution $v(\cdot, t)$ of Navier-Stokes equations such that

$v(\cdot, t)$ is a symmetric vector field for $t \geq 0$
$$|v(x,t)| \leq C(1 + |x|)^{-\gamma}, \qquad |v(x,t)| \leq C(1 + t)^{-\gamma/2}.$$

The conclusion still holds in the limit case $\gamma = 5$, if we assume

$$|e^{\tau \Delta} v_0(x)| \leq C_0(1 + |x|)^{-5}, \qquad \tau \geq 0. \tag{11.16}$$

If it is the case, we have

$$|e^{\tau \Delta} u(x,t)| \leq C(1 + |x|)^{-5} \qquad \text{for } t, \tau \geq 0, \tag{11.17}$$

and (11.16) is preserved by the nonlinear evolution.

The situation dramatically changes when the vorticity is concerned. Indeed the evolution equation which is satisfied by the vorticity reads

$$\begin{cases} \frac{\partial \omega}{\partial t} = \Delta \omega - \partial_j(v_j \omega - \omega_j v) \\ \nabla \cdot \omega = 0 \\ \omega(x, 0) = \omega_0 \\ v = K \star \omega, \ K = (K_1, K_2, K_3), \ K_j(x) = -\frac{1}{4\pi}\frac{x_j}{|x|^3} \end{cases} \tag{11.18}$$

This fourth equation is known as the Biot-Savart law. This relation was already used in the Giga and Miyakawa theorem. It means that the mapping $\omega \mapsto v$ is given by a pseudo-differential operator of order -1.

Using the heat semigroup $S(t) = \exp(t\Delta)$, (11.18) can be rewritten as

$$\omega(t) = S(t)\omega_0 - \int_0^t S(t - s)\partial_j(v_j \omega - \omega_j v)ds \tag{11.19}$$

which needs to be coupled with $v = K \star \omega$.

We would like to construct smooth, localized and oscillating solutions to (11.16). The smoothness and localization properties of ω will be specified in the following definition.

Definition 11.2. *Let A be the Banach algebra consisting of $f \in \mathcal{S}(\mathbf{R}^3)$ such that*

$$|f(x)| \le C \exp(-|x|), \qquad |\hat{f}(\xi)| \le C \exp(-|\xi|)(1 + |\xi|)^{-n-1} \qquad (11.20)$$

where $\hat{f}$ is the Fourier transform of f.

This functional space A is included in the Schwartz class.

We now arrive at the following theorem by L. Brandolese [13].

Theorem 11.3. *Let us assume the following conditions on the initial vorticity:*
$\omega_0 \in A, \operatorname{div}(\omega_0) = 0$ *and*

$$\int \omega_0(x)dx = \int x_j \omega_0(x)dx = 0, \qquad j = 1, 2, 3 \qquad (11.21)$$

Then there exists a $\tau > 0$ and a solution $v(x, t)$ of the Navier-Stokes equations such that the corresponding vorticity $\omega(x, t)$ belongs to the Banach algebra A when $0 \le t \le \tau$.
Moreover this vorticity $\omega(x, t)$ still satisfies the cancellation relations (11.21).

Roughly speaking, this theorem says that 'individual coherent structures' do exist. Here coherent structures are defined as localized and oscillating patterns. The continuation of this program consists in (a) proving that there exist coherent structures with explicit geometical properties and (b) letting several such coherent structures interact.

Let us assume that the initial vorticity ω_0 is a sum between two components $\lambda^2 w_0(\lambda(x - x_0))$ and $\lambda^2 w_1(\lambda(x - x_1))$. Here w_0 and w_1 are smooth, localized and satisfy the hypotheses of Brandolese's theorem and λ is a large parameter. We assume that w_0 and w_1 satisfy (11.21). Then we would like to know whether or not these two coherent structures will develop and eventually merge, as described in a scenario by Marie Farge. Then it would pave the way to M. Farge's program. It consists in expanding the vorticity field into an orthonormal basis of divergence-free vector wavelets in order to decouple the Navier-Stokes equations [38].

Let us mention that most numerical analysts compute the vorticity field. In contrast pure analysts focus on the velocity. There is still some hope that divergence-free wavelet bases might be efficient for speeding the computation of the vorticity field.

12 Large Time Behavior of Solutions to the Navier-Stokes Equations

This section is borrowed from L. Brandolese's Ph.D [13]. In this section the linear or nonlinear cancellations of the solutions to Navier-Stokes equations are related to the decay of the energy $\|v(\cdot,t)\|_2$ as the time variable t tends to infinity.

From the preceding chapters we know that the oscillating behavior of the initial condition plays an important role in the global existence of a solution.

Here we want to know whether the energy $\|u(\cdot,t)\|_2^2$ can have a fast decay as t tends to infinity. Here also the Besov space will play a key role. If one sticks to the linear evolution driven by the heat equation , we have the following result.

Indeed we have:

Lemma 12.1. *If $a \in L^2(\mathbf{R}^n)$, the following two properties are equivalent ones*

(a) $$\|e^{t\Delta}a\|_2 \leq C(1+t)^{-\alpha}, \qquad t \geq 0,$$

(b) $$a \in \dot{B}_2^{-2\alpha,\infty}(\mathbf{R}^n) \qquad (\alpha > 0).$$

Large time decay of $\|e^{t\Delta}a\|_2$ depends on the cancellation properties of a which are made precise in (b).

Let us first ask the following: does $\|v(\cdot,t)\|_2$ tends to 0 at infinity?

The first contribution to this issue is a fundamental theorem by T. Kato [62] which says the following

Theorem 12.1. *Let us assume that the initial condition v_0 belongs to $L^2(\mathbf{R}^3) \cap L^3(\mathbf{R}^3)$ and let $v(\cdot,t) \in \mathcal{C}([0,\infty); L^3(\mathbf{R}^3))$ be a mild solution to the Navier-Stokes equations, as given by Theorem 5.1. Then $\|v(\cdot,t)\|_2$ tends to 0 at infinity.*

An important remark in [62] is that a Leray solution will always become a Kato solution when the time variable will be large enough. This follows from the energy inequality

$$\|v(t)\|_2^2 + 2 \int_s^t \|\nabla v(\tau)\|_2^2 \, d\tau \leq \|v(s)\|_2^2, \qquad (12.1)$$

which holds for $s = 0$,and almost evrywhere in $s > 0$ for all $t \geq s$. More details are given in L. Brandolese's dissertation [13].

This led to relate the decay rate of $\|v(t)\|_2$ to some properties of the initial condition.

T. Kato's theorem implies

$$\|u(t)\|_2 \leq C(1+t)^{-\frac{3}{4}+\epsilon}, \quad \epsilon > 0 \qquad (12.2)$$

whenever $a \in L^1(\mathbf{R}^3)$. A main breakthrough was achieved by Maria Elena Schonbek who used an adapted partitioning of the frequency domain in [95] and in [96]. These new techniques were further improved by Kajikiya et Miyakawa [61]. They proved the following

$$\|v(t)\|_2 \leq C(1+t)^{-3/4} \tag{12.3}$$

$$\|v(t) - e^{t\Delta}v_0\|_2 = o(t^{-3/4}), \quad t \to \infty. \tag{12.4}$$

when $v_0 \in L^1(\mathbf{R}^3)$. The second estimate shows that $t^{-3/4}$ cannot be improved when v_0 is a generic function in $L^1(\mathbf{R}^3)$.

For some special initial velocities v_0, M. Wiegner obtained a sharper decay rate

Theorem 12.2. *Let $v_0 \in L^2(\mathbf{R}^3)$ and let v be a weak solution of the Navier-Stokes equations with $v(\cdot, 0) = v_0$. Let us assume that v satisfies the energy inequality (12.1). Then*

i) $\|v(\cdot, t)\|_2 \to 0$ when $t \to \infty$
ii) If the solution $S(t)v_0$ of the heat equation with initial condition v_0 has the decay rate $\|e^{t\Delta}v_0\|_2 \leq C(1+t)^{-5/4}$, then v satisfies

$$\|v(\cdot, t)\|_2 \leq C(1+t)^{-5/4} \tag{12.5}$$

If the initial condition v_0 is more oscillating which here means that the Fourier transform $\widehat{v}_0 \in L^2(\mathbf{R}^3)$ vanishes at 0 at a high order, then the decay of $\|e^{t\Delta}v_0\|_2$ is faster than $t^{-5/4}$. But this faster decay does not pay and the decay of $\|v(t)\|_2$ is not improved.

Using in a clever way her technique of Fourier splitting, Maria Elena Schonbek proved that the exponent $5/4$ is optimal. Indeed she obtained

$$\|v(t)\|_2 \geq C(1+t)^{-5/4} \tag{12.6}$$

for generic solutions.

This was improved by Miyakawa and Schonbek. These authors found a necessary and sufficient condition for getting a better decay estimate than (12.5). The Miyakawa–Schonbek theorem resembles Theorem 11.1. But this is accidental since this similarity disappears when one is looking for more decay. The Miyakawa–Schonbek theorem reads as follows

Theorem 12.3. *Let us assume $v_0 \in L^2(\mathbf{R}^3)$, $\operatorname{div} v_0 = 0$ and $\int |v_0(x)|(1 + |x|)\, dx < \infty$. Let v be a weak solution of the Navier-Stokes equations with initial condition v_0.*
We define $b_{hk} = \int x_h v_{0,k}(x)\, dx$ and $M_{hk} = \int_0^\infty \int (v_h v_k)(x, t)\, dx\, dt$. Then we have:

i) If $(b_{hk}) = 0$ and if M_{hk} is a multiple of the identity matrix, then
$\limsup_{t \to \infty} t^{\frac{5}{4}} \|v(t)\|_2 = 0$.
ii) Conversely if $(b_{hk}) \neq 0$ or if M_{hk} is not a multiple of the identity matrix, then

$$\liminf_{t \to \infty} t^{\frac{5}{4}} \|v(t)\|_2 > 0. \tag{12.7}$$

This theorem is striking but i) is useless since a full knowledge of the solution is needed to conclude. Condition i) does not concern the initial condition but the solution itself.

L. Brandolese achieved the construction of nontrivial solutions whose energy have a fast decay at infinity. He used the same tool he introduced for proving the existence of localized velocities. Indeed a simple computation shows that the symmetric solutions of Theorem 11.2 satisfy $\|v(\cdot,t)\|_2 \leq Ct^{-\gamma}$ for any $\gamma < 7/4$.

13 Improved Gagliardo-Nirenberg Inequalities

The properties of Littlewood-Paley expansions which will be used are given in Section 18. We begin with the definition of the Triebel-Lizorkin spaces. Let us start with the norm. The space will come later on, as the collection of tempered distributions for which the norm is finite.

Definition 13.1. *For $p, q \in (0, \infty), s \in \mathbf{R}$, we consider*

$$\|f\|_{F_p^{s,q}} = \|S_0(f)\|_p + \|\left(\sum_{j=0}^{+\infty} 2^{jsq}|\Delta_j(f)|^q(\cdot)\right)^{1/q}\|_p \qquad (13.1)$$

If $p = \infty$, this definition should be modified into the following one: one considers all possible expansions

$$f = S_0(h) + \sum_{j=0}^{+\infty} \Delta_j(h_j) \qquad (13.2)$$

where h and h_j are arbitrary functions and define the norm $\|f\|_{F_\infty^{s,q}}$ as the infimum of all the corresponding L^∞ norms

$$\|h\|_\infty + \left(\sum_{j=0}^{+\infty} 2^{jsq}|h_j|^q(\cdot)\right)\|_\infty \qquad (13.3)$$

If $p = q$, the Besov and Triebel-Lizorkin spaces are identical and we will be mainly interested in the case $p = q = \infty$. Then we recover the classical Hölder spaces.

Finally $F_p^{s,q}$ is the subspace of L^p consisting of all $f \in L^p$ for which the norm (13.1) is finite.

We now turn to the more interesting homogeneous Triebel-Lizorkin spaces $\dot{F}_p^{s,q}$. The definition of the norms are similar to (13.1) or (13.3) but the term $S_0(f)$ disappears while the index j runs from $-\infty$ to $+\infty$. The main difficulty concerns the a priori regularity assumption which should be imposed to f to belong to one of these homogeneous spaces. The reader is reffered to Section 18 but we would like to give a flavour of the problems one is facing. If $p =$

$q = 2$, the Triebel-Lizorkin space is the homogeneous Sobolev space $\dot{H}^s$. If $0 < s < n/2$, then $\dot{H}^s \subset L^p$ when $1/2 - 1/p = s/n$ and this inclusion is not true for other values of p. The a priori assumption is $f \in L^p$ in this case. When $s = n/2$, then $\dot{H}^s \subset BMO$ and f cannot be considered as a function, but instead as a function modulo constant functions. If $n/2 < s < n/2 + 1$, then $\dot{H}^s \subset \dot{\mathcal{C}}^r$ where $r = s - n/2$ and $\dot{\mathcal{C}}^r$ denotes the standard homogeneous Hölder space defined by

$$|g(x) - g(y))| \leq C|x - y|^r \tag{13.4}$$

We now arrive to the main theorem

Theorem 13.1. *Let us assume* $0 < p_1, p_2, p_3 \leq \infty, 1 \leq q_1, q_2, q_3 \leq \infty$ *and* $s \in \mathbf{R}$. *Let us denote by* $F^{(1)}, F^{(2)}, F^{(3)}$ *the Triebel-Lizorkin spaces* $\dot{F}^{s_1,q_1}_{p_1}, \dot{F}^{s_2,q_2}_{p_2}, \dot{F}^{s_3,q_3}_{p_3}$ *and let us denote by* $\|\cdot\|_{(m)}, m = 1, 2, 3$ *the three Triebel-Lizorkin norms. Let us assume that*

$$\begin{cases} 1/p_3 = (1 - \theta)/p_1 + \theta/p_2 \\ s_3 = (1 - \theta)s_1 + \theta s_2, s_1 \neq s_2 \end{cases} \tag{13.5}$$

Then we have

$$\|f\|_{(3)} \leq C\|f\|_{(1)}^{1-} \|f\|_{(2)} \tag{13.6}$$

This theorem is well known in some particular cases. For instance when we add $1/q_3 = (1 - \theta)/q_1 + \theta/q_2$ to the hypotheses of Theorem 13.1 or when $q_1 = q_2 = q_3, s_1 = s_2 = s_3$. In such settings, classical results in real or complex interpolation apply. Moreover if $q_1 = q_2 = q_3 = 2$, we recover the standard Gagliardo-Nirenberg inequalities.

The proof of Theorem 13.1 is surprisingly easy and should become a classic. Since the Triebel-Lizorkin spaces are increasing with q, when p is frozen, we can assume $q_3 < \infty$, $q_1 = q_2 = \infty$. Without loosing generality, we suppose $s_1 < s_3 < s_2$. The integer $N = N(x)$ will be fixed in a minute. We have
$F_3(x) = \left(\sum_{j=-\infty}^{+\infty} 2^{js_3 q_3} |\Delta_j(f)|^{q_3}(x) \right)^{1/q_3}$
$\leq \left(\sum_{j=-\infty}^{N} |2^{j(s_3-s_1)} 2^{js_1} \Delta_j(f)|^{q_3}(x) \right)^{1/q_3} +$
$\left(\sum_{j=N+1}^{+\infty} |2^{-j(s_2-s_3)} 2^{js_2} \Delta_j(f)|^{q_3}(x) \right)^{1/q_3}$
$\leq C\left[2^{N(s_3-s_1)} A_1(x) + 2^{-N(s_2-s_3)} A_2(x) \right]$ where $A_1(x) = \sup_{j \in \mathbf{Z}} |2^{js_1} \Delta_j(f)(x)|$
and $A_2(x)$ is defined similarly.
We now choose N such that $2^{N(s_2-s_1)} = \frac{A_2(x)}{A_1(x)}$ which means that the two terms $2^{N(s_3-s_1)} A_1(x)$ and $2^{-N(s_2-s_3)} A_2(x)$ have the same magnitude. Altogether it gives

$$F_3(x) \leq C(A_1(x))^{1-} (A_2(x)) \tag{13.7}$$

It now suffices to apply Hölder inequality to (13.7) to conclude.

We now give several examples of these improved Gagliardo-Nirenberg inequalities.

All these applications depend on the identification of the Triebel-Lizorkin spaces with usual functional spaces. As it is well known, these identifications rely on the standard Littlewood-Paley theory (see Section 18). This excludes the spaces $L^1, L^\infty, BV, \ldots$ and other spaces based upon these spaces. We will return to BV in the next section.

The first example improves on the famous Sobolev embedding theorem which in three dimensions reads

$$\|f\|_6 \leq C\|\nabla f\|_2 \tag{13.8}$$

The enemy in (13.8) is given by $f(x) = \exp(i\omega \cdot x)\phi(x)$ where $\omega = \lambda\nu, \nu \in S^2, \lambda \to \infty$. Then the left-hand side of (13.8) is $\|\phi\|_6$ while the right-hand side is asymptotic to $\lambda\|\phi\|_2$. Theorem 13.1 applies and yields

$$\|f\|_6 \leq C\|\nabla f\|_2^{1/3}\|f\|_{(-1/2)}^{2/3} \tag{13.9}$$

We have $p_3 = 6, p_1 = 2, p_2 = \infty, \theta = 2/3, s = 0, s_1 = 1, s_2 = -1/2$ which corresponds to the assumptions of Theorem 13.1.

Let us comment on (13.9). The norm in $\|\cdot\|_{(-1/2)}$ is the norm in the homogeneous Besov space $\dot{B}_\infty^{\infty,-1/2}$. Here we do not need the general theory since this Besov space and the corresponding norm will be defined in a simpler way. We start with $\dot{B}_\infty^{\infty,1/2}$ which is simply the homogeneous Hölder space $\dot{C}^{1/2}$ defined by $|f(x) - f(y)| \leq C|x - y|^{1/2}, x, y \in \mathbf{R}^3$. Then a function belongs to $\dot{B}_\infty^{\infty,-1/2}$ if and only if it is the divergence of a vector field in $\dot{C}^{1/2}$. For a better understanding of (13.9), one plays with dilations. If $f(x)$ is replaced by $f_\lambda(x) = \lambda^{1/2}f(\lambda x)$, then the three norms in (13.9) remain invariant. This shows *le bien fondé* of the exponent $-1/2$. We now consider $f(x) = \exp(i\omega \cdot x)\phi(x)$. Then the two norms in the right-hand side of (13.9) grow as $|\omega|$ and $|\omega|^{-1/2}$. This explains the role of the exponents $1/3$ and $2/3$.

A main difference between (13.9) and the usual Sobolev theorem is the following: the validity of (13.9) does not depend on the dimension. If this dimension is 3, then $L^6 \subset \dot{H}^1$ and (13.9) tells us a little more. If the dimension is not 3, then the finiteness of the Besov norm is playing an even more important role.

We now turn to the fundamental inequality

$$\|f\|_4^2 \leq C\|\nabla f\|_2\|f\|_B, \quad B = \dot{B}_\infty^{-1,\infty} \tag{13.10}$$

which was stated in Section 2. Returning to Theorem 13.1, (13.10) corresponds to $p_3 = 4, p_1 = 2, p_1 = \infty, \theta = 1/2, s_3 = 0, s_1 = 1, s_2 = -1$.

We now treat a last example which was used in Section 6. It reads

$$\|f\|_{\dot{H}^{1/2}} \leq C\|f\|_{\dot{H}^1}^{2/3}\|f\|_B^{1/3}, \quad B = B_2^{-1/2,\infty} \tag{13.11}$$

Here we have $p_1 = p_2 = p_3 = 2, s_3 = 1/2, s_1 = 1, s_2 = -1/2, \theta = 1/3$.

14 The Space   of Functions with Bounded Variation in the Plane

In studying the Banach space BV, we can assume that the dimension n is larger than 1. Indeed in the one-dimensional case, the space BV is trivial since it is isomorphic to the space of all bounded Borel measures on the line. Assuming $n \geq 2$, we say that a function $f(x)$ defined on $\mathbf{R}^n$ belongs to BV if (a) $f(x)$ vanishes at infinity in a weak sense and (b) the distributional gradient ∇f of $f(x)$ is a (vector valued) bounded Borel measure. The norm of $f \in BV$ is defined as the total mass of ∇f. A seemingly less demanding definition reads as follows: f belongs to BV if its distributional gradient is a (vector valued) bounded Borel measure. Then it is easily proved that $f = g + c$ where c is a constant and g tends to 0 at infinity in the weak sense.

The condition at infinity says that $f \star \varphi$ should tend to 0 at infinity whenever φ is a function in the Schwartz class. For instance, any function in $L^p(\mathbf{R}^n)$, $1 \leq p < \infty$, tends to 0 at infinity in the weak sense.

A crucial property of the space BV is the fact that it is a dual space. Let $\mathcal{C}_0(\mathbf{R}^n)$ denote the Banach space of all continuous functions on $\mathbf{R}^n$ which vanish at infinity, the norm of $f \in \mathcal{C}_0(\mathbf{R}^n)$ being $\|f\|_\infty$. We introduce the Banach space Γ consisting of all tempered distributions $f = \partial_1 g_1 + \partial_2 g_2 + ... + \partial_n g_n$ where $g_1, ..., g_n$ belong to the Banach space $\mathcal{C}_0(\mathbf{R}^n)$ and where the norm of $f \in \Gamma$ is the L^∞ norm of $\sqrt{|g_1|^2 + ... + |g_n|^2}$. Let us observe that the decomposition of f as the divergence of the vector field $(g_1, ..., g_n)$ is not unique. It implies that the norm of f is the infimum of all L^∞ norms of $\sqrt{|g_1|^2 + ... + |g_n|^2}$ where the infimum is computed over all such decompositions.

Then the Hahn-Banach theorem yields

Lemma 14.1. *The Banach space BV is the dual space of Γ and is equipped with the dual norm.*

This fundamental property implies weak-compactness. Indeed we have:

Lemma 14.2. *If f_j denotes a bounded sequence of functions in BV and if this sequence converges to f in the distributional sense, then f belongs to BV and $\|f\|_{BV} \leq \liminf \|f_j\|_{BV}$, $j \to \infty$.*

If the Banach space BV is equipped with its weak* topology, then test functions are dense in BV.

A second and equivalent definition of BV reads the following:

Definition 14.1. *A function $f(x)$ belongs to $BV(\mathbf{R}^n)$ if it vanishes at infinity in the weak sense and if there exists a constant C such that*

$$\int_{\mathbf{R}^n} |f(x+y) - f(x)|\, dx \leq C|y| \tag{14.1}$$

for each $y \in \mathbf{R}^n$.

Let us stress that this new definition does not yield the same norm as above.

From this second definition, it is immediately concluded that if a real valued function $f(x)$ belongs to BV, so do $f^+(x) = \sup(f(x), 0)$ and $f^-(x) = \sup(-f(x), 0)$. In other words, it is often sufficient to consider non-negative functions in BV.

The Banach space BV is contained in $L^{n/n-1}(\mathbf{R}^n)$ and this embedding is sharp: we cannot replace $p = n^* = \frac{n}{n-1}$ by a larger value of p. Here is a counter-example. Let us begin with a function $\theta(t)$ of the real variable t with the following properties: $\theta(t)$ is smooth on $(0, +\infty)$ and equal to $t^{(-n+1)}|\log(t)|^{-\gamma}$ whenever t is small or large (the trouble comes from $t = 1$). Let us assume $\gamma > 1$. We define $f(x)$ by $f(x) = \theta(|x|)$. Then $f(x)$ belongs to BV and $f(x)$ does not belong to L^p for $p \neq n^*$. This specific function is not the worst among BV functions. Indeed its gradient belongs to the Besov space $\dot{B}_1^{0,1}$ which is much smaller than L^1 and which will be defined now. In other words we have $f \in \dot{B}_1^{1,1}$ where $\dot{B}_1^{1,1}$ denotes the usual homogeneous Besov space. This Besov space is much smaller than BV and has a trivial wavelet characterization. Let $\psi_{j,k}(x) = 2^{jn/2}\psi(2^j x - k), \psi \in \{\psi_1, ..., \psi_N\}, N = 2^n - 1$, be an orthonormal basis consisting of smooth (for instance $\psi \in \mathcal{C}^2$ suffices) and compactly supported wavelets. Then the characterization of $\dot{B}_1^{1,1}$ reads

Definition 14.2. *The Banach space $\dot{B}_1^{1,1}$ is the subspace of $L^{\frac{n}{n-1}}(\mathbf{R}^n)$ defined by*

$$\sum_{j \in \mathbf{Z}} \sum_{k \in \mathbf{Z}^n} |c(j,k)| 2^{j(1-n/2)} < \infty \tag{14.2}$$

The dual space of $\dot{B}_1^{1,1}$ is the Banach space $\dot{B}_\infty^{-1,\infty}$.

It is clear that the left-hand side of (14.2) should concern the $N = 2^n - 1$ wavelets $\psi_1, ... \psi_N$. If we now return to the larger space BV, then l^1 will be replaced by weak l^1.

We have seen that a function in BV belongs to L^{n^*} which is sharp. However if we a priori know that a given function f both belongs to BV and to L^q for some $q > n^*$, then this f is more regular than a generic function in BV is expected to be. Indeed f belongs to the Sobolev space $L^{(n^*, s)}$ whenever $s = 1/n^* - 1/q$ (Theorem 15.4). The Sobolev space $L^{(n^*, s)}$ consists of functions whose fractional derivative of order s belongs to L^{n^*}.

If E is a measurable set in $\mathbf{R}^n$ and if χ_E denotes the indicator function of E, we would like to compute the BV norm of this indicator function χ_E whenever it is finite. If E is a domain delimited by a smooth boundary Γ, then the BV norm of χ_E is the total surface of Γ. However if E is an arbitrary open set, then the BV norm of χ_E is in general smaller than the $n - 1$-dimensional Hausdorff measure $\mathcal{H}^{n-1}(\Gamma)$ of its boundary $\Gamma = \partial E$. Indeed $\chi_E = \chi_F$ almost everywhere does not imply $H^{n-1}(\partial E) = \mathcal{H}^{n-1}(\partial F)$. That is why De Giorgi defined the reduced boundary $\partial^* E$ of a measurable set E and proved that the

BV norm of χ_E is $\mathcal{H}^{n-1}(\partial^* E)$. Here $\mathcal{H}^q$ denotes the q-dimensional Hausdorff measure.

For defining this reduced boundary, let us denote by $B(x, r)$ the ball centered at x with radius r. We then follow De Giorgi

Definition 14.3. *The reduced boundary $\partial^* E$ of E is the set of points x belonging the closed support of $\mu = \nabla \chi_E$ such that the following limit exists*

$$\lim_{r \to 0} \frac{\mu\{B(x, r)\}}{|\mu|\{B(x, r)\}} = \nu(x) \tag{14.3}$$

With these new notations the co-area identity reads as follows [2], [82].

Theorem 14.1. *Let $f(x)$ be a real valued measurable function defined on $\mathbf{R}^n$ and belonging to BV. Let us denote by Ω_t, $t \in \mathbf{R}$, the measurable set defined by*

$$\Omega_t = \{x \in \mathbf{R}^n \mid f(x) > t\} \tag{14.4}$$

Let $\partial^* \Omega_t$ be the reduced boundary of Ω_t and $l(t) = \mathcal{H}^{n-1}(\partial^* \Omega_t)$. Then the BV norm of f is given by the sum of the surfaces of its level sets. In other words

$$\|f\|_{BV} = \int_{-\infty}^{+\infty} l(t) dt \tag{14.5}$$

This identity needs to be completed with the following observation

$$f(x) = \lim_{m \to \infty} \Big[\int_{-m}^{\infty} \chi_{\Omega_t}(x) dt - m \Big] \tag{14.6}$$

where the limit is taken with respect to the weak* topology of BV.

A first approximation to this theorem was given in the pioneering work by Fleming and Rishel and Theorem 14.1 was completed by De Giorgi [2], [82].

Roughly speaking, we may interpret these two identities by saying that BV is a space with a remarkable atomic decomposition. Let us be more precise.

Definition 14.4. *A Banach space E is equipped with an atomic decomposition if there exists a bounded subset Λ of E and a constant C such that each x in E can be written as a convex combination of $\gamma_j \lambda_j$ where γ_j are some scalar factors of modulus not exceeding C and λ_j, $j \in \mathbf{N}$, are some atoms belonging to Λ. In other words this reads*

$$x = \sum_{j \in} \alpha_j \lambda_j \tag{14.7}$$

with

$$\sum_{j \in} |\alpha_j| \leq C \|x\|_E \tag{14.8}$$

Every Banach space E can be equipped with a trivial atomic decomposition where $\Lambda = E$ and the series (14.7) reduces to $x = x \in \Lambda$. When one is seeking an atomic decomposition, it means that the atoms $\lambda \in \Lambda$ are expected to be rather simple building blocks.

In the application to the Banach space BV, the series in (14.7) will be replaced by an integral representation of x as a convex combination $x = \int_\Omega \lambda(\omega) d\mu(\omega)$ where $\lambda(\omega) \in \Lambda$ and the total mass of μ does not exceed $C\|x\|_{BV}$. A more precise statement and the definition of the space Ω will be given in (14.9).

For instance the Hardy space $\mathcal{H}^1(\mathbf{R}^n)$ has an atomic decomposition in the strict sense given by (14.8). The atoms are the simple building blocks in $\mathcal{H}^1(\mathbf{R}^n)$ which are defined by the following properties:

(a) $a(x)$ is supported by a cube Q
(b) $\|a\|_\infty \leq 1/|Q|$
(c) $\int a(x)\, dx = 0$

Whenever a Banach space E is equipped with a nontrivial atomic decomposition, proving the continuity of a mapping $T : E \mapsto F$ is often an easy task since it suffices to prove a uniform estimate for the action of T on atoms. That is the approach which will be used for proving that BV is contained in the Lorentz space $L^{(2,1)}(\mathbf{R}^2)$. However proving the continuity of a mapping $F \mapsto E$ still remains difficult.

Returning to BV, the atoms $a_\Omega(x)$ which will be used are indicator functions χ_Ω of measurable sets Ω whose reduced boundaries $\partial^*\Omega$ have a finite surface. We know that the BV norm of such an atom χ_Ω is the surface of the reduced boundary $\partial^*\Omega$. Finally (14.4), (14.5) and (14.6) show that any function $f(x)$ in BV is a Bochner integral $\int_{-\infty}^{+\infty} a_t(x)dt$ of such atoms. More precisely we define Ω_t by (14.4) and write $a_t = \chi_{\Omega_t}$. Then (14.5) implies

$$\int_{-\infty}^{+\infty} \|a_t\|_{BV}\, dt = \|f\|_{BV} \tag{14.9}$$

For proving an embedding theorem of the type $BV \subset E$ where E is some Banach space, it suffices to check that there exists a constant C such that the E-norm of any atom $a_\Omega(x)$ in BV does not exceed C times the surface of $\partial^*\Omega$.

For instance in three dimensions we have

$$BV \subset L^{3/2,1} \tag{14.10}$$

Our last remark concerns the pointwise multipliers of BV. They are characterized by the following theorem

Theorem 14.2. *A function $m(x)$ is a pointwise multiplier of BV if and only if $m(x)$ belongs to L^∞ and if ∇m satisfies the following condition: there exists a constant C such that*

$$\int_{\Omega} |\nabla m|\, dx \leq C\mathcal{H}^{n-1}(\partial\Omega) \tag{14.11}$$

holds for all Ω.

This characterization is an easy corollary of the atomic decomposition. Indeed we need to bound $\int |\partial_j(mf)|\, dx$ for $f \in BV$. Leibnitz formula yields two terms. The first one is fine since we can multiply a measure with a bounded function (we are sketching a proof where measurability issues are not discussed). The second term is saddler. We need to estimate $\int |\nabla m||f|\, dx$ by $C\|f\|_{BV}$. Here the atomic decomposition permits to reduce this issue to the case of indicator functions. Let us mention for example that in three dimensions, any bounded function $m(x)$ such that $\nabla m \in L^3$ is a multiplier. Indeed Hölder inequality reads $\int_{\Omega} |\nabla m|\, dx \leq \|\nabla m\|_3 |\Omega|^{2/3}$ which is estimated by the isoperimetric inequality in $C\|\nabla m\|_3 \mathcal{H}^2(\partial\Omega)$.

15 Gagliardo-Nirenberg Inequalities and

Let us start with the twodimensional case and with the Sobolev embedding of $BV(\mathbf{R}^2)$ into $L^2(\mathbf{R}^2)$.

The fundamental estimate

$$\|f\|_2 \leq (2\sqrt{\pi})^{-1}\|f\|_{BV} \tag{15.1}$$

is obviously consistent with translations and dilations. Indeed, for any positive a and $f_a(x) = af(ax)$, we will have $\|f_a\|_2 = \|f\|_2$ and similarly $\|f_a\|_{BV} = \|f\|_{BV}$. But (15.1) is not consistent with modulations: if M_ω denotes the pointwise multiplication operator with $\exp(i\omega x)$, then M_ω acts isometrically on L^2 while $\|M_\omega f\|_{BV}$ blows up as $|\omega|$ tends to infinity.

For addressing this invariance through modulations, let us introduce an adapted Besov norm.

Definition 15.1. *Let B be the Banach space of all tempered distributions $f(x)$ for which there exists a constant C such that, for $g(x) = \exp(-|x|^2)$ and $g_{(a,b)} = ag(a(x - b))$, the following condition is satisfied:*
There exists a constant C such that, for any $(a,b) \in (0,\infty) \times \mathbf{R}^2$, we have

$$|<f, g_{(a,b)}>| \leq C \tag{15.2}$$

The infimum of these constants C is the norm of f in B and is denoted by $\|f\|_*$.

It is easily proved that this Banach space coincides with the space of second derivatives of functions in the Zygmund class. In other words B is the homogeneous Besov space $\dot{B}_\infty^{-1,\infty}$ of regularity index -1. The definition and the norm given by (15.2) will be crucial for trying to obtain estimates which do not depend on the dimension.

We then have

Theorem 15.1. *There exists a constant C such that*

$$\|f\|_2 \leq C\left(\|f\|_{BV}\|f\|_*\right)^{1/2}, \quad f \in BV(\mathbf{R}^2) \tag{15.3}$$

Moreover the norm $\|f\|_*$ cannot be replaced in (15.3) by a smaller norm which would satisfy the same scaling laws as the L^2 or BV norm.

To better understand this theorem, let us stress that we always have $\|f\|_* \leq \|f\|_{BV}$. The ratio $\|f\|_*/\|f\|_{BV}$ between these norms is then denoted by β and is expected to be small in general. With these new notations, (15.3) reads

$$\|f\|_2 \leq C\beta^{1/2}\|f\|_{BV} \tag{15.4}$$

which yields a sharp estimate of the ratio between the L^2 norm and the BV norm of f. Moreover $\beta^{1/2}$ in (15.4) is sharp as the example of $f(x)=\exp(i\omega x)w(x)$ shows. Indeed if $|\omega|$ tends to infinity and $w(x)$ belongs to the Schwartz class, then $\|f\|_2 = \|w\|_2$ is constant, $\|f\|_* \simeq |\omega|^{-1}\|w\|_\infty$, and finally $\|f\|_{BV} \simeq |\omega|\|w\|_1$. In this example β is of the order of magnitude of $|\omega|^{-2}$ which corresponds to $\beta^{1/2} \simeq |\omega|^{-1}$. If $|\omega|$ tends to infinity, (15.3) yields the classical estimate $\|w\|_2 \leq \sqrt{\|w\|_1\|w\|_\infty}$.

The proof of (15.3) relies on the following theorem

Theorem 15.2. *Let $\psi_\lambda, \lambda \in \Lambda$, be a two-dimensional orthonormal wavelet basis as described in Theorem 19.1. Then for every f in $BV(\mathbf{R}^2)$, the wavelet coefficients $c_\lambda =< f, \psi_\lambda >, \lambda \in \Lambda$, belong to weak $l^1(\Lambda)$.*

This theorem was proved by A.Cohen et al. [24] in the Haar system case. The general case was obtained by the author and a good reference is [82].

Theorem 15.2 says the following: if $c_\lambda =< f, \psi_\lambda >$ and if the $|c_\lambda|, \lambda \in \Lambda$,are sorted out by decreasing size, the rearranged sequence c_n^* satifies $c_n^* \leq C/n$ for $1 \leq n$.

One cannot replace the vector space weak $l^1(\Lambda)$ by $l^1(\Lambda)$ in Theorem 15.2. Indeed let $f(x)$ be the indicator function of any smooth domain Ω and let L be the length of the boundary of Ω. Then when $2^j L > 1$, the cardinality of the set of λ such that $2^{-j} < |c_\lambda| \leq 2^{-j+1}$ is precisely $2^j L$. This does not mean that Theorem 15.2 is optimal. Indeed BV is a Banach space which is the dual X^* of a separable Banach space X while weak l^1 does not have this property. Therefore BV and weak l^1 cannot be isomorphic. The Besov space $\dot{B}_1^{1,1}$ is included inside BV and is defined by the property that the wavelet coefficients of a function in $\dot{B}_1^{1,1}$ belongs to l^1. Therefore the vector space Y of wavelet coefficients of BV functions is sitting somewhere between l^1 and weak l^1.

Corollary 15.1. *Let $c_\lambda, \lambda \in \Lambda$, be a given sequence. Let us sort out $|c_\lambda|, \lambda \in \Lambda$, by decreasing size and let us denote the rearranged sequence by c_n^*, $n \geq 1$. Let us assume that $\sum_{n\geq 1} c_n^*/n$ is finite. Then $f = \sum_{\lambda \in \Lambda} c_\lambda \psi_\lambda$ belongs to $B_\infty^{-1,\infty}$.*

The proof of (15.3) is straightforward. One uses the following trivial estimate on sequences

$$\sum_{n=1}^{\infty} |c_n|^2 \leq 2\|c_n\|_\infty \|nc_n\|_\infty \tag{15.5}.$$

Then one applies Theorem 15.2 to an orthonormal wavelet basis of class $\mathcal{C}^2$. If c_n^* denotes the nonincreasing rearrangement of the wavelet coefficients $|c(\lambda)|, \lambda \in \Lambda$, then $\|nc_n^*\|_\infty$ is precisely the norm of $c(\lambda)$ in the space weak l^1.

Let us observe that (15.3) is, as announced in the title of this section, an improvement on the celebrated Gagliardo-Nirenberg estimates. Let us clarify this statement. Let p, q, r and σ be four exponents satisfying $1 \leq p, q, r < \infty$, $\sigma \in (0,1)$ and let j and m be two integers. One assumes $j \leq \sigma m$ and $1/p - j/n = \sigma(1/r - m/n) + (1-\sigma)/q$ where n is the dimension. Then the Gagliardo-Nirenberg inequalities read

$$\|D^j f\|_p \leq C\|D^m f\|_r^\sigma \|f\|_q^{1-\sigma} \tag{15.6}$$

The notation $\|D^j f\|_p$ means $\sup\{\|\partial^\alpha f\|_p; |\alpha| = j\}$. Gagliardo-Nirenberg estimates are related to the Sobolev embedding theorem. Indeed $\|D^j f\|_p \leq C\|D^{j'} f\|_{p'}$ whenever $p \geq p'$ and $1/p' - 1/p = (j' - j)/n$. This simple observation implies that the interesting case in (15.6) occurs when $j = \sigma m$, $\sigma \in (0,1)$, and $1/p = \sigma/r + (1-\sigma)/q$. Then the Gagliardo-Nirenberg inequalities simply read

$$\begin{cases} \|D^{\sigma m} f\|_p \leq C\|D^m f\|_r^\sigma \|f\|_q^{1-\sigma} \\ 1/p = \sigma/r + (1-\sigma)/q, \quad 1 \leq p, q, r < \infty \\ \sigma \in (0,1) \end{cases} \tag{15.7}$$

Let us compare (15.3) to the Gagliardo-Nirenberg estimate (15.6). If $p = 2, j = 1, m = 2, r = 1$ and $q = \infty$, then (15.6) can be improved into $\|\nabla f\|_2^2 \leq C\|\Delta f\|_1 \|f\|_\infty$. Unfortunately this estimate is trivial. Indeed integrating by parts yields $\int |\nabla f|^2 \, dx = -\int f\Delta f \, dx$ and the (L^1, L^∞) duality suffices to bound the last integral. However this simply minded estimate is not implied by (15.3). Indeed the Riesz transformations are not bounded on L^1 and $\Delta f \in L^1$ does not imply $\nabla f \in BV$. On the other hand, $f \in B_\infty^{(-1,\infty)}$ is much weaker than $\nabla f \in L^\infty$.

Albert Cohen, Wolfgang Dahmen, Ingrid Daubechies and Ron DeVore [25] proved the following theorem:

Theorem 15.3. *In any dimension $n \geq 1$, let us assume that a function f belongs both to BV and to the homogeneous space $\dot{B}_\infty^{-1,\infty}$. Then we have*

$$\|f\|_2 \leq C(\|f\|_{BV}\|f\|_*)^{1/2} \qquad (15.8)$$

where $\|f\|_*$ is norm of f in the Besov space $\dot{B}_\infty^{-1,\infty}$.

The norm of f in $\dot{B}_\infty^{-1,\infty}$ can be defined as the optimal constant C for which one has $| < f, g_{(a,b)} > | \leq Ca2^{-n}$, $(a,b) \in (0,\infty) \times \mathbf{R}^n$. The notations are the same as in Definition 14.4. Let us stress that BV is contained in L^2 if and only if $n = 2$. In other words when $n = 1$ or $n > 2$, the assumption $f \in \dot{B}_\infty^{-1,\infty}$ complements $f \in BV$ and both are needed to get an L^2 estimate.

Theorem 15.3 cannot be deduced from Theorem 15.2 when $n \neq 2$. New estimates on wavelet coefficients of functions with bounded variation will be needed. These estimates will be provided by Theorem 15.5.

Every function with a bounded variation belongs to $L^{\frac{n}{n-1}}$ and it is natural to ask the following question: what would happen if a function f belongs both to BV and to L^q for some $q > n^*$ where $n^* = \frac{n}{n-1}$?

The Lebesgue space L^q is contained inside the larger space $\mathcal{C}^{-\beta} = \dot{B}_\infty^{-\beta,\infty}$ when $\beta = n/q$.

Our problem will then be restated the following way. Let us assume that a function f belongs both to BV and to $\dot{B}_\infty^{-\beta,\infty}$. Is it then possible to say more? The following theorem settles this issue.

Theorem 15.4. *If $0 \leq s < 1/p$, $1 < p \leq 2$ and $\beta = \frac{1-sp}{p-1}$, then*

$$\|f\|_{L^{p,s}} \leq C\|f\|_{BV}^{1/p}\|f\|_{-\beta}^{1-1/p} \qquad (15.9)$$

where $\|.\|_{-\beta}$ stands for the norm of f in the homogeneous Besov space $\dot{B}_\infty^{-\beta,\infty}$.

Theorem 15.3 is a special case of Theorem 15.4 where $s = 0, p = 2$. Let us prove Theorem 15.4. It uses both the upcoming Theorem 15.5 and the characterization of the Sobolev spaces $L^{p,s}(\mathbf{R}^n)$ by estimates on wavelet coefficients [77].

The normalization of the wavelet coefficients of f is the same as in Theorem 15.5. Indeed one writes the wavelet expansion of f as

$$f(x) = \sum\sum \alpha(j,k)2^{j(n-1)}\psi(2^j x - k) \qquad (15.10)$$

Since $1 < p \leq 2$, the norm of f in $L^{(p,s)}$ is estimated by $\sum\|\Delta_j(f)\|_p^p 2^{sjp}$. Moreover using the heuristics given by $\Delta_j(f)(k2^{-j}) \simeq \alpha(j,k)2^{j(n-1)}$ we write

$$\|\Delta_j(f)\|_p^p \simeq \sum_k |\alpha(j,k)|^p 2^{jp(n-1)}2^{-nj} \qquad (15.11)$$

which yields

$$\sum \|\Delta_j\|_p^p 2^{sjp} \simeq \sum\sum |\alpha(j,k)|^p 2^{jp(n-1)} 2^{-nj} \qquad (15.12)$$

This being said we turn to the Besov norm σ of f. We have

$$\sigma = \|f\|_{-\beta} \simeq \sup\{2^{-\beta j} 2^{j(n-1)} |\alpha(j,k)|;\ (j,k) \in \mathbf{Z} \times \mathbf{Z}\ \} \qquad (15.13)$$

which implies

$$|\alpha(j,k)| \le \sigma 2^{\beta j} 2^{-j(n-1)} \qquad (15.14)$$

Let us write $\gamma = 1 - \frac{1}{n} - \frac{\beta}{n}$. Then the right-hand side of (15.14) is $\sigma 2^{-\gamma jn}$. We then define F_m, $m \in \mathbf{Z}$, as the collection of $(j,k) \in \mathbf{Z} \times \mathbf{Z}$ such that

$$2^{-m} 2^{-\gamma jn} \le |\alpha(j,k)| < 2^{(-m+1)} 2^{-\gamma jn} \qquad (15.15)$$

Since β is positive, we have $\gamma < 1 - \frac{1}{n}$. Then Theorem 15.5 can be applied and yields

$$\sum_{(j,k) \in F_m} 2^{-\gamma jn} \quad 2^m \|f\|_{BV} \qquad (15.16)$$

If $q \in \mathbf{Z}$ is defined by $2^{-q} \le \sigma < 2^{-q+1}$, we do have $m \ge q$.

We now return to (15.12) and split the sum over (j,k) into a triple sum. We use the partition of $\mathbf{Z} \times \mathbf{Z}$ into F_m.
We end with $\sum_{m \ge q} 2^{-mp} \sum_{(j,k) \in F_m} 2^{-\gamma jn}$ and the estimate given by (15.16) suffices to complete the proof.

As it was already mentioned, Albert Cohen, Wolfgang Dahmen, Ingrid Daubechies and Ron DeVore [25] were motivated by the conjectural estimate (15.8). The main tool in their approach is the following new estimate on wavelet coefficients of $f \in BV$.

Theorem 15.5. *In any dimension $n \ge 1$, let us assume $\gamma < 1 - \frac{1}{n}$ where γ is a real exponent. Then there exists a constant $C = C(\gamma, n)$ such that for $f \in BV(\mathbf{R}^n)$, $c(j,k) = \int_{\mathbf{R}^n} f(x) 2^j \psi(2^j x - k)\, dx$ and $\lambda > 0$, one has*

$$\sum_{\{|c(j,k)| > \lambda 2^{-nj}\ \}} 2^{-nj\gamma} \le C \frac{\|f\|_{BV}}{\lambda} \qquad (15.17)$$

Let us first observe that Theorem 15.5 improves on Theorem 15.2. Indeed if $\gamma = 0$ and $n \ge 2$, (15.17) implies that the wavelet coefficients $c(j,k) = \int_{\mathbf{R}^n} f(x) 2^j \psi(2^j x - k)\, dx$ of a function $f(x)$ in BV belong to weak-l^1. We observe that the corresponding wavelet expansion
$f(x) = \sum_j \sum_k c(j,k) 2^{(n-1)j} \psi(2^j x - k)$ is using some normalized wavelets where the normalization is adapted to the BV nrom.

It is easily seen that this estimate is false when $\gamma = 1 - \frac{1}{n}$ and it is not difficult to construct sequences $c(j,k)$ belonging to weak l^1 for which (15.17) is not fulfilled. Indeed both Theorem 15.2 and Theorem 15.5 tell us that the sorted wavelet coefficients c_m^* of a function in $BV(\mathbf{R}^n)$ decay as $1/m$. But these two theorems do not tell us the same story since this $1/m$ decay does not say anything about the way the dyadic boxes are filled. For instance let us assume that γ is a positive exponent. Then the sequence defined by $c(j,k) = 2^{-\gamma j}$ for $k \in F_j$ will belong to weak l^1 when the cardinality of F_j is $2^{\gamma j}$ but does not fulfil (15.17). However the sorted c_m^* are of the order of magnitude of $1/m$. In the opposite direction, let us denote by $\mathcal{P}_\gamma$ the conclusion of Theorem 15.5. It is interesting to observe that $\mathcal{P}_\gamma$, $\gamma > 0$, does not imply weak l^1 which is $\mathcal{P}_\gamma$ with $\gamma = 0$. A trivial counter-example is given by $c(j,k) = 1$, $k = 0$, $c(j,k) = 0$, $k \neq 0$. Telling the same story with other words, we may investigate the group $\mathcal{G}$ of permutations g of $\mathbf{Z} \times \mathbf{Z}^n$ for which the left-hand side of (15.17) is preserved. We observe that, for any frozen j, any permutation acting on $k \in \mathbf{Z}^n$ belongs to $\mathcal{G}$. But these permutations are inconsistent with the locations of the corresponding wavelets since, at scale 2^{-j}, the corresponding location is given by $x_{j,k} = k2^{-j}$. Therefore Theorem 15.5 does not throw any light on the interscale correlations which exist on wavelet coefficients of functions in BV. These correlations come from the fact that edges exist. Therefore one is likely to find chains of large wavelet coefficients located on edges.

16 Improved Poincaré Inequalities

In this section some remarkable improvements on Poincaré inequalities will be given. We first start with the following example in two dimensions. Then Poincaré inequality reads as follows:

If Ω is a connected bounded open set in the plane with a Lipschitz boundary $\partial\Omega$, then there exists a constant $C = C_\Omega$ such that for every f in $BV(\Omega)$ we have

$$\int_\Omega |f(x) - m_\Omega(f)|^2 \, dx \leq C\|f\|_{BV}^2 \tag{16.1}$$

Here $m_\Omega(f)$ denotes the mean value of the function f over Ω. Such an estimate cannot be true in $\mathbf{R}^n$ for $n > 2$ since BV is not locally embedded in L^2 if $n > 3$. However there exists a sharpening of this Poincaré inequality which is valid in any dimension. Let $\mathcal{C}^{-1}(\Omega)$ denote the Banach space of all distributions f on Ω which can be written as $f = \Delta F$ where F is the restriction to Ω of a function G belonging to the Zygmund class on $\mathbf{R}^n$. The Zygmund class is defined by the classical condition that a constant C exists such that

$$|G(x+y) + G(x-y) - 2G(x)| \leq C|y| \quad x,y \in \mathbf{R}^n \tag{16.2}$$

The norm of f in $\mathcal{C}^{-1}(\Omega)$ is denoted by $\|f\|_*$ and is defined as the infimum of these constants C. This infimum is computed over all extensions G of F such that $f = \Delta F$ on Ω.

Here are some examples of functions belonging to $\mathcal{C}^{-1}(\Omega)$. Let $\rho(x)$ denote the distance from $x \in \Omega$ to the boundary $\partial\Omega$. Let f be measurable on Ω. Then the following growth condition

$$|f(x)| \leq C/\rho(x) \tag{16.3}$$

implies $f \in C^{-1}(\Omega)$. Moreover this condition is necessary and sufficient for $f \in C^{-1}(\Omega)$ if f is harmonic on Ω. These examples show that $C^{-1}(\Omega)$ is not contained in $L^1(\Omega)$.

The theorem which follows is valid in any dimension. It will be proved in several geometrical contexts. In the simplest case, Ω is a domain in $\mathbf{R}^n$ with a smooth boundary $\partial \Omega$. In the second one Ω will be a Lipschitz domain.

Theorem 16.1. *Let Ω be a smooth domain in $\mathbf{R}^n$. Then there exists a constant $C = C_\Omega$ such that, for every $f \in BV(\Omega) \cap C^{-1}(\Omega)$, one has*

$$\int_\Omega |f(x) - m_\Omega(f)|^2 \, dx \leq C \|f\|_{BV} \|f\|_* \qquad (16.4).$$

Let us insist that no boundary conditions are imposed on f. Moreover $C_{\lambda\Omega} = C_\Omega$ for any positive λ and C_Ω is translation invariant.

This theorem will be deduced from Theorem 15.3. Let us use the following local coordinates on some neighborhood of the boundary $\partial \Omega$. The coordinate ρ is the signed distance to the boundary. Then for a small enough ϵ, (y, ρ), $y \in \partial \Omega$, $\rho \in [-\epsilon, \epsilon]$ are the local coordinates which will be used on a neighborhood of the boundary. Now the proof is extremely simple. One extends f across the boundary $\partial \Omega$ by imposing that the extended function F which agrees with f on Ω should be locally odd in ρ. Here locally odd means $F(y, \rho) = -F(y, -\rho)$ for $|\rho| \leq \epsilon$.

The key fact which enters into the proof of Theorem 16.1 is that this odd extension operator is both continuous with respect to the Besov norm and the BV norm. This needs to be proved but there are no difficulties. Both the Besov norm and the BV norm can be localized and the boundary of Ω can be locally mapped to a hyperplane by a suitable diffeomorphism. The case of the half space is now obvious.

An even extension operator would also be adequate for the BV norm but certainly not for our Besov norm. Finally one applies Theorem 15.3 to this new function F, once it has been cut by a convenient cut-off function.

This proof does not extend to non-smooth domains Ω. Here is another approach which is valid for any lipschitzian domain Ω. We first consider the special case of the unit ball $B(0, 1)$ centered at 0, with radius 1. Then our preceding proof applies. The same is true if $B(0, 1)$ is replaced by the unit cube Q_0. Let us now consider a general bounded domain Ω.

The proof runs as follows. We consider a Whitney covering of Ω by open dyadic cubes $Q \in \Lambda$. The distance form Q to the boundary is of the order of magnitude of the side length l_Q of Q. The sum of the indicator functions χ_Q does not exceed a constant C. The construction of this Whitney covering does not require any smoothness on the boundary $\partial \Omega$ of Ω.

We now consider a smooth partitioning $\varphi_Q(x)$ of the identity associated to these cubes Q. Each φ_Q is supported by the doubled cube $2Q$, we have $\sum_{Q \in \Lambda} \varphi_Q(x) = 1$ and $|\partial^\alpha \varphi_Q(x)| \leq C_\alpha l_Q^{-|\alpha|}$.

Let us assume that the BV norm of f is 1.

We begin with defining averages $\mu_Q(f) = \int f\Phi_Q$ where $\Phi_Q(x) = l_Q^{-n}\Phi((x - x_Q)/l_Q)$, Φ being a testing function supported by the unit cube and with integral 1.

Our given f is decomposed into a sum $f = g + h$ where $g(x) = \sum \mu_Q(f)\varphi_Q(x)$ and $h(x) = \sum(f(x) - \mu_Q(f))\varphi_Q(x)$.

A first remark concerns the BV norm of $h(x)$. We have

$$\|h\|_{BV} \le C\|f\|_{BV} \tag{16.5}$$

For proving this estimate, it suffices to show the following

$$\int_Q |f(x) - \mu_Q(f)|\, dx \le Cl_Q \int_Q |\nabla f(x)|\, dx \tag{16.6}$$

We first use the Sobolev embedding $BV \subset L_{n/(n-1)}$ written in a scale invariant way

$$\left(\int_Q |f(x) - \mu_Q(f)|^{n*}\, dx\right)^{1/n*} \le C \int_Q |\nabla f(x)|\, dx \tag{16.7}$$

The constant C does not depend on Q and $n^* = n/(n-1)$.

Then Hölder's inequality yields (16.6) and this, together with $|\nabla\varphi_Q| \le C/l_Q$, suffices for estimating the BV norm of h.

We just proved that h belongs to BV and this information will be also used for $g = f - h$.

What do we know of the L^2 norm of h? We use (16.4) on each of the doubled cube $2Q$. We know that (16.4) is scale invariant. We then have, by the first version of Theorem 16.1,

$$\int_Q |f(x) - \mu_Q(f)|^2\, dx \le C\|f\|_* \int_Q |\nabla f(x)|\, dx \tag{16.8}$$

It now suffices to add these estimates over all cubes $Q \in \Lambda$ to obtain $\int_\Omega |h|^2\, dx \le C\|f\|_{BV}\|f\|_*$.

Let us now estimate the L^2 norm of g.

We trivially have $|\mu_Q(f)| \le C/l_Q$ since f belongs to $C^{-1}(\Omega)$. Therefore

$$|g(x)| \le \gamma/\rho(x) \tag{16.9}$$

where γ is the norm of f in $C^{-1}(\Omega)$ and $\rho(x)$ denotes the distance from $x \in \Omega$ to the boundary $\partial\Omega$.

In local coordinates, Ω is the domain lying above the graph $\Gamma = \{x = (x', x_n);\ x_n > \theta(x')\}$ of a lipschitz function θ defined on $\mathbf{R}^{n-1}$. Then $\frac{\partial}{\partial x_n}g$ belongs to $L^1(\mathbf{R}^n)$. Therefore

$$\|g(x', \cdot)\|_\infty = \sigma(x') \in L^1(\mathbf{R}^{n-1}) \tag{16.10}$$

We then assume that the BV norm of f is 1 and write $\|g\|_2^2 = \int_A |g|^2\, dx + \int_B |g|^2\, dx = I_1 + I_2$ where $A = \{x; x_n \ge \theta(x') + \gamma/\sigma(x')\}$ and B is the complement of A in Ω. of A. For estimating the first integral, we simply use the pointwise estimate (16.9) and we obtain $I_1 \leq C\gamma$ as expected. For estimating the second integral, we use (16.10) and obtain

$$\int_{(x')}^{(x')+\gamma/\sigma(x')} |g(x',t)|^2 \, dt \leq C\gamma\sigma(x') \tag{16.11}$$

and it suffices to integrate (16.11) over x' to conclude.

There are other useful theorems along these lines. Let Ω be a connected bounded open set with a smooth boundary. Let $\|f\|_*$ be defined as above.

Theorem 16.2. *With the preceding notations, we have*

$$\Big(\int_\Omega |f(x) - m_\Omega(f)|^4 \, dx\Big)^{1/2} \leq C\|\nabla f\|_2 \|f\|_* \tag{16.12}$$

There exists a similar theorem where Ω is replaced by $\mathbf{R}^n$ and this case is trivial. Indeed one is using a simply minded Littlewood-Paley analysis and writes $f = \sum_{-\infty}^{+\infty} f_j$. Then $\|f\|_4^4 \quad \sum_{j,k} \int |f_j|^2 |f_k|^2 \, dx$ and it then suffices to estimate the L^2 norm of f_j by $2^{-j}\epsilon_j$ where ϵ_j is square integrable. Similarly the L^∞ norm of f_j is estimated by $\gamma 2^j$ where γ is the $\mathcal{C}^{-1}$ norm of f. However it is harder to obtain the localized result (16.12) from the global one since we cannot extend a function f in the Sobolev space $H^1(\mathbf{R}^n)$ by an odd symmetry across the boundary. This extension is valid only when f vanishes at the boundary. However the second proof of Theorem 16.1 can be adapted to this setting and the details can be found in [82].

A last example of an improved Poincaré inequality applies to the inclusion $BV \subset L^{3/2}$ in three dimensions. More precisely G. Gallavotti used the following estimate in his CIME lectures

$$\Big(\int_\Omega |f(x) - m_\Omega(f)|^{3/2}\Big)^{2/3} dx \leq C\|f\|_{BV} \tag{16.13}$$

But as before (16.13) can be improved. We have

$$\Big(\int_\Omega |f(x) - m_\Omega(f)|^{3/2}\Big)^{2/3} dx \leq C\|f\|_{BV}^{2/3} \|f\|_{**}^{1/3} \tag{16.14}$$

where the norm $\|f\|_{**}$ in the space $\mathcal{C}^{-2}(\Omega)$ is the restriction norm to Ω of functions or distributions in the homogeneous Besov space $B_\infty^{-2,\infty}$. This Besov space consists of third derivatives of functions in the Zygmund class.

In contrast with (16.13), this new estimate is valid in any dimension $n \geq 2$. As above, we begin with the global inequality where $\Omega = \mathbf{R}^n$ and $m_\Omega(f)$ is erased.

The proof of the global estimate directly follows from Theorem 15.4. In the three dimensional case it is so simple and elegant that one cannot resist providing the reader with more details. The proof relies on Theorem 15.5. Keeping the notations of Theorem 15.5, we know that the wavelet coefficients $c(j,k) = \int_{\mathbf{R}^n} f(x) 2^j \psi(2^j x - k)\, dx$ of a function in BV belong to weak l^1. The same wavelet coefficients of a distribution in $B_\infty^{-2,\infty}$ belong to l^∞. Finally one uses the Littlewood-Paley theory and obtains $\|f\|_{3/2} \leq C \left(\sum |c(j,k)|^{3/2} \right)^{2/3}$. Let c_m^* be the sorted wavelet coefficients $|c(j,k)|$ and $\| \cdot \|_{**}$ be the norm in $B_\infty^{-2,\infty}$. Then we have $c_m^* \leq A = (C_0/m)\|f\|_{BV}$ and $c_m^* \leq B = C_1\|f\|_*$ which imply

$$\sum (c_m^*)^{3/2} \leq A^{2/3} B^{1/3} = (C_0\|f\|_{BV})^{2/3}(C_1\|f\|_{**})^{1/3} \qquad (16.15)$$

as announced.

It remains to prove the local estimate. We proceed as above and use a Whitney decomposition of Ω to construct a version of an extension operator $T : \mathcal{C}^{-2}(\Omega) \mapsto B_\infty^{-2,\infty}$ which is also bounded on BV. More precisely we have with the same notations as above $f(x) = \sum f\varphi_Q = h(x) + g(x) = \sum (f\varphi_Q - \mu_Q\varphi_Q) + \sum \mu_Q\varphi_Q$.

As above $\mu_Q = \|\varphi_Q\|_1^{-1} \int f\varphi_Q = O(d_Q^{-2})$. Therefore the sum $g(x) = \sum \mu_Q\varphi_Q$ is $O(\operatorname{dist}(x, \partial\Omega))$ and the corresponding integral can be handled as above. We now turn to $h(x) = \sum h_Q$ where $h_Q = (f\varphi_Q - \mu_Q\varphi_Q)$. As above, we map the boundary $\partial\Omega$ to $\{x_3 = 0\}$. Once this function $h(x)$ is pulled back to the upper half space, the extension is an odd extension as above. A crucial role is played by $\int h_Q(x)\, dx = 0$ and the details are left to the reader.

17 A Direct Proof of Theorem 15.3

Michel Ledoux found a direct proof of Theorem 15.3. Let $S(t) = \exp(t\Delta)$ be the heat semigroup. The proof starts with the following lemma:

Lemma 17.1. *There exists a positive constant $C = C_n$ such that*

$$\|f - S(t)f\|_1 \leq C t^{1/2} \|f\|_{BV} \qquad (17.1)$$

The proof of (17.1) is almost trivial. Indeed $S(t)f - f = \int [f(x - y) - f(x)]g_t(y)dy$ which yields $\|S(t)f - f\|_1 \leq \int \|f(x - y) - f(x)\|_1 g_t(y)dy$ and it suffices to observe that $\|f(x - y) - f(x)\|_1 \leq |y|\|f\|_{BV}$. Here $g_t(x) = t^{(-n/2)}g(\frac{x}{\sqrt{t}})$ and $g(x)$ is the familiar Gaussian function.

We are now ready to prove Theorem 15.3. Let $\theta(t)$ be the odd function defined by $\theta(t) = 0$ if $0 \leq t \leq 1$, $\theta(t) = t - 1$ if $1 \leq t \leq M$, and $\theta(t) = M - 1$ if $t \geq M$. Here M is a constant ($M = 10$ will suffice). Next $\theta_\lambda = \lambda\theta(t/\lambda)$. Let us assume $\|f\|_* \leq 1$ which reads $\|S(t)f\|_\infty \leq t^{-1/2}$. We denote by $|E|$ the Lebesgue measure if a measurable set $|E|$. We obviously have $5^{-2}\|f\|_2^2 = \int_0^\infty |\{|f(x)| > 5\lambda\}| d(\lambda^2)$. That is why we want to estimate the Lebesgue measure $|E_\lambda|$ of $E_\lambda = \{|f(x)| > 5\lambda\}$. Our first observation is $|f(x)| > 5\lambda$ $|\theta_\lambda(f)| > 4\lambda$. Then for simplifying the notations let us write $f_\lambda = \theta_\lambda(f)$ and we split this function into the following three pieces:

$$f_\lambda = f_\lambda - S(\mu)f_\lambda + S(\mu)[f_\lambda - f] + S(\mu)f \tag{17.2}$$

with $\mu = \lambda^{-2}$.

Since we already have $\|S(\mu)f\|_\infty \leq \lambda$, $|f_\lambda(x)| > 4\lambda$ implies that one of the first two terms in (17.2) is large. In other words $E_\lambda \subset F_\lambda$ G_λ where

$$F_\lambda = \{x; |f_\lambda - S(\mu)f_\lambda(x)| > \lambda\} \tag{17.3}$$

and

$$G_\lambda = \{x; |S(\mu)(f_\lambda - f)(x)| > 2\lambda\} \tag{17.4}$$

We obviously have

$$|F_\lambda| \leq \lambda^{-1}\|f_\lambda - S(\mu)(f_\lambda)\|_1 \tag{17.5}$$

and (17.1) yields $|F_\lambda| \leq C\lambda^{-2}\|\nabla(f_\lambda)\|_1 = C\lambda^{-2}\|\chi_\lambda \nabla f\|_1$. Here χ_λ is the indicator function of $\lambda \leq |t| \leq M\lambda$. Finally

$$\int_0^\infty |F_\lambda| d(\lambda^2) \leq 2\int_0^\infty \int_{\mathbf{R}^n} \chi_\lambda(f)|\nabla f| \, dx \, d\lambda = 2\log M \int_{\mathbf{R}^n} |\nabla f| \, dx \tag{17.6}$$

We now treat $|G_\lambda|$. At each point x we obviously have $|S(\mu)(f_\lambda - f)| \leq S(\mu)|f_\lambda - f|$ and $|f_\lambda - f| \leq \lambda + |f|\mathbf{1}_{\{|f|>M\lambda\}}$. Altogether it yields

$$|S(\mu)(f_\lambda - f)| \leq \lambda + S(\mu)(|f|\mathbf{1}_{\{|f|>M\lambda\}}) \tag{17.7}$$

Finally G_λ is contained in the set Ω_λ of points x such that

$$S(\mu)(|f|\mathbf{1}_{\{|f|>M\lambda\}})(x) > \lambda.$$

We obviously have

$$|\Omega_\lambda| \leq \lambda^{-1} \int S(\mu)(|f|\mathbf{1}_{\{|f|>M\lambda\}}) \, dx \leq \lambda^{-1} \int |f|\mathbf{1}_{\{|f|>M\lambda\}} \, dx.$$

Then

$$2\int_0^\infty |\Omega_\lambda|\lambda \, d\lambda \leq 2\int_0^\infty \int_{\mathbf{R}^n} |f|\mathbf{1}_{\{|f|>M\lambda\}}\lambda \, d\lambda = M^{-2}\|f\|_2^2 \tag{17.8}$$

We proved that

$$5^{-2}\|f\|_2^2 \leq M^{-2}\|f\|_2^2 + \log M\|\nabla f\|_1 \tag{17.9}$$

We choose $M = 6$ and we assume that all norms are finite. Then (17.9) ends the proof.

It now remains to get rid of the a priori assumption that the L^2 norm of f is finite. We are indeed assuming that f both belongs to BV and to the Besov space $B_\infty^{-1,\infty}$. We then know that f tends to 0 at infinity in a weak sense. Therefore f belongs to $L^{n/(n-1)}$. Assuming $n \geq 2$, it implies that, for any positive ϵ, $f_\epsilon = S(\epsilon)f$ belongs to L^2. We then apply (17.9) to f_ϵ and then let ϵ tend to 0. The case $n = 1$ is left to the reader.

18 Littlewood-Paley Analysis

Littlewood-Paley theory was introduced in the thirties by Littlewood and Paley in order to deal with some nonlinearities for which plain Fourier methods do not work. In the late seventies, the author of these notes successfully used Littlewood-Paley analysis to prove L^2 estimates when the symbol of a pseudodifferential operator is a 'rough' function [27]. In this situation usual tools as Cotlar's lemma do not work, since they are too demanding in term of regularity. Soon after J-M. Bony created the theory of para-differential operators for which the Littlewood-Paley theory is playing a key role again. J-M. Bony applied these new tools to non-linear PDE's and Jean-Yves Chemin could prove his famous theorem on the regularity of the boundary of vortex patches for the 2-D Euler equation. This proof was later simplified by P. Constantin. But let us return to the thirties.

Non-linear analysis heavily relies on the resources of L^p spaces where $p \neq 2$. While the Fourier transformation is unitary on the Hilbert space L^2, L^p cannot be characterized by size estimates on Fourier coefficients. To overcome this drawback, Littlewood and Paley got the idea of grouping together the dyadic blocks of the Fourier expansion $\sum_{-\infty}^{+\infty} c_k \exp(ikx)$ of a 2π-periodic function f. They obtained

$$\Delta_j(f) = \sum_{2^j \leq |k| < 2^{j+1}} c_k \exp(ikx), \quad j \geq 0 \qquad (18.1)$$

The Fourier expansion of $f(x)$ reads

$$f(x) = c_0 + \sum_0^\infty \Delta_j(f)(x) \qquad (18.2)$$

If $1 < p < \infty$, Littlewood and Paley proved that $f \in L^p$ is characterized by size estimates on these dyadic blocks. Indeed, if $c_0 = 0$, the L^p-norm of f is equivalent to the L^p norm of the Littlewood-Paley square function defined by

$$S(f)(x) = \left(\sum_0^\infty |\Delta_j(f)(x)|^2 \right)^{1/2} \qquad (18.3)$$

and this beautiful discovery opened a new chapter in analysis.

Extending Littlewood-Paley theory from 2π-periodic functions to functions defined on $\mathbf{R}^n$ was not a trivial task since most of the proofs relied on complex analysis. In dimensions larger than 1 and if $p \neq 2$, Charles Fefferman proved that the indicator (or characteristic) function of the unit ball is not a multiplier for the Fourier transforms of L^p functions This implies that the recipe for defining the dyadic blocks should be modified.

This was achieved in the late fifties by Alberto Calderón and Antoni Zygmund and later improved by Elias Stein. We now follow Stein's presentation of Littlewood-Paley analysis.

One starts with a function φ in the Schwartz class $\mathcal{S}(\mathbf{R}^n)$ with the following two properties

(a) the Fourier transform $\widehat{\varphi}(\xi)$ of $\varphi(x)$ vanishes outside the unit ball $|\xi| < 1$
(b) $\widehat{\varphi}(\xi) = 1$ on $\quad |\xi| \leq \frac{1}{2}$.

In some applications, one might wish to assume that $\varphi(x)$ is a radial function, a property which is not needed in what follows.

For $j \in \mathbf{N}$, we introduce the dilated function $\varphi_j(x) = 2^{nj}\varphi(2^j x)$ and denote by S_j the convolution operator with φ_j. In other words, $S_j(f) = f \star \varphi_j$. Using the Fourier transformation we obtain

$$[S_j(f)] \ (\xi) = \widehat{\varphi}(2^{-j}\xi)\hat{f}(\xi) \tag{18.4}$$

In signal processing, S_j is called a low-pass filter since it retains only those frequencies which are less than a cut-off frequency which is here 2^j. For mathematicians S_j is a mollifier or an approximation to the identity.

The speed of convergence of $S_j(f)$ towards f is related to the smoothness of f and the goal of the *Littlewood-Paley theory* is to give a precise meaning to this remark. For studying this speed of convergence, one is led to use series expansions instead of sequences. Therefore we introduce

$$\Delta_j = S_{j+1} - S_j \tag{18.5}$$

and observe that

$$I = S_0 + \sum_0^\infty \Delta_j \tag{18.6}$$

This identity can be applied to any tempered distribution f :

$$f = S_0(f) + \sum_0^\infty \Delta_j(f) \tag{18.7}$$

The function $\Delta_j(f)$ belongs to $C^\infty(\mathbf{R}^n)$ and is named a dyadic block. Indeed the Fourier transform of $\Delta_j(f)$ is supported by the dyadic annulus Γ_j defined by $2^{j-1} \leq |\xi| \leq 2^{j+1}$. In signal processing, this operator Δ_j is called a 'band-pass filter'. It retains only the frequencies which belong to the dyadic annulus Γ_j. These dyadic annuli are named channels in signal processing. The expansion of f given by (18.6) can be viewed as a *decoupling* of this function. Indeed the dyadic blocks $\Delta_j(f)$ are (almost) orthogonal. More precisely, if $|j' - j| \geq 2$, $\Delta_{j'}(f)$ is orthogonal to $\Delta_j(f)$ in L^2. When Littlewood-Paley analysis is applied to L^p, *decoupling* means that one can freely multiply the dyadic blocks by a bounded sequence of real numbers and the new expansion still is an L^p function. More pecisely we have

$$C_0(p)\|f\|_p \leq \|S(f)\|_p \leq C_1(p)\|f\|_p \qquad (18.8)$$

when $1 < p < \infty$ and when $S(f)(x) = \left(|S_0(f)(x)|^2 + \sum_0^\infty |\Delta_j(f)(x)|^2\right)^{1/2}$ as above.

The dyadic blocks $\Delta_j(f)$ are well localized in the frequency domain. However they are not localized with respect to the variable x and the goal of *wavelet analysis* is to impose both localizations. This would contradict Heisenberg's uncertainty principle, unless the localization with respect to the space variable reads $x \in Q$ where Q is a dyadic cube with side length $d \simeq 2^{-j}$. This is exactly the localization which will be imposed to wavelets in the next section.

A *Littlewood-Paley analysis* of f is given by the right-hand side of (18.7). This decomposition obviously depends on the choice of φ but the properties we want to discuss do not.

The function which plays the key role in a Littlewood-Paley analysis is not $\varphi(x)$ but $\psi(x) = 2^n\varphi(2x) - \varphi(x)$.

Let us stress a few properties satisfied by ψ

(a) ψ belongs to $\mathcal{S}_0(\mathbf{R}^n)$
(b) the Fourier transform $\hat\psi(\xi)$ of ψ is supported by the annulus $1/2 \leq |\xi| \leq 2$
(c) $\hat\psi(\xi) + \hat\psi(2\xi) = 1$ if $\frac{1}{2} \leq |\xi| \leq 1$
(d) $\sum_{-\infty}^\infty \hat\psi(2^j\xi) = 1$ if $\xi \neq 0$
(e) $\Delta_j(f) = f * \psi_j$ where $\psi_j(x) = 2^{nj}\psi(2^j x)$.

In (a) $\mathcal{S}_0(\mathbf{R}^n)$ denotes the subspace of the Schartz class $\mathcal{S}(\mathbf{R}^n)$ defined by the condition

$$\int x^\alpha f(x)\, dx = 0, \qquad \alpha \in \mathbf{N}^n \qquad (18.9)$$

The function ψ is the prototype of a wavelet and Littlewood-Paley analysis is the first step towards *wavelet analysis*. Indeed the analyzing wavelet ψ satisfies the localization, smoothness and the vanishing moment conditions which will be imposed on an analyzing wavelet. Let us however stress a few differences between ψ and the mother wavelet of an orthonormal wavelet analysis. For obtaining an orthonormal wavelet basis, conditions (c) and (d) should be replaced by

(c') $|\hat\psi(\xi)|^2 + |\hat\psi(2\xi)|^2 = 1$
(d') $\sum_{-\infty}^\infty |\hat\psi(2^j\xi)|^2 = 1$ if $\xi \neq 0$.

Thes remarks will be further detailed in the next section.

There is a second and more involved version of the Littlewood-Paley analysis where (18.6) is replaced by a summation running from $-\infty$ to $+\infty$. Before being more specific, let us give three examples where this new brand of a Littlewood-Paley analysis is crucially needed.

In physics, this definition is relevant whenever there is no reference scale, in other words, when large scales are as important as small scales. A specific example is given by Navier-Stokes equations or turbulence. Many scientists think that one of the main problems in Navier-Stokes equations or turbulence is the coupling between small and large scales. Such couplings between scales might be related to the elusive 'butterfly effect' in atmospheric turbulence.

A similar problem concerns wavelet analysis. Should one use wavelet analysis to investigate large scale behavior? This problem was often raised by Jean Morlet. Jean Morlet thought that wavelet analysis should only be used for analyzing behaviors at small scales.

There are many other examples where large scales are as important as the small scales. For instance, let us consider the Brownian motion or the fractional Brownian motion. If the standard Brownian motion is excepted, the fractional Brownian motions or more generally the '1/f' processes, have a remarkable property named 'long memory'. This long memory concerns the large scale behavior and is related to some 'infrared divergence' of the spectral density. The fractional Brownian motion is an example of Gaussian processes with stationary increments. A standard Fourier analysis does not make any sense in this context, as it was stressed by B. Picinbono. The correct tool for analyzing these processes is a Littlewood-Paley analysis.

A third example is given by homogeneous function spaces. Let us be more precise. A function space is defined as a Banach space E which is continuously embedded inside $\mathcal{S}'(\mathbf{R}^n)$. A function space E is homogeneous if its norm $\|.\|_E$ is translation invariant and satisfies a scaling law.

Definition 18.1. *The norm $\|.\|_E$ satisfies a scaling law if there exists an exponent α such that for every f in E and every positive λ we have*

$$\|f(\lambda \cdot)\|_E = \lambda^\alpha \|f\|_E \tag{18.10}$$

Many function spaces do have this property. Let us list Lebesgue L^p spaces, Hardy spaces, homogeneous Besov spaces etc. However many mathematicians prefer non-homogeneous space since homogeneous spaces are often quotient spaces.

Let us now return to the second form of a Littlewood-Paley analysis.

We still define Δ_j by (18.4) but now negative values of j are allowed. Similarly S_j is still defined by (18.3) with $j \in \mathbf{Z}$.

Then we still have $\Delta_j = S_{j+1} - S_j$ and we would like to know if the following identity is true

$$f = \sum_{-\infty}^{\infty} \Delta_j(f). \tag{18.11}$$

It is obvious that (18.11) cannot be true for any $f \in \mathcal{S}'(\mathbf{R}^n)$. Indeed any polynomial P is a counter-example since $\Delta_j(P) = 0$ identically. However there are many functional spaces $E \subset \mathcal{S}'(\mathbf{R}^n)$ for which (18.11) is valid : when $f \in E$, the right-hand side of (18.11) converges to f in E.

This problem is related to the properties of "an infra-red cut-off" which deletes all low frequencies components in (18.11). Let us be more specific and give a precise definition.

Let $\theta(\xi)$ be a function in $C^\infty(\mathbf{R}^n)$ such that

(a) $\theta(\xi) = 0$ if $|\xi| \leq 1$
(b) $\theta(\xi) = 1$ if $|\xi| \geq 2$.

This function is defined in the Fourier domain and the problem raised by the convergence of the right-hand side of (18.11) can be reformulated with the following definition.

Definition 18.2. *An infra-red cut-off is a sequence of linear mappings defined on tempered distributions f by*

$$\widehat{f_q}(\xi) = \theta(2^q \xi) \, \widehat{f}(\xi) \,, \qquad q \in \mathbf{N} \tag{18.12}$$

The convergence issue in (18.11) is related to an infra-red cut-off. Indeed $\sum_{-N}^{N-1} \Delta_j(f) = S_N(f) - S_{-N}(f)$. Since $\varphi \in \mathcal{S}(\mathbf{R}^n)$ and $\int \varphi(x)\,dx = 1$, we obviously have $S_N(f) \to f$ in $\mathcal{S}'(\mathbf{R}^n)$. We need to know whether $S_{-N}(f) \to 0$ in $\mathcal{S}'(\mathbf{R}^n)$. Using the Fourier transformation, this can be rewritten $\widehat{\varphi}(2^N \xi)\,\widehat{f}(\xi) \to 0$ $(N \to +\infty)$.

We now define $\theta(\xi) = 1 - \widehat{\varphi}(\xi)$ and our problem can be restated as

$$\theta(2^q \xi)\,\widehat{f}(\xi) = \widehat{f_q}(\xi) \to \widehat{f}(\xi) \,, \qquad q \to +\infty. \tag{18.13}$$

If $\mathcal{S}_0(\mathbf{R}^n)$ is dense in E, then $\|f_q - f\|_E \to 0$ $(q \to +\infty)$. then (18.11) holds. This applies to $E = L^p(\mathbf{R}^n)$ when $1 < p < \infty$ but not to $E = L^1(\mathbf{R}^n)$ or to $E = L^\infty(\mathbf{R}^n)$. In both cases (18.11) does not hold. Before leaving these observations, let us give a quantitaive version of (18.11) when $E = L^p(\mathbf{R}^n)$. We define $S(f)(x) = \left(\sum_{-\infty}^{\infty} |\Delta_j(f)(x)|^2 \right)^{1/2}$ and we have, as in (18.8),

$$C_0(p)\|f\|_p \leq \|S(f)\|_p \leq C_1(p)\|f\|_p \tag{18.14}$$

We now return to the space $\mathcal{S}'(\mathbf{R}^n)$ of tempered distributions and give another solution to the problem of the infra-red cut-off. This approach does not yield (18.11) but instead provides us with a weaker conclusion.

Proposition 18.1. *If $f(x)$ is any tempered distribution, there exists an integer N and a sequence $P_q(x)$ of polynomials of degrees less than or equal to N such that*

$$f(x) = \lim_{q \to +\infty} \left\{ \sum_{-q}^{0} \Delta_j(f) - P_q(x) \right\} + \sum_{1}^{\infty} \Delta_j(f) \tag{18.15}$$

Let us observe that (18.15) is consistent with the fact that Littlewood-Paley analysis is blind to polynomials. It implies that some 'floating plolynomials' are needed to fix the algorithms.

This proposition will be illustrated with the example of the homogeneous Besov spaces $\dot{B}_\infty^{s,\infty}$ when the regularity exponent s is a real number. Applying a wavelet analysis or a Littlewood-Paley decomposition will raise the same issues which concern the role of floating polynomials in (18.15).

These are characterized by

$$\|\Delta_j(f)\|_\infty \leq C2^{-js}, \quad j \in \mathbf{Z} \tag{18.16}$$

If the regularity exponent s is positive, then this Besov space $\dot{B}_\infty^{s,\infty}$ is usually denoted by $\dot{C}^s$ and named a Hölder space. It is defined modulo polynomials of degrees not exceeding s. Let us observe that any polynomial whatever his degree satisfies (18.16). If (18.16) was exactly defining $\dot{C}^s$, we would face the following problem: all polynomials, whatever be their degree would belong to $\dot{C}^s$ while only polynomials with degrees not exceeding s do belong to $\dot{C}^s$. In other words, in Proposition 18.1, the degree of the floating polynomial is related to the order of the distribution f. The same relation holds in the case of Hölder spaces. Let us observe that if s is negative, there are no floating polynomials.

We would like to end this section by defining other Besov spaces which have been used in a systematic way throughout this book. These spaces depend on three indices; the first one measures the smoothness and is denoted by s, the second one gives a small improvement on the regularity measurement and the third one tells us what is the functional norm which is used for these regularity measurements. Moreover these spaces exist as homogeneous spaces or non-homogeneous spaces. Finally they are either spaces of functions (or generalized functions) or spaces of tempered distributions.

Let us begin with the simpler definition of the non-homogeneous Besov spaces.

Definition 18.3. *Let s be a real number and p, q be two exponents belonging to $[1, \infty]$. Then a tempered distribution $f \in \mathcal{S}'(\mathbf{R}^n)$ belongs to $B_p^{s,q}(\mathbf{R}^n)$ if and only if the following two conditions are satisfied*

$$\|\Delta_j(f)\|_p \leq \epsilon_j 2^{-js}, \quad j \in \mathbf{Z} \quad \epsilon_j \in l^q(\mathbf{N}) \tag{18.17}$$

and

$$\|S_0(f)\|_p \leq C \tag{18.18}$$

The norm of f in $B_p^{s,q}(\mathbf{R}^n)$ is the sum between C and the $l^q(\mathbf{Z})$ norm of ϵ_j.

If $p = q = \infty$, this Besov space coincides with the non-homogeneous Besov space $\mathcal{C}^s$. If $p = q = 2$, we find the standard Sobolev space.

A second definition will provide us with homogeneous spaces. We erase condition (18.18) and impose (18.17) on all $j \in \mathbf{Z}$. But this definition immediately raises a difficult problem. The homogeneous Besov space $\dot{B}_p^{s,q}(\mathbf{R}^n)$ is no more embedded inside $\mathcal{S}'(\mathbf{R}^n)$ in general. Indeed this embedding is equivalent to $s < n/p$ if $1 < p \leq \infty$ or $1 < q \leq \infty$ and to $s \leq n$ if $1 = p = q$. When $s \geq n/p$ and when either p or q is larger than 1, then $\dot{B}_q^{s,p}(\mathbf{R}^n)$ is a space of functions modulo polynomials. The degrees of these floating polynomials does not exceed $s - n/p$.

19 Littlewood-Paley Analysis and Wavelet Analysis

Wavelet analysis is an alternative to the standard Fourier analysis. This alternative is aimed at improving the representation and processing of some large classes of signals, images, functions or operators. Wavelets should provide us with fast algorithms which yield (1) *accurate approximations* and (2) *sparse and robust expansions*.

The first example which illustrates these remarks concerns *accurate approximations*. The story begins in the nineteen century with the following unpleasant fact: the Fourier series expansion of a continuous function may diverge at a given point. This counter-example raised the following problem: could an orthonormal basis $h_n(x)$, $n \in \mathbf{N}$, of $L^2[0,1]$ have the property that the expansion of any continuous function $f(x) \in \mathcal{C}[0,1]$ converges uniformly to $f(x)$? In his dissertation submitted in July 1909 at the University of Göttingen, A. Haar constructed what we today call the *Haar basis* and proved that this basis is a solution to the problem. The structure of the Haar basis is identical to the general structure of a wavelet basis. It is given by $h_n(x) = 2^{j/2}h(2^j x - k)$, $n = 2^j + k$, $0 \leq k \leq 2^j - 1$, $j \geq 0$. Here $h(x) = 1$ on $[0, 1/2)$, -1 on $[1/2, 1)$ and $h(x) = 0$ outside $[0, 1)$. The approximation to $f(x)$ which given by its expansion in the Haar basis coincides with the ordinary approximation of a continuous function $f(x)$ by piecewise constant functions $f_j(x)$ with jump discontinuities at dyadic points $k2^{-j}$, $0 \leq k \leq 2^j - 1$.

A similar question can be raised when $f(x)$ has m continuous derivatives. We then write $f \in \mathcal{C}^m[0,1]$. Does the corresponding expansion in the Haar basis converges to f in the $\mathcal{C}^m$ norm? The answer is obviously no and adapted bases were constructed by Ph. Franklin (1927). These new bases heavily depend on m and their algorithmic structure is not as simple as in the Haar basis. New orthonormal wavelet bases were constructed by Ingrid Daubechies (1987) and provide a much more interesting solution. Moreover these bases come equipped with fast algorithms. These fast algorithms have been previously discovered in signal and image processing (1982) and are called *subband coding*.

The second example concerns *robust expansions*. In the abstract theory of Banach spaces *robust expansions* are related to *unconditional bases*. An unconditional basis e_n, $n \in \mathbf{N}$, of a Banach space E is defined by the following two properties: (a) each $x \in E$ has a unique expansion as a series $x = \sum_0^\infty \alpha_n e_n$ and (b) this expansion still converges to x whatever be the permutation imposed to the terms. In (a) and (b) convergence has the strongest meaning which is defined by the norm in E. Condition (b) is then equivalent to the seemingly stronger property (c): if $(\lambda_n) \in l^\infty(\mathbf{N})$ is any bounded sequence and if $x = \sum_0^\infty \alpha_n e_n$, then the series $\sum_0^\infty \lambda_n \alpha_n e_n$ still converges in E to $y = M(x)$. This leads us to another equivalent condition (d): for any multiplier sequence $(\lambda_n) \in l^\infty(\mathbf{N})$, the operator M defined by $M(e_n) = \lambda_n e_n$, $n \in \mathbf{N}$, is bounded on E.

In 1938, Marcinkiewicz proved that the Haar system is an unconditional basis for $L^p[0, 1]$ when $1 < p < \infty$.

We now turn to operator theory. Given a class $\mathcal{T}$ of operators acting on some Hilbert space $\mathcal{H}$, we wish to represent these operators in the simplest possible way. If the collection $\mathcal{T}$ consists of self-adjoint operators which commute, then one can find an orthonormal basis in which these operators are simultaneously diagonalized. If these conditions are not fulfilled, we instead look after an orthonormal basis in which the operators $T \in \mathcal{T}$ are represented by banded matrices or matrices with a strong off-diagonal decay. If it is possible, it paves the way to the existence of fast algorithms for computing the product between two operators in $\mathcal{T}$. This program can be completed when $\mathcal{T}$ consists of singular integral operators of Calderón-Zygmund type [26].

What can be said about *sparsity* or *best bases*? This fundamental issue will be illustrated by a simple example. Let us consider a Hilbert space $\mathcal{H}$ and a compact subset $\mathcal{B} \subset \mathcal{H}$. We would like to find an orthonormal basis $e_n, n \in \mathbf{N}$, such that, for each $n \geq 0$, the errors $\rho_n = \sup_{\{x \in \mathcal{B}\}} \|x - x_n\|$ have the strongest decay rate. Here x_n is the best approximation to x by a linear combination of at most n terms among the basis sequence (these n terms are not the first n terms in general). Compression algorithms for signals or images are related issues.

In many interesting cases a near optimal solution to these four problems is provided by an orthonormal wavelet basis. Moreover wavelets come along with fast and efficient algorithms which implement Littlewood-Paley analysis. Let us explain this connection.

There exists a 1-D orthonormal wavelet basis where both the 'mother wavelet' ψ and the 'scaling function' φ belong to the Schartz class $\mathcal{S}(\mathbf{R})$. Moreover the scaling function and the corresponding wavelet have the following properties

(a) $\hat{\varphi}(\xi) = 1$, $|\xi| \leq 2\pi/3$
(b) $\hat{\varphi}(\xi) = 0$, $|\xi| \geq 4\pi/3$
(c) $\sum_{-\infty}^{+\infty} |\hat{\varphi}(\xi + 2k\pi)|^2 = 1$, $-\infty < \xi < \infty$

and these properties are close to the corresponding ones in a Littlewood-Paley analysis. Let S_j denote the convolution operator with $2^j \varphi(2^j x)$ and similarly let Δ_j denote the convolution operator with $2^j \psi(2^j x)$. In an orthonormal wavelet analysis one insists on getting $S_j^2 + \Delta_j^2 = S_{j+1}^2$ while in a Littlewood-Paley analysis, one is satisfied with $S_j + \Delta_j = S_{j+1}$. The first requirement implies that energy estimates are preserved since the operators S_j are selfadjoint.

A second remark concerns the definition of wavelet coefficients. If ψ is a mother wavelet in some orthonormal wavelet basis, then the wavelet coefficients of a function f and the dyadic blocks of the corresponding Littlewood-Paley expansion are related by

$$\Delta_j[f](k2^{-j}) = 2^{j/2} c_{j,k} - 2^{j/2} <f, \psi_{j,k}> \qquad (19.1)$$

In other words, once the dyadic blocks have been computed, then the wavelet coefficients are obtained by a down-sampling which, waving a hand, is consistent with Shannon's theorem. Here we have been sloppy since Shannon's theorem implies that the fine grid $\Gamma_j = 2^{-j}\mathbf{Z}$ should be replaced by a finer one given by $(3/8)\Gamma_j$. In other words, Shannon's theorem does not explain why a function can be recovered from its wavelet coefficients. To be more precise, this does not hold for a given j. Information which is not provided by a given j is coming from $j \pm 1$.

This connection between wavelet analysis and signal processing is not surprising, since most of the ideas and tools which have been developed in wavelet based algorithms were indeed implicit in signal processing. For instance multiresolution analysis mirrors the pyramidal algorithms of Burt and Adelson. Similarly quadrature mirror filters or subband coding are almost identical to an orthonormal wavelet analysis. However many scientists belonging to the signal processing community are still shocked by the fact that wavelet analysis does not respect Shannon's theorem *stricto sensu*.

Wavelet analysis is almost identical to Littlewood-Paley theory. It implies that all the well known characterizations of function spaces in terms of size estimates on dyadic blocks immediately translate into characterizations in terms of size estimates on wavelet coefficients. As an obvious consequence, wavelet bases are unconditional bases for such function spaces.

This section will end with describing the three-dimensional wavelets which are generated by the above mentioned basis. For constructing these 3-D wavelets, we need both the one-dimensional wavelet ψ and the corresponding scaling function φ. Then the three 3-D mother wavelets are

$$\begin{cases} \psi_1(x_1, x_2, x_3) = \psi(x_1)\varphi(x_2)\varphi(x_3) \\ \psi_2(x_1, x_2, x_3) = \varphi(x_1)\psi(x_2)\varphi(x_3) \\ \psi_3(x_1, x_2, x_3) = \varphi(x_1)\varphi(x_2)\psi(x_3) \\ \psi_4(x_1, x_2, x_3) = \psi(x_1)\psi(x_2)\varphi(x_3) \\ \psi_5(x_1, x_2, x_3) = \varphi(x_1)\psi(x_2)\psi(x_3) \\ \psi_6(x_1, x_2, x_3) = \psi(x_1)\varphi(x_2)\psi(x_3) \\ \psi_7(x_1, x_2, x_3) = \psi(x_1)\psi(x_2)\psi(x_3) \end{cases} \qquad (19.2)$$

The 3-D wavelet analysis is described by the following theorem:

Theorem 19.1. *Keeping the same notations as in (19.2), we have:*

(a) each $\psi_m(x_1, x_2, x_3)$ belongs to the Schwartz class $S(\mathbf{R}^3)$ and all moments vanish: $\int x^\alpha \psi_m(x)\,dx = 0$ $(\alpha \in \mathbf{N}^3)$

(b) $2^{3j/2}\psi_m(2^j x_1 - k_1, 2^j x_2 - k_2, 2^j x_3 - k_3)$, $j \in \mathbf{Z}$, $k = (k_1, k_2, k_3) \in \mathbf{Z}^3$, $m = 1, 2, 3, 4, 5, 6, 7$, is an orthonormal basis for $L^2(\mathbf{R}^3)$.

This basis is an unconditional basis for all function spaces if one excepts those which are built on L^1 or L^∞. The latter spaces do not admit an unconditional basis. For simplifying the notations, the orthonormal wavelet basis $2^{3j/2}\psi_m(2^j x - k)$, $m = 1, 2, 3$, $j \in \mathbf{Z}$, $k \in \mathbf{Z}^3$, of Theorem 16.2 will be indexed by the corresponding dyadic cube $Q = \{x;\ 2^j x - k \in [0, 1]^3\}$ and written in short $\psi_Q(x)$. Let   be the collection of all such dyadic cubes Q. The index m will be forgotten in the following theorem. If the mother wavelet ψ was replaced by the indicator function of $[0, 1]^3$, then the support of $2^{3j/2}\psi_m(2^j x - k)$ would be the dyadic cube Q and the actual wavelet ψ_Q is almost supported by Q. Most of the classical functional spaces are characterized by size estimates on wavelet coefficients. Fot instance the space BMO admits the following characterization.

Theorem 19.2. *Let $f(x) \in S'(\mathbf{R}^3)$ be a tempered distribution. Then f belongs to BMO if and only if its wavelet coefficients $\alpha_Q = <f, \psi_Q>$ satisfy the following size condition: there exists a constant C such that for each dyadic cube R we have*

$$\sum_{Q \subset R} |\alpha_Q|^2 \le C|R| \tag{19.3}$$

Moreover f belongs to the Kochd Tataru space if and only if

$$\sum_{Q \subset R} 2^{-2j} |\alpha_Q|^2 \le C|R| \tag{19.4}$$

This theorem is attractive and one cannot hope for a simpler analysis of BMO anf of the Koch&Tataru space. Let us mention that Theorem 19.1 would not true if our wavelet basis was replaced by the Haar basis.

The characterization of the homogeneous Besov spaces $\dot{B}_q^{-(1-3/q),\infty}$, $3 \le q \le \infty$ is as simple. It is here convenient to change the normalization of the wavelets and to write $\beta(j, k) = \int f(x) 2^{3j} \psi(2^j x - k)\,dx$. We then have

Theorem 19.3. *Let $f(x) \in S'(\mathbf{R}^3)$ be a tempered distribution. Then f belongs to $\dot{B}_q^{-(1-3/q),\infty}$ if and only if there exists a constant C such that, for each $j \in \mathbf{Z}$, its normalized wavelet coefficients $\beta(j, k)$ satisfy*

$$\left(\sum_k |\beta(j, k)|^q\right)^{1/q} \le C\, 2^j \tag{19.5}$$

It is then obvious that $\dot{B}_q^{-(1-3/q),\infty} = B_q \subset B_r$ when $q \le r$. When $q = \infty$, (19.5) simply reads $|\beta(j, k)| \le C\, 2^j$, $j \in \mathbf{Z}$.

This section indicates that wavelet bases are a marvellous tool for facing theoretical problems but might be clumsy in practice, as the lengthy list (19.2) shows.

References

1. S. Alinhac and P. Gérard, Opérateurs Pseudo-différentiels et Théorème de Nash-Moser. Savoirs Actuels. Editions du CNRS (1991).
2. L. Ambrosio. *Corso introduttivo alla teoria geometrica della Misura ed alle Superfici Minime*. Scuola Normale Superiore Lecture Notes, Pisa, (1997).
3. C. Amrouche, V. Girault, M. E. Schonbek, T. P. Schonbek *Pointwise Decay of Solutions and of higher derivatives to Navier–Stokes Equations*, SIAM J. Math. Anal., **31**, N. 4, 740–753, (2000).
4. J. M. Ball, *Remarks on blow-up and nonexistence theorems for non-linear parabolic evolution equations*, Quart J. Math. Oxford Ser. **28** (1977) 473-486.
5. G. Battle, P. Federbush, *Divergence-free vector wavelets,* Mich. Math. J., **40**, 181–195 (1993).
6. G. Berkooz, P. Holmes, J. L. Lumley, *The proper orthogonal decomposition in the analysis of turbulent flows*, Annu. Rev. Fluid Mech., **25**, 539–575 (1993).
7. J-M. Bony, *Calcul symbolique et propagation des singularités pour les equations aux dérivées partielles non linéaires*. Ann. Sci. ENS (1981) 209-246.
8. G. Bourdaud, *Réalisation des espaces de Besov homogènes*, Ark. Mat., **26**, N. 1, 41–54 (1988).
9. G. Bourdaud, *Localisation et multiplicateurs des espaces de Sobolev homogènes* Manuscripta Mathematica, **60**, 93–130 (1988).
10. G. Bourdaud, *Ondelettes et espaces de Besov*, Rev. Mat. Iberoamericana, **11**, N. 3, 477–512 (1995).
11. A. Braides. *Approximation of free discontinuity problems*. Lecture Notes in Mathematics no. 1694, Springer (1991).
12. L. Brandolese, *On the Localization of Symmetric and Asymmetric Solutions of the Navier–Stokes Equations dans* **R**^{n}, C. R. de l'Acad. de Sciences de Paris, **t. 332**, Série I, 125–130 (2001).
13. L. Brandolese, Ph.D. Dissertation (to be asked to ymeyer@cmla.ens-cachan.fr).
14. L. Brandolese, Y. Meyer, *On the instantaneous spreading for the Navier–Stokes system in the whole space*, A tribute to J. L. Lions. ESAIM Control Optim. Calc. Var. 8 (2002), 273–285 (electronic).
15. L. Caffarelli, R. Kohn, L. Nirenberg, *Partial regularity of suitable weak solutions of the Navier–Stokes equations*, Comm. Pure Appl. Math, **25**, 771–831 (1982).
16. C.P. Calderón, Existence of weak solutions for the Navier-Stokes equations with initial data in L^p. Trans. AMS **318** (1990) 179-200.
17. M. Cannone, *Ondelettes, paraproduits et Navier–Stokes*, Diderot Éditeur, (1995).
18. M. Cannone, F. Planchon, *On the regularity of the bilinear term for solutions to the incompressible Navier–Stokes equations,* Rev. Mat. Iberoamericana, **16**, N. 1, 1–16, (2000).
19. A. Carpio, *Large time behavior in the incompressible Navier–Stokes equations*, SIAM J. Math. Anal., **27**, N. 2, 449–475 (1996).

20. Th. Cazenave and F. Weissler. *The Cauchy problem for the critical nonlinear Schrödinger equation in H·*. Nonlinear Anal. T.M.A. **14** (1990) 807-836.
21. Th. Cazenave and F. Weissler. *Asymptotically self-similar global solutions of the nonlinear Schrödinger and heat equations*, Math. Zeit. 228 (1998), 83-120.
22. J.-Y. Chemin, *Fluides parfaits incompressibles*, Astérisque, **230** (1995).
23. A. Cohen. *Numerical analysis of wavelet methods*. Handbook of numerical analysis, P.G.Ciarlet and J.L.Lions eds. (1999).
24. A. Cohen, R. DeVore, P. Petrushev and H. Xu. *Nonlinear approximation and the space $BV(R^2)$*, American Journal of Mathematics, 121 (1999) 587-628.
25. A. Cohen, W. Dahmen, I. Daubechies and R. DeVore. *Harmonic Analysis of the space BV*. (September 25, 2000)
26. R. R. Coifman *Adapted multiresolution analysis, Computation, Signal processing, and Operator theory*. Proceedings of the ICM, Kyoto (1990) pp. 879-888, Springer-Verlag.
27. R. R. Coifman, Y. Meyer, *Au delà des opérateurs pseudo-différentiels*, Astérisque, **57**, seconde édition (1978).
28. R. R. Coifman, G. Weiss, *Extentions of Hardy Spaces and their use in Analysis*, Bull. Amer. Mat. Soc., **83**, 569–645 (1977).
29. P. Constantin, C. Foias, *Navier-Stokes equations*, The University of Chicago Press, Chicago, 1988.
30. Y. Couder, C. Basdevant, *Experimental and numerical study of vortex couples in two-dimensional flows*, J. Fluid Mech., **173**, 225–251.
31. R. Danchin, *Analyse numérique et harmonique d'un problème de mécanique des fluides*, Thèse de doctorat, École Polytechnique, Palaiseau, France, Décembre (1996).
32. I. Daubechies. *Ten lectures on wavelets*. SIAM Philadelphia (1992).
33. R. DeVore. *Nonlinear approximation*. Acta Numerica (1998) pp. 51-150.
34. R. DeVore, B. Jawerth and V. Popov. *Compression of wavelet decompositions*. American Journal of Mathematics 114 (1992) pp. 737-785.
35. S. Y. Dobrokhotov, A. I. Shafarevich, *Parametrix and the Asymptotics of Localized Solutions of the Navier–Stokes Equations, Linearized on a smooth Flow*, Institute for problems of Mechanics, Academy of Sciences of USSR. Translated from Matematicheskie Zametki, **51**, N. 1, 72–82 (1992).
36. S. Y. Dobrokhotov, A. I. Shafarevich, *Some Integral Identities and Remarks on the Decay at Infinity of the Solutions to the Navier–Stokes Equations in the Entire Space*, Russ. J. Math. Phys. Vol. 2, N. 1, 133–135 (1994).
37. D. Donoho *Abstract statistical estimation and modern harmonic analysis*. Proceedings of the ICM, Zürich (1994) pp. 997-1005 Birkäuser.
38. D. Donoho, R. DeVore, I. Daubechies and M. Vetterli. *Data compression and harmonic analysis*, IEEE Trans. on Information Theory, Vol. 44, No 6, October 1998, 2435-2476.
39. M. Farge, *Transformée en ondelettes continue et application à la turbulence*, Soc. Math. de France, 5 mai (1984).
40. M. Farge, M. Holschneider, J. F. Colonna, *Wavelet Analysis of Coherent Structures in Two-Dimensional Turbulent Flows*, Topological Fluid Mech., Ed. Keith Moffat, Cambridge University Press.
41. P. Federbush. *Navier and Stokes meet the wavelet*. Communications in Mathematical Physics 155 (1993) ·219-248.
42. C. L. Fefferman, *Existence and Smoothness of the Navier–Stokes Equation*, Princeton, NJ 08544-1000, May 1 (2000).

43. M. Frazier, B. Jawerth, G. Weiss, *The Littlewood–Paley Theory and the study of Function spaces*, CBMS Regional Conf. Series in Math., **79**, Amer. Math. Soc., Providence, (1991).

44. P. Frick, V. Zimin, *Hierarchical models of turbulence. Wavelets, Fractals and Fourier Transforms*, Farge et al. eds. New York, Clarendon, 265–283, (1995).

45. Y. Fujigaki, T. Miyakawa, *Asymptotic profiles of non stationary incompressible Navier–Stokes flows in* $\mathbf{R}^n$ *and* $\mathbf{R}^n_+$. Mathematical analysis in fluid and gas dynamics (Japanese) (Kyoto, 2000). Surikaisekikenkyusho Kokyuroku No. 1225 (2001), 14–33

46. H. Fujita and T. Kato. *On the Navier-Stokes initial value problem I*, Arch. Rational Mech. Anal. **16** (1964) 269-315.

47. G. Furioli, P-G. Lemarié-Rieusset and E. Terraneo. *Sur l'unicité dans* $L^3(\mathbf{R}^3)$ *des solutions 'mild' des équations de Navier-Stokes*. C. R. Acad. Sciences Paris, Série 1 (1997) 1253-1256.

48. G. Furioli, P-G. Lemarié-Rieusset, E. Zahrouni and A. Zhioua. *Un théorème de persistance de la régularité en norme d'espaces de Besov pour les solutions de H. Koch et D. Tataru des équations de Navier-Stokes dans* $\mathbf{R}^3$, CRAS Paris, t. 330, Série I (2000)

49. G. Furioli, P.G. Lemarié-Rieusset et E. Terraneo, *Unicité* $L^3(\mathbf{R}^3)$ *et d'autres espaces fonctionels limites pour Navier–Stokes*, Rev. Mat. Iberoamericana, **16**, N. 3, 605–667 (2000).

50. G. Furioli, *Applications de l'analyse harmonique réelle à l'étude des équations de Navier–Stokes et de Schrödinger non linéaires*, Thèse, Univérsité d'Orsay (1999).

51. G. Furioli, E. Terraneo, *Molecules of the Hardy space and the Navier–Stokes equations*, Funkcial Ekvac., 45 (2002), no. 1, 141–160.

52. T. Gallay, C. E. Wayne, *Long-time asymptotics of the Navier-Stokes and vorticity equations on* $\mathbf{R}^3$, Recent developments in the mathematical theory of water waves (Oberwolfach, 2001). R. Soc. Lond. Philos. Trans. Ser. A Math. Phys. Eng. Sci. 360 (2002), no. 1799, 2155–2188.

53. T. Gallay, C. E. Wayne, *Invariant manifolds and the long-time asymptotics of the Navier–Stokes equations on* $\mathbf{R}^2$, Arch. Ration. Mech. Anal. 163 (2002), no. 3, 209–258.

54. Y. Giga, T. Miyakawa, *Navier–Stokes Flow in* $\mathbf{R}^3$ *with Measures as Initial Vorticity and Morrey Spaces*, Commun. in Partial Diff. Equations, **14**, 5, 577–618 (1989).

55. Y. Giga, T. Miyakawa, H. Osada, *Two dimensional Navier–Stokes flow with measures as initial vorticity*, Arch. Rational Mech., **104**, N. 3, 223–250 (1988).

56. R. Hanks, *Interpolation by the Real Method between* BMO, $L^\alpha(0 < \alpha < \infty)$, *and* $H^\alpha(0 < \alpha < \infty)$, Indiana Univ. Math. J., **26**, N. 4 (1977).

57. C. He, Z. Xin, *On the decay properties of Solutions to the nonstationary Navier-Stokes Equations in* $\mathbf{R}^3$, Proc. Roy. Edinbourgh Soc. Sect. A, **131**, N. 3, 597–619 (2001).

58. S. Jaffard, Y. Meyer, *Wavelet Methods for Pointwise Regularity and Local Oscillations of Functions*, Memoirs Amer. Math. Soc., **123**, N. 587 (1996).

59. S. Jaffard, Y. Meyer, *On the pointwise regularity of functions in critical Besov Spaces*, J. of Funct. Anal., **175**, 415–434 (2000).

60. J-P. Kahane and P-G. Lemarié-Rieusset. Fourier *Series and Wavelets*. Gordon and Breach Science Publishers (1996).

61. R. Kajikiya, T. Miyakawa, *On L^2 Decay of Weak Solutions of the Navier–Stokes Equations in* $\mathbf{R}^n$, Math Z., **192**, 135–148 (1986).
62. T. Kato, *Strong L^p-Solutions of the Navier–Stokes Equations in* $\mathbf{R}$ *, with applications to weak solutions*, Math. Z., **187**, 471–480 (1984).
63. T. Kato, *The Navier–Stokes equations for an incompressible fluid in* $\mathbf{R}^2$ *with a measure as initial velocity*, Diff. and Int. Equations, **7**, 949–966 (1994).
64. T. Kato, *Strong Solutions to the Navier–Stokes Equations in Morrey Spaces*, Bol. Soc. Bras. Mat., **22**, N. 2, 127–155 (1992).
65. T. Kato, H. Fujita, *On the non-stationary Navier–Stokes system*, Semin. Mat. Univ. Padova, **32**, 243–260 (1962).
66. T. Kato and G. Ponce, *Commutator estimates and the Euler and Navier-Stokes equations*. Comm. in PDE **41** (1988) 891-907.
67. H. Koch and D. Tataru. *Well-posedness for the Navier-Stokes equations*, Advances in Mathematics, 157 (2001) 22-35.
68. O. Ladyzenskaija, *The mathematical theory of viscous incompressible flow*, Gordon and Breach, New York, English translation, Second edition (1969).
69. P-G. Lemarié-Rieusset. *Recent developments in the Navier-Stokes problem* Capmant Hall, Research Notes in Mathematics 431.
70. P.G. Lemarié-Rieusset, *Analyses multirésolutions non-orthogonales, commutation entre proœcteurs et dérivation et ondelettes vecteurs à divergence nulle*, Rev. Mat. Iberoamericana, **8**, 221–236 (1992).
71. P.G. Lemarié-Rieusset, *Ondelettes et poids de Muckenoupt*, Studia Math., **108**, N. 2, 127–147 (1994).
72. J. Leray, *Étude de diverses équations intégrales non linéaires et des quelques problèmes que pose l'Hydrodynamique*, J. .Math. Pures et Appl., **12**, 1–82 (1933).
73. J. Leray. *Sur le mouvement d'un liquide visqueux emplissant l'espace*. Acta Mathematica 63 (1934) 193-248.
74. P-L. Lions, N. Masmoudi *Unicité des solutions faibles de Navier-Stokes dans* L (Ω). CRAS Paris, 327 (1998) 491-496.
75. Lu Ting, *On the application of the integral invariants and decay laws of vorticity distributions*, J. Fluid Mech. **127**, 497–506 (1983).
76. S. Mallat. *A Wavelet Tour of Signal Processing*. Academic Press (1998).
77. Y. Meyer. *Wavelets and Operators*. Cambridge studies in advanced mathematics 37 CUP (1992).
78. Y. Meyer. *Wavelets, paraproducts and Navier-Stokes equations*, Current Developments in Mathematics, 1996, pp 105-212, Ed. Raoul Bott, Arthur Jaffe, S.T. Yau, David Jerison, George Lusztig, Isadore Singer. International Press.
79. Y. Meyer. *Wavelets, vibrations and scalings*, CRM Monograph Series, Vol.9 (1998).
80. Y. Meyer. *Large time behavior and self-similar solutions of some semi-linear diffusion equations*. Essays in Honor of Alberto Calderón, University of Chicago Press (1999).
81. Y. Meyer. *Wavelets and functions with bounded variations from image processing to pure mathematics*. Mathematics towards the Third Millennium. Atti della Accademia Nazionale dei Lincei, ROMA (2000).
82. Y. Meyer, *Oscillating patterns in image processing and nonlinear evolution equations. The Fifteen Dean daqueline B. Lewis Memorial Lectures*. Univeristy Lecture Series **22** AMS (2001).

83. Y. Meyer and T. Rivière, *A partial regularity result for a class of stationary Yang-Mills fields in high dimension.* Revista Mathemática Iberoamericana, **19**, (2003) 195-219.

84. T. Miyakawa, *Hardy spaces of soleinoidal vector fields, with applications to the Navier–Stokes equations,* Kyushu J. Math., **50**, 1–64 (1996).

85. T. Miyakawa, *Application of Hardy Space Techniques to the Time-Decay Problem for Incompressible Navier–Stokes Flows in $\mathbf{R}^n$,* Funkcial. Ekvac., **41**, 383–434 (1998).

86. T. Miyakawa, *On space time decay properties of nonstationary incompressible Navier–Stokes flows in $\mathbf{R}^n$,* Funkcial. Ekvac., **32**, N. 2, 541–557 (2000).

87. T. Miyakawa, M. E. Schonbek, *On Optimal Decay Rates for Weak Solutions to the Navier–Stokes Equations in $\mathbf{R}^n$,* Proceedings of Partial Differential Equations and Applications (Olomouc, 1999). Math. Bohem. 126 (2001), no. 2, 443–455.

88. S. Monniaux. *Uniqueness of mild solutions to the Navier-Stokes equation and maximal L^p-regularity,* CRAS. Paris, t. 328, Série I (1999) 663-668.

89. B. Muckenhoupt, *Wheighted norm inequalities hor the Hardy Maximal function,* Trans. Amer. Soc., **165**, 207–226 (1972).

90. F. Oru. *Le rôle des oscillations dans quelques problèmes d'analyse non-linéaire.* Thèse. CMLA. ENS-Cachan (9 Juin 1998).

91. J. Peetre. *New thoughts on Besov Spaces.* Duke Univ. Math. Series (1976).

92. F. Planchon. *On the Cauchy problem in Besov spaces for a nonlinear Schrödinger equation,* Communications in Contemporary Mathematics (2000), 243–254.

93. F. Planchon, *Solutions globales et comportement asymptotique pour les équations de Navier–Stokes,* Thèse, École Polytechnique (1996).

94. E. Sawyer, *Multipliers of Besov and power-weighted L^2-spaces,* Indiana, Univ. Math. J. **33**, 353–366 (1984).

95. M. E. Schonbek, *L^2 decay for Weak Solutions of the Navier–Stokes Equations,* Arch. Rat. Mech. Anal., **88**, 209–222 (1985).

96. M. E. Schonbek, *Large Time Behavior of Solutions to the Navier–Stokes Equations,* Comm. in Partial Diff. Equations, **11**, N. 7, 733–763 (1986).

97. M. E. Schonbek, *Lower Bounds of Rates of Decay for Solutions to the Navier-Stokes Equations,* J. of the Amer. Math. Soc., **4**, N. 3, 423–449 (1991).

98. M. E. Schonbek, *Some results on the Asymptotic Behaviour of Solutions to the Navier–Stokes Equations,* dans *The Navier–Stokes Equations II — Theory and Numerical Methods,* Proceedings, Oberwolfach 1991, Lecture Notes in Mathematics, 1530, 146–159.

99. M. E. Schonbek, *Asymptotic Behaviour of Solutions to Three-Dimentional Navier-Stokes Equations,* Indiana Univ. Math. J., **41**, N. 3, 809–823 (1992).

100. M. E. Schonbek, T. P. Schonbek, *On the boundedness and decay of moments of solutions of the Navier–Stokes equations,* Advances in Diff. Eq., 7–9 July-September (2000).

101. M. E. Schonbek, T. P. Schonbek, E Süli, *Decay results to the Magneto Hydrodynamics equations,* Mathematische Annalem, **304**, N. 4, 717–756 (1996).

102. J. Serrin, *The initial value problem for the Navier–Stokes equation,* Nonlinear Problems, Proc. Symp. MRC Univ. Wisconsin, Madison, 69–98 (1963).

103. K. Schneider, M. Farge, *Numerical Simulation of a mixing layer in an adaptive wavelet basis,* C. R. Acad. Sci. Paris, **t. 328**, Série II, *b*, Mécanique de fluides, 263–269 (2000).

104. H. Sohr, W. von Wahl, M. Wiegner, *hur Asymptotik der i leichungen von Navier–Stokes*, Nachr. Acad. Wiss. Göttingen (1987).

105. E. M. Stein. *Singular integrals and differentiability properties of functions.* Princeton University Press, Princeton, NJ (1970).

106. E. M. Stein. *Harmonic Analysisj Real-Variable methods, orthogonality and oscillatory integrals.* Princeton University Press, Princeton, NJ (1993).

107. S. Takahashi, *A wheighted equation approch to decay rate estimates for the Navier–Stokes equations*, Nonlinear Analysis, **37**, 751–789 (1999).

108. M. E. Taylor, *Pseudodifferential operators and nonlinear PDE.* Birkhäuser (1991).

109. R. Temam, *Navier–Stokes equations. Theory and numerical analysis,* third edition, Studies in Mathematics and its applications, vol. 2, North Holland Publishing Co., Amsterdam-New York (1984).

110. R. Temam, *Some developements on Navier–Stokes equations in the second half of the kl th century,* dans "Développements des Mathématiques au cours de la seconde moitié du XXème siècle", J.P. Pier Editor (1998).

111. E. Terraneo. *Sur la non-unicité des solutions faibles de l'équation de la chaleur non linéaire avec non-linéarité u^3,* CRAS. Paris t. 328, Série I (1999) 759-762.

112. E. Terraneo, *Applications de certains espaces de l'analyse harmonique aux équations de Navier–Stokes et de la chaleur non linéaires,* Thèse, Université d'Evry–Val d'Essonne (1999).

113. H. Triebel, *Theory of Function Spaces,* Birkhäuser Verlag, Basel-Boston, (1977).

114. F. Weissler. *Local existence and nonexistence for semilinear parabolic equations in L^p.* Indiana Univ. Math. J. **29** (1980) 79-102 J. (1980) 79-102.

115. M. Wiegner, *Decay Results for Weak Solutions of the Navier–Stokes Equations on $\mathbf{R}^n$,* J. London Math. Soc., **2**, N. 35, 303–313 (1987).

116. M. Wiegner, *Decay of the L_∞-norm of solutions of the Navier-Stokes equations in unbounded domains,* Acta Appl. Math., **37**, 215–219 (1994).

117. M. Yamada, K. Ohkitani, *Orthonormal Wavelet Expansion and its Application to Turbulence,* Dicaster Prevention Research Institute Kyoto University, Uji 611, Japan ; Departement of Physics, Kyoto University, Kyoto 606, Japan.

118. A. Youssfi, *Localisation des espaces de Besov homogènes,* Indiana Univ. Math. J., **37**, 565–587 (1988).

119. V. Yudovich, *Écoulement non-stationnaire d'un liquide parfait incompressible,* Zhurn. Vich. Mat., **3**, 1032–1066 (1963) (en Russe).

120. W. P. Ziemer, *Weakly differentiable functions. Sobolev spaces and functions of bounded variation,* N. 120, Springer-Verlag (1989).

121. V.D. Zimin, *Hierarchic Model of Turbulence,* Izvestiya, Atmospheric and Oceanic Physics, **17**, N. 12, UDC 532.517.4, 941–946 (1981).

Asymptotic Analysis of Fluid Equations

Seiji Ukai

Department of Applied Mathematics
Yokohama National University
79-5 Tokiwadai, Hodogaya-ku, Yokohama 240-8501, Japan
ukai mathlab.sci.ynu.ac.φ

1 Introduction

Nonlinear partial differential equations describing the fluid motion make a long list, most of which date back to the 19th century or earlier. Among them are the Boltzmann equation, the Navier-Stokes and Euler equations, compressible and incompressible, to mention a few. The Newton equation should be also included in the list as a microscopic fluid equation which consider the fluid to be a many-particle system. On the other hand, the Navier-Stokes and Euler equations are macroscopic fluid equations regarding the fluid as a continuum, and the Boltzmann equation is in-between.

Apart from the Newton equation, they are all nonlinear partial differential equations, but belong to quite different classes from each other both in the type as partial differential equation and in the structure of nonlinearity. This has been raising a variety of challenging and interesting mathematical problems with important and fruitful results on the theory of nonlinear partial differential equations. The fluid equations have been and will remain to be a rich source for the progress of this field.

Obviously, any of the equations mentioned above can be applied to one and the same fluid. In other words, they must be closely interrelated to one another. Mathematically, this is a problem of the asymptotic analysis or of the theory of singular perturbation. Since all the equations are of different type and nature, this provides a number of highly sophisticated mathematical problems of asymptotic analysis of nonlinear partial differential equations. It should be stressed that it is only in these decades that the theory and method have been developed on the asymptotic analysis of fluid equations with mathematical rigor. However, the progress was remarkable and results are fruitful. The aim of the present article is to review this progress and results. More precisely, we will discuss the mathematical theory and method on the fluid equations and their asymptotic relations depicted in the following two figures.

: YV() *Asymptotic Analysis of Fluid Equations+, - .   . 0   +189I250 (2006)*
    "   "   ⑨ : ;) =   A=B A C)   Æ =2006

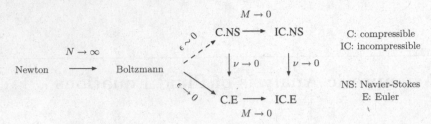

Fig. 1. N: number of particles, ϵ: mean free path, ν: viscosity, M: Mach number

The first figure is classical. In physics, the following asymptotic diagram has been believed to be valid for a long time. The explicit forms of the fluid equations mentioned in this figure will be presented in §2. The diagram given here implies that any fluid equations should originate in the Newton equation and bifurcate through various level of limits. This is in the spirit of the classical mechanics that all phenomena should obey the Newton mechanics. The arrows and physical parameters appearing in Figure 1 indicate which is the nearest ancestor of each fluid equation and in which physical regime this fluid equation is valid. For example, it says that the C.E (compressible Euler equation) is derived from the Boltzmann equation in the zero limit of the mean free path ϵ or the *Knudsen number* witch is the dimensionless distance between collisions of gas particles., and therefore that it must be valid in the physical sense in the regime of small ϵ. To prove this mathematically is our aim.

The second figure we target is a recent one. In the 1990's, there has been a remarkable progress of the multi-scale analysis which made possible to establish other kinds of links that have not been known in the classical diagram of Figure 1, that is, the direct links (not via compressible macroscopic fluid equations) between the Boltzmann equations and the incompressible macroscopic (Euler and Navier-Stokes) equations. This is indicated in Figure 2. The explicit formula of the equations appearing in this figure will be given in §2.5, together with their formal derivation.

Remark 1.1. The reader might notice that the naming in Figure 2 is somewhat inappropriate. Indeed, it is true for the equations named with 'L'. The L.C.E (linear compressible Euler equation) has the formal name 'the equation of acoustic wave' and L.IC.NS (linear incompressible Navier-Stokes equation) 'the Stokes equation', while L.IC.E (linear incompressible Euler equation) is nothing but a trivial equation describing a constant state (see Remark 2.6). Although the present naming is adopted to make explicit the original equations from which they come by linearization, it is the naming only within this article.

As stated already, we now have mathematical theories and methods, though not necessarily enough, on all the asymptotic relations indicated in Figures 1

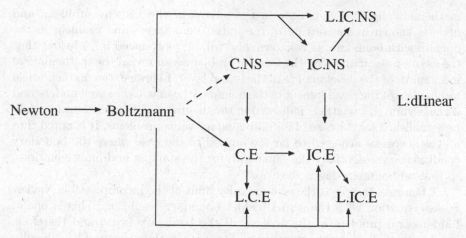

Fig. 2. Multi-scale Analysis

and 2. The aim of this article is to present some of their main ideas, but the full proof will be given only for some topics.

At this point, we mention two related open problems which are important both in mathematics and physics. Checking the above two diagrams, one might notice that there are some missing rings. They are the links which connect the Newton equation and macroscopic fluid equations directly, not via the Boltzmann equation;

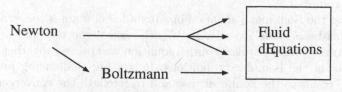

Fig. 3. Missing rings

Verifying the links in this diagram would serve, together with those in Figure 1 and 2, to establish a unified mathematical picture of a family of fluid equations. Therefore, many attempts have been done so far also on Figure 3, but without any essential success up to the present. This is now understood to be one of big open issues in the mathematical theory of fluid equations. Note that some results have been established by replacing the Newton equation by stochastic models of particles, see e.g. [30]. From the physical view point, however, deterministic models like the Newton equation seems more natural than stochastic models for the microscopic dynamical law of fluid motion.

Another big open issue is concerned with the boundary layer. As stated above, all of arrows in Figures 1 and 2 have been more or less verified mathematically. But it is only in the setting of the Cauchy problem, and little is known in the setting of the initial boundary value problem in the domain with boundary. It is known, and will be reproduced in §5 below, that the asymptotic analysis on the Cauchy problems can reveal the mathematical mechanism of the development of the initial layer. Likewise, the mathematical mechanism of the development of the boundary layer could be well understood if the asymptotic relations indicated in the diagrams of Figures 1 and 2 could be established for the case of initial-boundary value problems. It is noted that all the successes achieved so far are limited to the case where the boundary conditions are essentially the same both for the starting and limit equations, so that no boundary layers develop.

A famous example is the zero viscosity limit of the incompressible Navier-Stokes equation with the zero Dirichlet boundary condition. This is one of fundamental problem in the analysis of the boundary layer, and therefore, has been attacked by many authors, but with little success. The difficulty lies in the fact that the limit equation, the incompressible Euler equation, is well-posed under the boundary condition of zero normal component, but ill-posed (overdetermined) under the Dirichlet boundary condition which is the well-posed boundary condition for the Navier-Stokes equation. Thus, the solutions are expected to develop singularities near the boundary as the viscosity coefficient ν becomes small. This is now known as one of the most difficult and challenging mathematical problems in the theory of nonlinear partial differential equations, as well as the existence problem of smooth solutions to the Navier-Stokes equation raised as 'the Millennium Prize'. The difficulty disappears as soon as the Dirichlet boundary condition is replaced by the slip boundary condition.

As far as the Boltzmann equation in a bounded domain is concerned, some progress has been made recently. In [37], the convergence of the Boltzmann equation to the (linear) Stokes-Fourier equation was proved together with the convergence of the boundary conditions. It is a big challenging problem to extend the result to the nonlinear case and to strength the convergence so as to make visible the boundary layer. The structure of the boundary layer of the Boltzmann equation has been also studied. The results for the linear case were given in 1980's in [34], [35], [36], and recently for the nonlinear case in [38], [39]. This topics is, however, out of the scope of this article. Throughout this article, only the case of the Cauchy problem will be discussed.

Further, we will discuss only the asymptotic relations involving the Boltzmann equation. Thus, we will not discuss interrelations between macroscopic fluid equations shown in Figure 1, but most of our ideas work for the proof of their asymptotic relations as well.

It should be also mentioned that we do not discuss the compressible Navier-Stokes equation (C.NS). The classical Chapman-Enskog expansion [22] gives a link between this and the Boltzmann equations, as indicated by the dotted arrows in Figures 1 and 2, but is not an asymptotic expansion. The compressible Navier-Stokes equation is an approximate equation to the Boltzmann equation, but cannot be characterized as a reduced equation of any limit of the Boltzmann equation. In fact, the compressible Navier-Stokes equation derived by the Chapman-Enskog expansion has the viscosity coefficient and the thermal diffusivity proportional to the mean free path ϵ .

We close this introduction with our basic setting of the problem. Mathematically, our problem is stated as follows: Given a family of equations

$$A_n[u] = 0,$$

parametrized with some parameter(s) μ near a certain value μ_*, we want to show that

(i) the solution $u = u_n$ converges to a limit u_* as $\mu \to \mu_*$,

$$u_n \to u_* \quad (\mu \to \mu_*)$$

and

(ii) the limit u_* solves some equation, say,

$$A_*[u_*] = 0.$$

This limit equation is called the reduced equation.

To prove (i) and (ii), we need to establish the following steps.

(1) Existence of local or global solutions u_n for each μ near μ_*.
(2) Uniform estimates in μ near μ_* of solutions u_n, for life spans (local solutions), size of initial data (global solutions), etc.
(3) Convergence of u_n, or equivalently, existence of a limit u_*, in the limit $\mu \to \mu_*$.
(4) Derivation of the limit (reduced) equation which u_* solves.

The most essential step is the step (2). The aim of this paper is to present main mathematical ingredients for establishing the steps (1)-(4) for verifying the asymptotic relations given in the diagrams in Figures 1 and 2. In particular, it will be seen that abstract versions of the Cauchy - Kovalevskaya theorem which is the most fundamental theorem in the theory of partial differential equations provide a simple but powerful tool for the step (2) in most of cases in the diagrams of Figures 1 and 2.

The present paper is planned as follows. In the next section, we reproduce explicit forms of all the fluid equations mentioned in Figures 1 and 2, in presenting the schemes which are adopted to prove the asymptotic relations between them. Two schemes will be given. The first scheme is given in Figure 4 in §2.1, which is for establishing the convergence of the Newton equation to the Boltzmann equation stated in Figure 1. This scheme is now classic, called the Boltzmann - Grad limit.

The second scheme is for the multi-scale analysis connecting the Boltzmann equation and macroscopic fluid equations. It is in Figure 5 in §2.4, which shows that introducing different scalings into the Boltzmann equation leads to different macroscopic fluid equations.

In §3, we introduce the abstract Cauchy Kovalevskaya theorem which will be useful to establish uniform estimates stated in the step (2) above. As application, two examples will be presented. The situation we meet in these examples are those we will meet in §4 and §5. We shall substantiate the diagram of Figure 4 in §4, and that of Figure 5 in §5.

2 Schemes for Establishing Asymptotic Relations

2.1 From Newton Equation to Boltzmann Equation: Boltzmann–Grad Limit

The validity of the Boltzmann equation [1] from the view point of the Newtonian mechanics is now well-established. The scheme for proving it is illustrated in the following diagram. This diagram goes back to Grad [5] and Cercignani [2], and was proved with a mathematical rigor by Lanford [8]. This verifies Boltzmann's derivation of his equation mathematically.

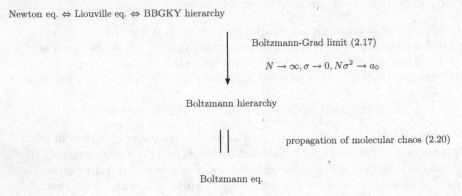

Newton eq. ⇔ Liouville eq. ⇔ BBGKY hierarchy

Boltzmann-Grad limit (2.17)

$$N \to \infty, \sigma \to 0, N\sigma^2 \to a_0$$

Boltzmann hierarchy

propagation of molecular chaos (2.20)

Boltzmann eq.

Fig. 4. Boltzmann - Grad Limit

A history and story associated with this diagram is nicely summarized in [2]. The existing proofs for this diagram are just only for two cases, i.e. for the hard spheher gas and a modified soft potential. Other cases are open big problems. In this paper, we shall discuss only the hard sphere case.

There are two keys for the mathematical verification of the link between the Newton and Boltzmann equation: The Boltzmann - Grad limit and the ansatz of propagation of molecular chaos. Figure 4 illustrates how they play central roles in this scheme. Their definitions and all the equations appearing in Figure 4 are reproduced in the below.

Newton Equation – Hard Sphere Gas

We consider a dynamical system of a finite number of hard spheres moving in the whole space according to the Newtonian mechanics without external forces. We assume that all spheres are identical.

Let N denote the number of hard spheres and σ the diameter of the ball. We assume the mass of the hard sphere $m = 1$ without loss of generality, and introduce the following notation.

$$x_i = (x_{i1}, x_{i2}, x_{i3}), \quad v_i = (v_{i1}, v_{i2}, v_{i3}) \in \mathbb{R}^3 :$$

the position and velocity of the center of the ball i,

$$z_i = (x_i, v_i) \in \mathbb{R}^6 \quad (i = 1, 2, \ldots, N), \tag{2.1}$$

$$z^{(N)} = (z_1, \ldots, z_N) \in \mathbb{R}^{6N} : \quad \text{the phase point of the hard-ball system.}$$

Then, the phase space of our system is

$$\Lambda_{N,\sigma} = \{ z^{(N)} \in \mathbb{R}^{6N} | \ |x_i - x_j| > \sigma, i \neq j \}, \tag{2.2}$$

and the Newton equation for our system is

$$\frac{dx_i}{dt} = v_i, \quad \frac{dv_i}{dt} = 0 \quad (i = 1, \ldots, N), \quad z^{(N)} \in \Lambda_{N,\sigma}. \tag{2.3}$$

To complete the description of the motion of our system, (2.3) should be supplemented with the boundary condition due to mutual collisions of hard spheres. The boundary of the phase space $\Lambda_{N,\sigma}$ is,

$$\partial \Lambda_{N,\sigma} = \{ z^{(N)} \in \mathbb{R}^{6N} | \ |x_i - x_j| = \sigma, i \neq j \}.$$

on which the collision occurs. The collision of hard spheres is elastic. The most frequent collision is the binary collision (collision of two balls), but the multiple collision (collision of more than two balls) can also occur. Here we describe the boundary condition for the binary collision.

A collision between the balls i and j is described by the relations

$$x_j = x_i - \sigma\omega, \tag{2.4}$$

$$v_i' = v_i - \big((v_i - v_j) \cdot \omega\big)\omega, \qquad v_j' = v_j + \big((v_i - v_j) \cdot \omega\big)\omega, \tag{2.5}$$

where $\cdot$ is for the inner product of $\mathbb{R}^3$, whereas $\omega \in \mathbf{S}^2$ is the unit vector in the direction from the center of the ball j to that of i. (2.4) indicates that the two balls cannot penetrate into each other. In (2.5), v_i' and v_j' are the velocities after collision. Since the collision is elastic, the conservation laws of momentum and energy hold throughout the collision:

$$\begin{cases} v_i + v_j = v_i' + v_j', & \text{(momentum)} \\ |v_i|^2 + |v_j|^2 = |v_i'|^2 + |v_j'|^2, & \text{(energy)} \end{cases} \tag{2.6}$$

Actually, any solution of (2.6) is given by (2.5) for some $\omega \in S^2$.

The description of the boundary condition due to the multiple collision is more complicated and omitted here.

(2.3) is also supplemented with the initial condition specifying the starting point of the system. Hence, the solution of (2.3) satisfying both the initial and boundary conditions thus specified describes a trajectory of our system in the phase space $\Lambda_{N,\sigma}$.

Liouville Equation

We now rewrite the Newton equation (2.3) to an equivalent equation called the Liouville equation. Its unknown is the N-particle probability density function

$$P_{N,\sigma} = P_{N,\sigma}(t, z^{(N)}). \tag{2.7}$$

Since all hard spheres are assumed to be identical, $P_{N,\sigma}$ is symmetric with respect to $z_1, \ldots, z_N$ of (2.1). The Liouville equation is given by

$$\frac{\partial P_{N,\sigma}}{\partial t} + \sum_{i=1}^{N} v_i \cdot \nabla_{x_i} P_{N,\sigma} = 0, \quad t > 0, \quad z^{(N)} \in \Lambda_{N,\sigma}, \tag{2.8}$$

where

$$v_i \cdot \nabla_{x_i} = \sum_{j=1}^{3} v_{ij} \frac{\partial}{\partial x_{ij}}, \quad x_i = (x_{i1}, x_{i2}, x_{i3}), \quad v_i = (v_{i1}, v_{i2}, v_{i3}).$$

Of course, (2.8) should be supplemented with the boundary condition on $\partial \Lambda_{N,\sigma}$ due to the elastic collisions described in the above, and also with the initial condition

$$P_{N,\sigma}(0, z^{(N)}) = P_{N,\sigma,0}(z^{(N)}). \tag{2.9}$$

The Newton equation (2.3) and the Liouville equation (2.8) are equivalent to each other in the sense that the ordinary differential equations (2.3) is the characteristic equation for the first order partial differential equation (2.8) while a solution to the above mentioned initial boundary value problem for (2.8) gives a trajectory of (2.3) if the initial data $P_{N,\sigma,0}$ is a delta function.

BBGKY Hierarchy

We introduce the k-particle marginal (joint) probability density function of $P_{N,\sigma}$. Put

$$z^{(k)} = (z_1, z_2, \cdots, z_k) \in \Lambda_{k,\sigma}, \qquad (k = 1, 2, \cdots, N).$$

Assuming $P_{N,\sigma} = 0$ for $z^{(N)} \notin \Lambda_{N,\sigma}$, the k-particle marginal probability density function is defined by

$$P_{N,\sigma}^{(k)} = P_{N,\sigma}^{(k)}(t, z^{(k)}) = \int_{\mathbb{R}^{6(N-k)}} P_{N,\sigma}(t, z_1, \cdots, z_N) dz_{k+1} \ldots dz_N. \quad (2.10)$$

By definition,

$$P_{N,\sigma}^{(N)} = P_{N,\sigma}. \quad (2.11)$$

The BBGKY hierarchy is a set of equations for $P_{N,\sigma}^{(k)}$ which are obtained by integrating (2.8) by parts with respect to $z_{k+1}, \cdots, z_N$,

$$\frac{\partial P_{N,\sigma}^{(k)}}{\partial t} + \sum_{i=1}^{k} v_i \cdot \nabla_{x_i} P_{N,\sigma}^{(k)} = (N-k)\sigma^2 Q_\sigma^{(k+1)}(P_{N,\sigma}^{(k+1)}) \quad \text{in } \Lambda_{k,\sigma}, \quad k = 1, \cdots, N.$$

$$(2.12)$$

Here, $Q_\sigma^{(k+1)}$, called the *collision operator*, is a linear integral operator defined by

$$Q_\sigma^{(k+1)}(g) = \sum_{i=1}^{k} \int_{\mathbb{R}^3 \times S^2} |(v_i - v_*) \cdot \omega|(g_i' - g_i) dv_* d\omega, \quad (2.13)$$

where $g = g(z_1, \cdots, z_k, z_{k+1})$, and with $z_i = (x_i, v_i)$,

$$g_i = g(z_1, \cdots, z_i, \cdots, z_k, x_i - \sigma\omega, v_*), \quad (2.14)$$

$$g_i' = g(z_1, \cdots, x_i, v_i', \cdots, z_k, x_i + \sigma\omega, v_*'),$$

$$v_i' = v_i - ((v_i - v_*) \cdot \omega)\omega, \qquad v_*' = v_* + ((v_i - v_*) \cdot \omega)\omega. \quad (2.15)$$

This is the equation satisfied by the k−particle distribution function as a consequence of the Liouville equation (2.8). The derivation of (2.12) is rather delicate, see [2]. The collision oerator (2.13) describes collisions between the first k particles with the remaining $N - k$ particles. Apart from notations, (2.15) is just (2.5) coming from the conservation laws (2.6). Note also that the operator $Q_\sigma^{(k+1)}$ depends on σ but not on N.

Remark 2.1. The form (2.13) is a traditional form given in terms of the gain term G and loss term L as

$$Q_\sigma^{(k+1)} = G - L,$$

so that they can lead to the same terms of the Boltzmann equation (see §2.1.5). Here G (resp. L) is the part of the integral (2.13) including g' (resp. g), and physically, it gives the rate of production (resp. loss) of particles in the k−particle system by collisions with the remainig $N - k$ particles. Note that this splitting form is obtained for $P^{(N)}$ satisfying the elastic boundary condition applied to particles after collision. If the same boundary condition is used for particles before collision, we will have the minus sign in front of Q and hence similarly for the Boltzmann equation. This fact is related to the 'ansatz of molecular chaos' and also to the time irreversibility of the Boltzmann equation. The ansatz of molecular chaos will be introduced in the next subsection (see (2.20)) while the time ireversibility will be touched in Remarks 2.2 and 4.6.

The boundary condition for (2.12) is the same as for the Liouville equation (2.8) but on the boundary $\partial \Lambda_{k,\sigma}$, which describes mutual collisions inside the $k-$ particle system. The initial condition is

$$P_{k,\sigma}(0, z^{(k)}) = P_{k,\sigma,0}(z^{(k)}). \tag{2.16}$$

Since the BBGKY hierarchy (2.12) for $k = N$ is nothing but the Liouville equation (2.8), it is clear that both equations are mutually equivalent. Whereas the Liouville equation is a single equation for each N, the number of equations composing the BBGKY hierarchy goes to ∞ with N.

Boltzmann Hierarchy

Now, we take the limit of (2.12) by letting

$$N \to \infty, \quad \sigma \to 0, \quad N\sigma^2 \to a_0, \tag{2.17}$$

for some positive constant a_0. This limit is now called the Boltzmann - Grad limit. Later, we shall show that the limit equation is an equation for the rarefied gas.

Under the last condition of (2.17), we see that $k\sigma^2 = (k/N)N\sigma^2 \to 0$ for each fixed k, and thereby, if $f^{(k)} = f^{(k)}(t, z^{(k)})$ denotes the (formal) limit of $P_{N,\sigma}^{(k)}$, the limit equation of (2.12) is

$$\frac{\partial f^{(k)}}{\partial t} + \sum_{i=1}^{k} v_i \cdot \nabla_{x_i} f^{(k)} = a_0 Q_0^{(k+1)}(f^{(k+1)}), \quad k \in \underline{\nu}. \tag{2.18}$$

Obviously, this equation is defined for each $k \in \underline{\nu}$ in the phase space $\Lambda_{k,0}$ and the boundary condition is to be imposed on the boundary $\partial \Lambda_{k,0}$ that is not empty but consists of the "diagonals $x_i = x_j$". According to [2], however, the distributions leading to pathologies due to this boundary condition is negligible (see Remark 4.4). This means that we may discard this boundary condition and consider (2.18) in the whole space $\mathbb{R}^{6k}$ instead of $\Lambda_{k,0}$. Thus, the left hand side of (2.18) describes the collision-free motion of a system of k hard spheres. (2.18) is to be supplemented by the initial condition

$$f^{(k)}(0, z^{(k)}) = f_0^{(k)}(z^{(k)}) \quad \text{in } \mathbb{R}^{6k}, \quad k \in \underline{\nu}. \tag{2.19}$$

The set of infinitely many equations (2.18) is called the Boltzmann hierarchy.

Boltzmann Equation

In order to derive his equation, Boltzmann [1] introduced an ansatz called "Propagation of Molecular Chaos". This is the hypothesis that the hard spheres are independent of each other stochastically, or more precisely, that the marginal probability density $f^{(k)}$ can be factorized with the 1-particle density $f^{(1)}$. Thus, writing $f = f^{(1)}$, we assume that

$$f^{(k)}(t, z^k) = \prod_{i=1}^{k} f(t, x_i, v_i), \qquad (2.20)$$

holds for all $k \in \underline{\nu}$". Under this hypothesis, it is easy to see that any of equations (2.18) reduces to one and the same nonlinear equation for f. It is the Boltzmann equation;

$$\frac{\partial f}{\partial t} + v \cdot \nabla_x f = \frac{1}{\epsilon} Q(f), \quad t > 0, \quad (x, v) \in \mathbb{R}^3 \times \mathbb{R}^3, \qquad (2.21)$$

where Q is the nonlinear *collision operator* defined by

$$Q(f) = \int_{\mathbb{R}^3 \times S^2} |(v - v_*) \cdot \omega|(f' f'_* - f f_*) dv_* d\omega, \qquad (2.22)$$

$$f = f(t, x, v), \quad f' = f(t, x, v'), \quad f'_* = f(t, x, v'_*), \quad f_* = f(t, x, v_*),$$

$$v' = v - ((v - v_*) \cdot \omega)\omega, \quad v'_* = v_* + ((v - v_*) \cdot \omega)\omega,$$

while

$$\epsilon = \frac{1}{a_0} \qquad (2.23)$$

is a number proportional to the *mean free path (mean free time)* of gas particles between collisions or *Knudsen number* that represents the ratio of the mean free path to some characteristic length of the domain containig the gas. In Figure 5 in the next subsection, this number plays an essential role as the parameter linking the Boltzmann equation to the macroscopic fluid equations.

The initial condition for (2.21) is

$$f(0, x, v) = f_0(x, v), \qquad (x, v) \in \mathbb{R}^6. \qquad (2.24)$$

Boltzmann's original derivation of (2.21) is based on the physical intuition but uses both the limit (2.17) and the ansatz (2.20) though just for $k = 2$. In order to complete the mathematical justification, we need (2.20) for all k.

The limit (2.17) means that the Boltzmann equation (2.21) is an equation of motion of a rarefied gas. For, $N\sigma^2$ is a multiple of the total surface area of our N hard spheres and hence proportional to the total collision cross section, while $N\sigma^3$ is a multiple of the total volume or mass of our hard spheres. Hence, under the limit (2.17), our hard sphere system tends to a gas with zero volume (mass) but with effective collisions. Such a gas is called the rarefied gas.

The fluid, on the other hand, must have non-zero volume and mass, and hence, the diagram of Figure 3 indicating the direct link between the Newton equation and macroscopic fluid equation should be substantiated under the assumption

$$N\sigma^3 \to b_0,$$

for some constant $b_0 > 0$ which is a multiple of the total volume or mass of the limit system. However, this means $N\sigma^2 \to \infty$ and thus the left hand side of (2.12) apparently diverges. This is a basic difficulty in proving the diagram of Figure 3.

Collision Operator Q

The multi-scale analysis shown in Figure 5 below establishes the links between the Boltzmann equation and various macroscopic fluid equations. The mathematical mechanism which makes this possible relies of course on properties of the collision operator Q. Most of them were deduced by Boltzmann himself. Here, we present three of them. Define the inner product

$$< f, g > = \int_{\mathbb{R}^3} f(v)g(v)dv \tag{2.25}$$

A function $\varphi(v)$ is said a *collision invariant* if

$$< \varphi, Q[g] > = 0 \quad (\forall g \in C_0^\infty(\mathbb{R}_v^3, \mathbb{R}+)). \tag{2.26}$$

The first property of Q is

[Q1] *Q has five collision invariants,*

$$\varphi_0(v) = 1, \quad \varphi_i(v) = v_i \ (i = 1, 2, 3), \quad \varphi_4(v) = \frac{1}{2}|v|^2. \tag{2.27}$$

This leads to the conservation laws of the Boltzmann equation (2.21). First, if $f(t, x, v)$ is a 1-particle probability density function of a gas in the phase space of x and v at time t, then its moments with respect to v give rise to corresponding macroscopic fluid quantities;

$$\begin{aligned}
\rho &= < \varphi_0, f(t, x, \cdot) >, \\
\rho u_i &= < \varphi_i, f(t, x, \cdot) > \ (i = 1, 2, 3), \\
\rho E &= < \varphi_4, f(t, x, \cdot) >,
\end{aligned} \tag{2.28}$$

where $\rho, u = (u_1, u_2, u_3)$ and E are functions of (t, x) and give the macroscopic mass density, bulk velocity, and energy density of the gas (fluid), respectively. The temperature T and the pressure p are related with E by

$$E = \frac{1}{2}|u|^2 + \frac{3}{2}T. \qquad p = R\rho T, \tag{2.29}$$

with R being the gas constant (the Boltzmann constant divided by the mass of the hard sphere). Throughout this note, we take $R = 1$. The last equation of (2.29) is called the equation of state for the case of the hard sphere (monoatomic) gas.

Now, let f be a smooth solution to (2.21) which vanishes sufficiently rapidly with (x, v). Then, by virtue of (2.27) and by integration by parts, we see that the quantities

$$\int_{\mathbb{R}^6} \varphi_i(v) f(t, x, v) dx dv, \quad i = 0, 1, 2, 3, 4, \tag{2.30}$$

do not change with time t. In view of (2.28), these are the conservation laws of total mass ($i = 0$), total momenta ($i = 1, 2, 3$) and total energy ($i = 4$) of the gas.

The second property to be mentioned is

[**Q2**] $< \log g, Q[g] > \le 0$ $(\forall g \in C_0^\infty(\mathbb{R}_v^3, \mathbb{R}+))$.

Let f be a solution to (2.21) with the same property as above. It is nonnegative since it is a density function, and hence we can define the *H-function*,

$$H = H(t) = \int_{\mathbb{R}^6} f \log f dx dv, \tag{2.31}$$

which is believed to give the minus of the entropy of the gas. Multiply (2.21) by $\log f$ and integrate over $\mathbb{R}^3$ by parts. By virtue of [Q2], Boltzmann himself found that

$$\frac{dH}{dt} \le 0, \tag{2.32}$$

holds and that the equality follows if and only if f is a Maxwellian. This is the celebrated *H-theorem*. Boltzmann then declared that he constructed the foundation of the second law of the thermodynamics, that is, the law of entropy, based on the classical mechanics. Although lots of objections were raised by his contemporaries, the controversies motivated the later development of various ergodic theorems and finally, Lanford [8] endorsed him 103 years later.

Remark 2.2. The law of entropy contradicts to the Newton mechanics because the mechanical law is time reversible. This time reversibility is phrased for our N-particle system as follows. If we reverse the velocities of all particles at time $t = t_0$ and follow their evolution, we find that all the particles return to their original positions (the positions at $t = 0$) with the reverse velocities after a lapse of time t_0. This would be rephrased for the Boltmann equation as follows. Given a solution $f(x, v, t)$, solve the Boltzmann equation with $f(x, -v, t_0)$ as the initial data. Denote this solution by $g(x, v, t)$. Then, $g(x, v, t_0) = f(x, -v, 0)$ would hold. Of course, this is false because it contradicts to the H theorem applied to both f and g:

$$H(f(0)) = H(g(t_0)) < H(g(0)) = H(f(t_0)) < H(f(0)).$$

as long as f, g are not Maxwellian. As we have shown above (formally), the Newton equation converges to the Boltmann equation, but the time reversibility is lost. This was the main point of the controversy mentioned above. It is not a surprise that Lanford's theorem gives a mathematical reasoning of this emergency of time irreversibility, see Remark 4.6.

Note from Remark 2.1 that if the ansatz of the molecular chaos is applied to the particles before collision, the time reversibility is lost backward in time.

The final property of Q is

[Q3] $Q[g] = 0 \Leftrightarrow <\log g, Q[g]> = 0 \Leftrightarrow g = M(v)$, *where*

$$M(v) = M[\rho, u, T](v) = \frac{\rho}{(2\pi RT)^{3/2}} \exp\left(-\frac{|v - u|^2}{2RT}\right) \quad (R = 1). \quad (2.33)$$

$M(v)$ is called a *Maxwellian* and is known to describe the velocity distribution of a gas in an equilibrium state with the mass density $\rho > 0$, bulk velocity $u = (u_1, u_2, u_3) \in \mathbb{R}^3$, and temperature $T > 0$. Here, (ρ, u, T) are taken to be parameters, and if they are constants, M is called a global (absolute) Maxwellian while if they are functions of (x, t), it is called a local Maxwellian. Evidently, the global Maxwellian is a stationary solution of (2.21).

2.2 From Boltzmann Equation to Fluid Equations – Multi-Scale Analysis

The macroscopic limits of the Boltzmann equation are obtained when the gas becomes dense enough that particles undergo sufficiently many collisions and that the gas becomes locally in equilibrium, which is one of the basic properties of the fluid. This situation is realized when the Knudsen number ϵ defined by (2.23) becomes sufficiently small. In fact, in this regime of ϵ, the solution $f = f^\epsilon$ of the Boltzmann equation can is expected to be close to a local Maxwellian which presents a local equilibrium state. For, writing (2.21) as

$$\epsilon\left(\partial_t f^\epsilon + v \cdot \nabla_x f^\epsilon\right) = Q(f^\epsilon), \quad (2.34)$$

and assuming that f^ϵ are sufficiently smooth and converge to a limit, say f^0, in a sufficiently strong norm, one reaches $0 = Q(f^0)$ in the limit $\epsilon \to 0$. By [Q3], then, the limit must be a Maxwellian, namely,

$$f^0 = M[\rho, u, T], \quad (2.35)$$

for some (ρ, u, T) which may be functions of t and x.

In order to obtain the incompressible limit suggested in the diagram of Figure 2, we shall further look at the case where this local Maxwellian is sufficiently close to an absolute Maxwellian in the whole space. One reason is that, as noticed in [14], if f^ϵ vanishes at $x = \infty$ (a gas in vacuum), all gas particles eventually will ran away (disappear) to infinity and any macroscopic (fluid dynamical) behaviors will not be observed. Another reason is that in view of the relation

$$\epsilon = \frac{Ma}{Re},$$

where Ma is the Mach number (ratio of the bulk velocity to the sound speed) and Re is the Reynolds number (dimensionless reciprocal viscosity), the macroscopic limit with a finite, non-zero Reynolds number will follow only when the Mach number vanishes. And the small Mach number is realized if f^ϵ is close to an absolute Maxwellian taken in a proper choice of the Galilean frame and dimensional units. Without loss of generality, we take the standard Maxwellian

$$M_0(v) = M[1,0,1](v) = \frac{1}{(2\pi)^{3/2}} \exp\left(-\frac{|v|^2}{2}\right). \qquad (2.36)$$

The distance to this abolute Maxwellian will be scaled in the unit of the Knudsen number ϵ as

$$f^\epsilon(t,x,v) = M_0(v) + \epsilon^\beta M_0^{1/2}(v) g^\epsilon(t',x,v), \qquad (2.37)$$

where $\beta \geq 0$ is the strength parameter of the scaling. It is easy to see that this ensures a infinitesimal small Mach number and non-zero Reynolds number for infinitesimally small ϵ, if $\beta \neq 0$.

The variable t' in g^ϵ above is the scaled time variable defined by

$$t = \frac{1}{\epsilon^\alpha} t', \qquad (2.38)$$

with the strength parameter $\alpha \geq 0$. This time scaling which is related to the Strouhal number (ratio of the oscillation frequency to the bulk velocity) is introduced to suppress, when $\alpha \neq 0$, the acoustic modes varying in a faster timescale than rotational modes of the fluid.

Varying the strengths of these two scalings yields different limits as $\epsilon \to 0$. Plug the scalings (2.37) and (2.38) into the Boltzmann equation and drop $'$ from t', to deduce the governing equation of the new unknown g^ϵ:

$$\epsilon^\alpha \frac{\partial g^\epsilon}{\partial t} + v \cdot \nabla_x g^\epsilon = \frac{1}{\epsilon} L g^\epsilon + \frac{1}{\epsilon^{1-\beta}} \Gamma(g^\epsilon), \qquad (2.39)$$

where L and Γ are defined by

$$Lg = 2M_0^{-1/2} Q(M_0, M_0^{1/2} g), \qquad (2.40)$$

$$\Gamma(g) = M_0^{-1/2} Q(M_0^{1/2} g, M_0^{1/2} g), \qquad (2.41)$$

$Q(f,g)$ being the bilinear symmetric operator associated with the quadratic operator (2.22) and defined by (3.28) in §3. L is the linearized operator of Q around the standard Maxwellian M_0 and hence called the linearized collision operator, while Γ is its remainder and is a quadratic operator.

The initial condition to be considered is

$$g^\epsilon|_{t=0} = g_0^\epsilon(x,v), \qquad (2.42)$$

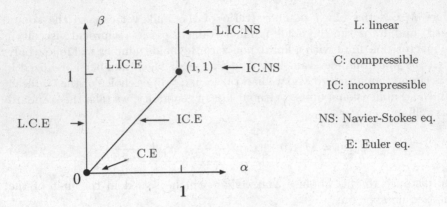

Fig. 5. Multi-scale Analysis

but for simplicity, we deal only with the case where initials are independent of ϵ, $g_0^\epsilon = g_0$. Obviously, the same conclusions will follow if $g_0^\epsilon \to g_0$ as $\epsilon \to 0$ in a sufficiently strong norm. The aim of this subsection is to present that different choices of the scaling parameters (α, β) in (2.39) invoke different fluid equations in the limit $\epsilon \to 0$. This is the multi-scale analysis of the Boltzmann equation and Figure 5.4 indicates its main results. The idea goes back to [18] and [23], where the formal derivations of the various limit equations indicated in Figure 5 are also nicely presented. The rest of this subsection is devoted to reproduce it.

The Case $(\alpha, \beta) = (0, 0)$: Compressible Euler Equation (C.E.)

This case corresponds to the famous Hilbert expansion. More precisely, we will show

Theorem 2.1. *Assume that f^ϵ be smooth enough and converge to a limit, say f^0, in a sufficiently strong norm. Then, when $\alpha = \beta = 0$, the limit is a local Maxwellian*

$$f^0 = M[\rho, u, T], \tag{2.43}$$

and the fluid quantities (ρ, u, T) are governed by the compressible Euler equation

$$\begin{cases} \dfrac{\partial \rho}{\partial t} + \operatorname{div}(\rho u) = 0, \\ \dfrac{\partial (\rho u)}{\partial t} + \operatorname{div}(\rho u \otimes u) + \nabla p = 0, \\ \dfrac{\partial (\rho E)}{\partial t} + \operatorname{div}((\rho E + p)u) = 0, \end{cases} \tag{2.44}$$

where

$$E = \frac{1}{2}|u|^2 + \frac{3}{2}T, \qquad p = R\rho T \qquad (R = 1),$$

E being the energy density and p the pressure, and

$$\nabla = (\frac{\partial}{\partial x_1}, \frac{\partial}{\partial x_2}, \frac{\partial}{\partial x_3}), \qquad \text{div } u = \nabla \cdot u = \sum_{i=1}^{3} \frac{\partial u_i}{\partial x_i}, \quad u \otimes u = (u_i u_j)_{i,j=1,2,3}.$$

Proof. (2.43) was already proved in the begining of this subsection. Take the inner product $<,>$ of (2.34) with the collision invariants φ_j of (2.27), to deduce, by the help of [Q1],

$$\frac{\partial}{\partial t} < \varphi_i, f^\epsilon > + \text{div} \ < v\varphi_i, f^\epsilon >= 0, \quad i = 0, 1, 2, 3, 4,$$

for any $\epsilon > 0$, which becomes in the limit $\epsilon \to 0$,

$$\frac{\partial}{\partial t} < \varphi_i, M[\rho, u, T] > + \text{div} \ < v\varphi_i, M[\rho, u, T] >= 0.$$

Computing these inner products explicitly by (2.33) leads to (2.44).

The Case α $0, \beta = 0$

This is the case where only the time variable is scaled. Consequently, the above argument applies exactly in the same way with the only replacement of ∂_t by $\epsilon^\alpha \partial_t$, and so, the limit equation is (2.44) with all the time derivatives removed, that is, the stationary compressible Euler equation.

The Case $\alpha \geq 0, \beta$ 0

Put

$$\begin{cases} \psi_0(v) = M_0^{1/2}(v), \\ \psi_i(v) = v_i M_0^{1/2}(v) \quad (i = 1, 2, 3), \\ \psi_4(v) = \frac{1}{2}(|v|^2 - 3)M_0^{1/2}(v). \end{cases} \qquad (2.45)$$

and define

$$\mathcal{N} = \text{span}\Big\{\psi_0, \cdots, \psi_4\Big\}. \qquad (2.46)$$

(2.45) is an L^2-orthogonal (ψ_4 not being normalized) basis of the space $\mathcal{N}$ spanned by the collision invariants (2.27) weighted by $M_0^{1/2}$.

Theorem 2.2. *Suppose that the solutions g^ϵ be smooth enough and converge to a limit, say g^0, in a sufficiently strong norm. If $\beta > 0$, it follows that $g^0 \in \mathcal{N}$. Write this in the form*

$$g^0 = \eta\psi_0 + \sum_{i=1}^{3} u_i\psi_i + \theta\psi_4. \qquad (2.47)$$

Depending on the choice of (α, β), the coefficients $\eta, u = (u_1, u_2, u_3), \theta$, being functions of t and x solve one of the macroscopic fluid equations specified in Figure 5, as follows.

1. **The case $\alpha = 0, \beta > 0$, Linear Compressible Euler Equation (L.C.E):**

$$
\begin{cases}
\dfrac{\partial \eta}{\partial t} + \operatorname{div} u \quad = 0, \\
\dfrac{\partial u}{\partial t} + \nabla(\eta + \theta) = 0, \\
\dfrac{3}{2}\dfrac{\partial \theta}{\partial t} + \operatorname{div} u \quad = 0.
\end{cases}
\tag{2.48}
$$

2. **The case $\alpha > 0, \beta \geq 0$.**
   In this case, u always satisfies the divergence-free condition,

$$
\operatorname{div} u = 0,
\tag{2.49}
$$

   while η, θ always satisfy the Boussinesque relation,

$$
\nabla(\eta + \theta) = 0.
\tag{2.50}
$$

   On the other hand, the evolution equations they satisfy differ with the values of α, β:

   a) **The case $0 < \alpha = \beta < 1$, Incompressible Euler Equation (IC.E):**

$$
\begin{cases}
\dfrac{\partial u}{\partial t} + u \cdot \nabla u + \nabla p = 0, \\
\dfrac{\partial \theta}{\partial t} + u \cdot \nabla \theta = 0.
\end{cases}
\tag{2.51}
$$

   Here, p is the pressure determined through the divergence-free condition (2.49) as is known in the classical theory of Navier-Stokes equation, and

$$
u \cdot \nabla = \sum_{i=1}^{3} u_i \frac{\partial}{\partial x_i}.
$$

   b) **The case $0 < \alpha < 1$, $\alpha < \beta$, Linear Incompressible Euler Equation (L.IC.E):**

$$
\begin{cases}
\dfrac{\partial u}{\partial t} + \nabla p = 0, \\
\dfrac{\partial \theta}{\partial t} = 0,
\end{cases}
\tag{2.52}
$$

   p being the same as above.
   c) **The case $(\alpha, \beta) = (1, 1)$, Incompressible Navier-Stokes equation (IC.NS):**

$$\begin{cases} \dfrac{\partial u}{\partial t} + u \cdot \nabla u + \nabla p = \nu \triangle u, \\ \dfrac{\partial \theta}{\partial t} + u \cdot \nabla \theta = \mu \triangle \theta. \end{cases} \tag{2.53}$$

Here, p is the pressure as above while ν (viscosity) and μ (thermal diffusivity) are the positive constants defined by (2.62) in the below. Further,

$$\triangle = \nabla \cdot \nabla = \frac{\partial^2}{\partial x_1^2} + \frac{\partial^2}{\partial x_2^2} + \frac{\partial^2}{\partial x_3^2}$$

is the Laplacian.

d) **The case** $\alpha = 1$, $\beta > 1$, **Linear Incompressible Navier-Stokes Equation (L.IC.NS):**

$$\begin{cases} \dfrac{\partial u}{\partial t} + \nabla p = \nu \triangle u \quad \text{div } u = 0, \\ \dfrac{\partial \theta}{\partial t} = \mu \triangle \theta, \end{cases} \tag{2.54}$$

where μ and ν are the same constants as above.

e) **The cases** $0 < \beta < \alpha$: $u = 0$, and either $\eta = \theta = 0$ or (2.50) remains nontrivial..

f) **The cases** $\alpha > 1, \beta > 0$: $g^0 = 0$.

Remark 2.3. The formula on the right hand side of (2.47) is called the infinitesimal Maxwellian because it gives the infinitesimal fluctuation of Maxwellian $M[\rho, u, T]$ from the standard Maxwellian $M[1.0.1]$. This is seen from

$$\frac{d}{ds} M[1 + s\eta, su, 1 + s\theta]|_{s=0} = g_0 M_0^{1/2},$$

where s may be taken to be ϵ^β in the context of the scaling (2.37).

Remark 2.4. Suppose that g^0 be a 'smooth' solution to the Cauchy problem of (2.39) with the initial condition (2.42), and that the initial data g_0 be such that $g_0 \notin \mathcal{N}$. Then, the above theorem says that the convergence fails at $t = 0$ and hence cannot be a uniform convergence nearby. This explains how the initial layer occurs. Note that if the solution is not 'smooth', the initial layer is not visible.

Remark 2.5. It is easy to check that (2.48) (L.C.E) is nothing but the linearized equation of (2.44) around the stationary state $(\rho, u, T) = (1, 0, 1)$. (2.48) is called the acoustic equation and is reduced to the classical wave equation: For $\varphi = \eta + \theta$ or $\varphi = \text{div} u$,

$$\frac{\partial^2 \varphi}{\partial t^2} - c^2 \triangle \varphi = 0, \tag{2.55}$$

where

$$c = \sqrt{5RT/3} \qquad (T = R = 1) \tag{2.56}$$

is the sound speed of the mono-atomic (hard sphere) gas in equilibrium with the Maxwellian M_0.

Remark 2.6. Due to the classical orthogonality of solenoidal vectors against gradients of scalar functions, (2.52) is a trivial equation splitting into

$$\frac{\partial u}{\partial t} = 0, \quad \nabla p = 0, \quad \frac{\partial \theta}{\partial t} = 0,$$

showing that u, θ and hence also η due to (2.50), are all constant in t. However, it deserves its name, because it coincides with the linearized equation of ICE around $u = 0$.

The proof of Theorem 2.2 requires some classical properties of the operator L and Γ of (2.40). See for proof, e.g. [13], [2].

Proposition 2.1.

(i) *L is non positive self adjoint on the space L_v^2 and its null space is the space $\mathcal{N}$ of (2.46).*

(ii) *Let*

$$P^0 \; : \; L_v^2 \to \mathcal{N} \tag{2.57}$$

be the orthogonal projection. Then,

$$P^0 L g = 0, \qquad P^0 \Gamma(g) = 0, \tag{2.58}$$

for any g sufficiently smooth.

(iii) *For any $g^0 \in \mathcal{N}$, it holds that*

$$\Gamma(g^0) = -L(g^0)^2. \tag{2.59}$$

An elegant proof of (iii) is found in [18], but a direct computation with the expression (2.47) works as well.

A simple consequence of (i) is that the inverse L^{-1} exists and is a bounded negative definite self adjoint operator on the space $\mathcal{N}^\perp$. On the other hand, it is easy to check that the *Burnett* functions defined by

$$A(v) = (A_i(v)) = \frac{1}{2}\psi_0(v)(|v|^2 - 5)v, \quad B(v) = (B_{i,j}) = \psi_0(v)\left(v \otimes v - \frac{1}{3}|v|^2 I\right), \tag{2.60}$$

I being the 3-dimensional unit matrix, are in $\mathcal{N}^\perp$ componentwise, so that the functions

$$A' = L^{-1}A, \qquad B' = L^{-1}B, \tag{2.61}$$

are well-defined and belong to $\mathcal{N}^\perp$. It is known ([18]) that

Proposition 2.2.

(i) *There exist positive functions $a(r)$ and $b(r)$ on $[0, \infty)$ such that*

$$A'(v) = -a(|v|)A(v), \qquad B'(v) = -b(|v|)B(v).$$

(ii) *The following properties hold.*
  - $- < A_i, A_i' >$ *is positive and independent of i.*
  - $< A_i, A_j' >= 0$ *for any $i \neq j$.*
  - $< A_i, B_{j,k}' >= 0$ *for any i, j, k.*
  - $< B_{i,j}, B_{k,\ell}' >=< B_{k,\ell}, B_{i,j}' >=< B_{j,i}, B_{k,\ell}' >$ *holds and is independent of i, j for any fixed k, ℓ.*
  - $- < B_{i,j}, B_{i,j}' >$ *is positive and independent of i, j when $(i \neq j)$.*
  - $- < B_{i,i}, B_{j,j}' >$ *is positive and independent of i, j when $(i \neq j)$.*
  - $- < B_{i,i}, B_{i,i}' >$ *is positive and independent of i.*
  - $< B_{i,j}, B_{k,\ell}' >= 0$ *unless $(i, j) = (k, \ell), (\ell, k)$.*
  - $< B_{i,i}, B_{i,i}' > - < B_{i,i}, B_{j,j}' >= 2 < B_{i,j}, B_{i,j}' >$ *holds for any $i \neq j$*

The proof of (ii) follows from (i) and symmetry. A tedious computation is required to prove the last relation.

The thermal diffusivity and viscosity coefficient appearing in (2.53) and (2.54) are given by

$$\mu = - < A_i, A_i' >= \frac{1}{6\sqrt{2\pi}} \int_0^\infty a(r)(r^2 - 5)^2 r^4 e^{-r^2/2} dr, \tag{2.62}$$

$$\nu = - < B_{i,j}, B_{i,j}' >= \frac{1}{15\sqrt{2\pi}} \int_0^\infty b(r) r^6 e^{-r^2/2} dr \quad (i \neq j), \tag{2.63}$$

respectively. See also the proof of Theorem 5.8 in §5.

Proof of Theorem 2.2.

1. Proof of (2.47): Write (2.39) as

$$\epsilon \left(\epsilon^\alpha \frac{\partial g^\epsilon}{\partial t} + v \cdot \nabla_x g^\epsilon \right) = Lg^\epsilon + \epsilon^\beta \Gamma(g^\epsilon). \tag{2.64}$$

Going to the limit under the assumption of our theorem, we have $Lg = 0$ if $\beta > 0$, whence (2.47) follows from Proposition 2.1.

2. Proof of (2.48). Recall (ii) of Proposition 2.1. Then the inner product of (2.64) with ψ_j gives ($\alpha = 0$),

$$\frac{\partial}{\partial t} < g^\epsilon, \psi_j > + \nabla_x \cdot < vg^\epsilon, \psi_j >= 0 \qquad (j = 0, \cdots, 4). \tag{2.65}$$

Clearly, the limit equation for g^0 has the same form. Computing the appearing inner products explicitly with the expression (2.47) leads to (2.48).

3. Proof of (2.49) and (2.50). In this case, we obtain, instead of (2.65),

$$\epsilon^\alpha \frac{\partial}{\partial t} < g^\epsilon, \psi_j > + \nabla_x \cdot < vg, \psi_j >= 0 \qquad (j = 0, \cdots, 4), \qquad (2.66)$$

and since $\alpha > 0$, we have in the limit,

$$\nabla_x \cdot < vg^0, \psi_j >= 0 \qquad (j = 0, \cdots, 4).$$

(2.49) follows from $j = 0, 4$ (the same equation), and (2.50) from $j = 1, 2, 3$.

4. Proof of (a)-(f) of Theorem (2.2). It is convenient to rewrite the limit g^0 as

$$g^0 = \theta \Psi_0 + u \cdot \Psi_1,$$

where (2.50) was used and

$$\Psi_0 \equiv (\psi_4 - \psi_0) = \frac{1}{2}(|v|^2 - 5)\psi_0, \qquad \Psi_1 \equiv (\psi_1, \psi_2.\psi_3) = v\psi_0.$$

Observe that

$$\Psi_0 v = A(v), \qquad v \otimes \Psi_1 = B(v) + \frac{1}{3}|v|^2 \psi_0 I,$$

where A, B are the Burnett functions, and

$$< \Psi_0, \Psi_0 >= \frac{5}{2}, \qquad < \Psi_1, \Psi_1 >= I, \qquad < \Psi_0, \Psi_1 >= 0,$$

component wise. Write (2.66) in the form

$$\begin{aligned} \frac{\partial}{\partial t} < g^\epsilon, \Psi_0 > + \frac{1}{\epsilon^\alpha} \nabla_x \cdot < g^\epsilon, A >= 0, \\ \frac{\partial}{\partial t} < g^\epsilon, \Psi_1 > + \frac{1}{\epsilon^\alpha} \nabla_x \cdot < g^\epsilon, B > + \nabla_x p^\epsilon = 0, \end{aligned} \qquad (2.67)$$

where

$$p^\epsilon = \frac{1}{3\epsilon^\alpha}|v|^2 g^\epsilon.$$

The term $\nabla_x p^\epsilon$ is not dangerous because it is eliminated upon projecting the equation onto the divergence free space. Also, the time derivatives in (2.67) are not dangerous under the assumption of our theorem, as the direct computation leads to

$$\partial_t < g^\epsilon, \Psi_0 > \rightarrow \partial_t < g^0, \Psi_0 >= \partial_t < \theta \Psi_0 + u \cdot \Psi_1, \Psi_0 >= \frac{5}{2}\partial_t \theta,$$

$$\partial_t < g^\epsilon, \Psi_1 > \rightarrow \partial_t < g^0, \Psi_1 >= \partial_t < \theta \Psi_0 + u \cdot \Psi_1, \Psi_1 >= \partial_t u.$$

The two remaining terms are difficult terms because of the presence of the diverging factor $\epsilon^{-\alpha}$. To control it, (2.64) is used once again, but in this time, to obtain the expression for Lg. Thus, recalling $A' = L^{-1}A$, $B' = L^{-1}B$ of (2.61), we have,

$$\epsilon^{-\alpha} < g^\epsilon, A > = \epsilon^{-\alpha} < Lg^\epsilon, L^{-1}A >$$
$$= \epsilon < \partial_t g^\epsilon, A' > + \epsilon^{1-\alpha} < v \cdot \nabla_x g^\epsilon, A' > - \epsilon^{\beta-\alpha} < \Gamma(g^\epsilon), A' >,$$
$$\epsilon^{-\alpha} < g^\epsilon, B > = \epsilon^{-\alpha} < Lg^\epsilon, L^{-1}B >$$
$$= \epsilon < \partial_t g^\epsilon, B' > + \epsilon^{1-\alpha} < v \cdot \nabla_x g^\epsilon, B' > - \epsilon^{\beta-\alpha} < \Gamma(g^\epsilon), B' > .$$

Go to the limit $\epsilon \to 0$. The first terms on the last expressions vanish under our assumption. Similarly, the second term vanishes if $\alpha < 1$ and the third term if $\beta > \alpha$. This proves (2.52).

There remains the case $\alpha \geq 1$. We shall compute the limit of the inner products appearing in the relevant terms. One is, using the summention convention,

$$\nabla \cdot < v \cdot \nabla g^0, A' > = \nabla \cdot < v \cdot \nabla(\theta \Psi_0 + u \cdot \Psi_1), A' >$$
$$= \partial_i \partial_j \theta < A_j, A'_i > + \partial_i \partial_j u_k < B_{jk}, A'_i >$$
$$= -\partial_i \partial_j \theta(\mu \delta_{ij})$$
$$= -\mu \triangle \theta,$$

where Proposition 2.2 (ii) was used and $\partial_j = \partial/\partial x_j$. Similarly, for each fixed k,

$$\nabla \cdot < v \otimes \nabla g^0, B'_k > = \nabla \cdot < v \cdot \nabla(\theta \Psi_0 + u \cdot \Psi_1), B'_k >$$
$$= \partial_i \partial_j \theta < A_j, B'_{ik} > + \partial_i \partial_j u_m < B_{jm}, B'_{ik} >$$
$$= \sum_i \partial_i \left(\sum_{j=m} \partial_j u_m < B_{jm}, B'_{ik} > + \sum_j \partial_j u_j < B_{jj}, B'_{ik} > \right)$$
$$= \sum_{i=k} \partial_i(\partial_i u_k + \partial_k u_i) < B_{ik}, B'_{ik} >$$
$$= -\mu \triangle u_k.$$

Here, the divergence free condition (2.49) was used twice as well as Proposition 2.2 (ii).

Now, appealing to Proposition 2.1(iii), we get

$$< \Gamma(g^\epsilon), A' > \to < \Gamma(g^0), A' > = - < L(g^0)^2, A' > = - < (g^0)^2, A > = -5\theta u,$$
$$< \Gamma(g^\epsilon), B' > \to < \Gamma(g^0), B' > = - < L(g^0)^2, B' > = - < (g^0)^2, B > = -B(u),$$

and in view of the divergence free condition again,

$$\nabla \cdot (\theta u) = u \cdot \nabla \theta, \quad \nabla \cdot B(u) = u \cdot \nabla u - \nabla \frac{|u|^2}{3}.$$

The very last term is to be absorbed into the pressure term ∇p. We have now finished the proof of the theorem for the case $\alpha = 1 \leq \beta$.

On the other hand, when $\beta < \alpha < 1$, it is easy to check that these terms should vanish, so that $u = 0$, but $\nabla\theta$ may be or or may not be zero. This proves (e).

When $\alpha > 1$, the factor $\epsilon^{1-\alpha}$ diverges. However, we have again $g^0 = 0$. To see this, multiply (2.67) by $\epsilon^{\alpha-1}$ and then go to the limit. Applying the above results on various limit forms, one obtain whenever $\beta > 0$,

$$\triangle\theta = 0, \qquad \triangle u = 0.$$

Hence, if θ, u are in a Soblev space, say H_x^1, they must vanish identically. This is (f), and the proof of Theorem 2.2 is now complete.

3 Abstract Cauchy-Kovalevskaya Theorem

The main step for establishing the asymptotic relations discussed so far is the step 2 of §1, the uniform estimates for relevant solutions. In some parts of the diagrams in Figures 4 and 5, a difficulty is the built-in infinitude: In the Boltzmann-Grad limit of Figure 4, infinitely many equations come from the BBGKY hierarchy (2.12), whereas if $\beta > \alpha$ in Figure 5, the factor $\epsilon^{\alpha-\beta}$ appearing in (2.39) goes to infinity as ϵ tends to 0.

A very simple but powerful tool to overcome this difficulty is the abstract Cauchy-Kovalevskaya Theorem. The classical Cauchy-Kovalevskaya theorem, which is the most fundamental theorem in the theory of PDE's, provides the local existence and uniform estimate of solutions for general first order partial differential equations. Likewise, the abstract version does the same but for more general abstract evolution equations of the form

$$\begin{cases} \dfrac{du}{dt} = A[u], & t > 0, \\ u(+0) = u_0, \end{cases} \tag{3.1}$$

where A is a nonlinear 'unbounded' map. Actually, we will meet this situation repeatedly in the sequel.

The main ingredient of the abstract Cauchy-Kovalevskaya theorem is to introduce a suitable Banach scale and to take advantage of the smoothing effect of the integration in t appearing in the integral form of (3.1),

$$u(t) = u_0 + \int_0^t A[u(s)]ds. \tag{3.2}$$

Various versions of the abstract Cauchy-Kovalevskaya theorem have been developed by many authors, see e.g. [31], [27], [28]. The setting given below is a rather simple one. Let $\rho_0 > 0$ and put $I_0 = [0, \rho_0]$. Consider a one-parameter family of Banach spaces $\{(X_\rho, ||\cdot||_\rho)\}_{\rho\in I_0}$, and assume the following for the evolution equation (3.1).

(X) For any $\rho, \rho' \in I_0$ with $\rho' < \rho$, it holds that

$$X_\rho \subset X_{\rho'}, \qquad ||\cdot||_{\rho'} \leq ||\cdot||_\rho.$$

(A) A is well-defined as a map from X_ρ into $X_{\rho'}$ for any $\rho, \rho' \in I_0$ with $\rho' < \rho$, satisfying
  (A1) $A[0] = 0$.
  (A2) There exist a positive number $\sigma \in [0,1]$ and a nonnegative function $C(p,q)$ which is increasing both in $p, q \geq 0$, and for any $u, v \in X_\rho$,

$$||A[u] - A[v]||_{\rho'} \leq \frac{C(||u||_\rho, ||v||_\rho)}{(\rho - \rho')^\sigma} ||u - v||_\rho$$

  holds whenever $\rho, \rho' \in I_0$ with $\rho' < \rho$.

The family of Banach spaces satisfying the condition (X) is called a Banach scale. The condition (A) requires that the map A is a Lipschitz continuous map from X_ρ into $X_{\rho'}$, but this is required only for $\rho' < \rho$. For $\rho' \geq \rho$, A need not to be defined, or if defined, it is admitted to be "unbounded" (the denominator in (A1) is 0 for $\rho = \rho'$). This is just the same situation we meet in the classical Cauchy-Kovalevskaya theorem.

For any positive number T, denote the set of continuous functions on the interval $[0,T]$ with value in the space X_ρ by $C([0,T]; X_\rho)$. We have the following existence theorem for (3.2).

Theorem 3.1. *Assume* (X) *and* (A). *Then, for any initial $u_0 \in X_{\rho_0}$ and for any positive number $a > 1$, there exists a positive number γ such that, putting*

$$T = \rho_0/\gamma, \qquad \rho(t) = \rho_0 - \gamma t, \tag{3.3}$$

(3.2) has a unique solution $u = u(t)$ on the time interval $[0,T]$ satisfying

$$u \in C^0([0,T]; X_0), \qquad u(t) \in X_{\rho(t)} \quad (t \in [0,T]), \tag{3.4}$$

and

$$\sup_{t \in [0,T]} ||u(t)||_{\rho(t)} \leq a||u_0||_{\rho_0}. \tag{3.5}$$

Proof. The proof is very simple if $\sigma \in [0,1)$. First, with a positive constant γ to be determined later, we define a constant T and a function $\rho(t)$ by (3.3). Then, $\rho(t)$ is decreasing and

$$0 = \rho(T) \leq \rho(t) \leq \rho(0) = \rho_0, \qquad t \in [0,T].$$

With this choice, introduce the norm

$$|||u||| = \sup_{t \in [0,T]} ||u(t)||_{\rho(t)}, \tag{3.6}$$

and for any positive number $r > 0$, define a space by

$$Y_r = \left\{ u \mid u \text{ is in the class } (3.4), \; |||u||| \le r \right\}. \tag{3.7}$$

Obviously, Y_r is a complete metric space with the distance induced by the norm $||| \cdot |||$.

Next, we write the right hand side of (3.2) as $N[u]$;

$$N[u](t) = u_0 + \int_0^t A[u(s)] ds. \tag{3.8}$$

This defines a map, and u is a solution to (3.2) if and only if u is a fixed point of the map N. We shall show that N is a contraction map on the complete metric space Y_r with a suitable choice of r.

So, let u be in the class (3.4) with $(T, \rho(t))$ fixed as in (3.3). Then, by virtue of the conditions (X) and (A), we get for $\rho' < \rho$,

$$\begin{aligned} ||N[u](t)||_{\rho'} &\le ||u_0||_{\rho'} + \int_0^t ||A[u(s)]||_{\rho'} ds \\ &\le ||u_0||_{\rho_0} + \int_0^t \frac{C_\rho ||u(s)||_\rho}{(\rho - \rho')^\sigma} ds, \end{aligned}$$

where $C_\rho = C(||u||_\rho, 0)$. Choose

$$\rho' = \rho(t), \qquad \rho = \rho(s). \tag{3.9}$$

Since $\rho(t) < \rho(s)$ for $s < t$. we can deduce for any $u \in Y_r$ and $t \in [0, T]$,

$$\begin{aligned} ||N[u](t)||_{\rho(t)} &\le ||u_0||_{\rho_0} + \int_0^t \frac{C_{\rho(s)} ||u(s)||_{\rho(s)}}{(\rho(s) - \rho(t))^\sigma} ds \tag{3.10} \\ &\le ||u_0||_{\rho_0} + \frac{c_r}{\gamma^\sigma} \int_0^t \frac{1}{(t - s)^\sigma} ds \\ &= ||u_0||_{\rho_0} + \frac{c_r T^{1-\sigma}}{(1-\sigma)\gamma^\sigma}, \end{aligned}$$

with $c_r = C(r, 0)r$. Note from (3.3),

$$\frac{T^{1-\sigma}}{\gamma^\sigma} = \frac{\rho_0^{1-\sigma}}{\gamma}.$$

By a similar computation, we have, for any $u, v \in Y_r$,

$$||N[u](t) - N[v]||_{\rho(t)} \le \frac{d_r T^{1-\sigma}}{(1-\sigma)\gamma^\sigma} |||u - v|||, \tag{3.11}$$

with $d_r = C(r.r)$.

Finally, we take $r = a||u_0||_{\rho_0}$ with some $a > 1$ and choose γ to be so large that

$$\frac{c_r T^{1-\sigma}}{(1-\sigma)\gamma^\sigma} \leq (a-1)\|u_0\|_{\rho_0}, \qquad \mu \equiv \frac{d_r T^{1-\sigma}}{(1-\sigma)\gamma^\sigma} < 1,$$

can hold. Then, (3.10) and (3.12) yield, for any $u, v \in Y_r$,

$$\||N[u]\|| \leq r, \qquad \||N[u] - N[v]\|| \leq \mu\||u - v\||, \qquad (3.12)$$

which shows that N is a contraction map on Y_r. It is clear that the fixed point u belongs to the class (3.4). Thus the theorem follows for $\sigma < 1$.

In the above, a suitable choice of the constant a may lead to an optimal life span T of the solution u, but of course it depends on the growth order of c_r, d_r with r.

Actually, the solution $u(t)$ is more smooth than stated in Theorem 3.1. In fact, for any $\tau \in [0, T]$, it follows from (A) that with $t, t' \in [0, \tau]$, $t > t'$,

$$\|u(t) - u(t')\|_{\rho(t)} \leq \int_{t'}^t \|A[u(s)]\|_{\rho(t)} ds$$

$$\leq c_r \int_{t'}^t \frac{1}{(\rho(s) - \rho(t))^\sigma} ds \leq \frac{c_r}{(1-\sigma)\gamma^\sigma} |t - t'|^{1-\sigma},$$

$$\|\frac{u(t) - u(t')}{t - t'} - A[u(t)]\|_{\rho(\tau)} \leq \frac{1}{|t - t'|} \int_{t'}^t \|A[u(s)] - A[u(t)]\|_{\rho(\tau)} ds$$

$$\leq \frac{d_r}{(\rho(t) - \rho(\tau))^\sigma} \sup_{t' < s < t} \|u(s) - u(t)\|_{\rho(t)},$$

where c_r is as in (3.10) and d_r as in (3.11). This implies

Theorem 3.2. *The solution* $u = u(t)$ *given in Theorem 3.1 satisfies*

$$u \in C^0([0, \tau], X_{\rho(\tau)}) \cap C^1([0, \tau), X_{\rho(\tau)}), \qquad (3.13)$$

for any $\tau \in [0, T]$, *and solves the Cauchy problem* (3.1) *on the time interval* $[0, T)$.

It is easy to see that the same holds for the case $\sigma = 1$ with $C^0([0, \tau]; X_\rho(\tau))$ of (3.13) replaced by $C^0([0, \tau); X_{\rho(\tau)})$.

Theorem 3.1 is valid also for the case $\sigma = 1$, which corresponds to the situation of the classical Cauchy-Kovalevskaya theorem (cf. Example 1 below), and is the most important case in application, but its proof is much more complicated under the general setting (X) and (A) (see e.g. [29]).

However, there are some special situations for which the proof becomes much simpler. Here, we give two such examples. Actually, the situation of Example 1 below is an situation we will meet in §5 and that of Example 2 in §4. And, Example 2 establishes the existence theorem of local (in time) solution to the Cauchy problem for the Boltzmann equation (2.21).

3.1 Example 1: Pseudo Differential Equation

Our first example is a simple nonlinear first order pseudo differential equation. This example serves as a preparation for §5, and also gives a flavor of the classical Cauchy-Kovalevskaya theorem.

Let $u(x)$ be a function defined on $\mathbb{R}^n$ and let

$$\mathcal{F}(u)(\xi) = \hat{u}(\xi) = (2\pi)^{-n/2} \int_{\mathbb{R}^n} e^{-ix\cdot\xi} u(x) dx, \quad \xi \in \mathbb{R}^n, \quad (i = \sqrt{-1}), \quad (3.14)$$

be its Fourier transform. Define the norm

$$||u||_\rho = \sup_{\xi \in \mathbb{R}^n} \left\{ w_\rho(\xi) |\hat{u}(\xi)| \right\} = ||\hat{u}||_{L^\infty(\mathbb{R}^n, w_\rho(\xi)d\xi)}, \quad (3.15)$$

with a weight function

$$w_\rho(\xi) = e^{\rho(1+|\xi|)}(1 + |\xi|)^s, \quad s > n. \quad (3.16)$$

Then, the space

$$X_\rho = \{u| \; ||u||_\rho < \infty\}, \quad \rho \in \mathbb{R}, \quad (3.17)$$

meets the condition (X). The above norm and space should have been indexed, in addition, by the parameter s. Its role is only to make the space a Banach algebra in order to control the mulitiplication of functions. For this, it suffices to choose $s > n$. Thus fixed, we drop it from the index for simplifying the notation in the sequel.

We state two properties of the space X_ρ. First, if $\rho > 0$, $u(x) \in X_\rho$ is a real analytic function of x in the whole space $\mathbb{R}^n$ with the common convergence radius ρ. For simplicity, we show this for the case $n = 1$, but the proof is the same for $n \geq 2$. For any $k \in \mathbb{N}$, we have

$$u^{(k)}(x) = (2\pi)^{-1/2} \int_{-\infty}^{\infty} (i\xi)^k \hat{u}(\xi) e^{i\xi x} d\xi,$$

whence

$$|u^{(k)}(x)| \leq (2\pi)^{-1/2} \int_{-\infty}^{\infty} |\xi|^k |\hat{u}(\xi)| d\xi$$

$$\leq (2\pi)^{-1/2} \int_{-\infty}^{\infty} \frac{|\xi|^k e^{-\rho|\xi|}}{(1 + |\xi|)^s} d\xi ||u||_\rho$$

$$\leq C \frac{k^k e^{-k}}{\rho^k} ||u||_\rho,$$

where we have used the inequality $x^k e^{-x} \leq k^k e^{-k}$ for $x \geq 0$. Now, by Stirling's formula, we see that $u(x)$ has the convergent Taylor expansion at any point $a \in \mathbb{R}$ with the convergence radius ρ independent of a.

The second property of the space X_ρ is that if $s > n$ in (3.16), the space X_ρ becomes a Banach algebra as long as $\rho \geq 0$. For this, denote the convolution integral on $\mathbb{R}^n$ by $*$ and note that

$$|\mathcal{F}(uv)(\xi)| = |\hat{u} * \hat{v}(\xi)| = \left|\int_{\mathbb{R}^n} \hat{u}(\xi - \eta)\hat{v}(\eta)d\eta\right| \qquad (3.18)$$

$$\leq \int_{\mathbb{R}^n} \{w_\rho(\xi - \eta)w_\rho(\eta)\}^{-1}d\eta ||u||_\rho ||v||_\rho$$

$$\leq \frac{c_0 e^{-\rho}}{w_\rho(\xi)}||u||_\rho ||v||_\rho,$$

with some constant $c_0 > 0$. Indeed, by the help of the triangle inequality $|\xi - \eta| + |\eta| \geq |\xi|$, we have,

$$\int_{\mathbb{R}^n} \{w_\rho(\xi - \eta)w_\rho(\eta)\}^{-1}d\eta = \int_{\mathbb{R}^n} \frac{e^{-\rho\{(2+|\xi-\eta|)+|\eta|\}}}{(1+|\xi-\eta|)^s(1+|\eta|)^s}d\eta$$

$$\leq e^{-\rho(2+|\xi|)}\int_{\mathbb{R}^n} \frac{1}{(1+|\xi-\eta|)^s(1+|\eta|)^s}d\eta$$

$$\leq \frac{c_0 e^{-\rho}}{w_\rho(\xi)}.$$

This proves

$$||uv||_\rho \leq c||u||_\rho ||v||_\rho, \qquad (3.19)$$

which is what was desired.

Fix a positive number ρ_0 arbitrarily, and consider only $\rho \in I_0 = [0, \rho_0]$ for (3.17). Then, similarly as before, with a positive constant γ to be determined later, we define T and $\rho(t)$ by (3.3), and then with these, the norm $|||\cdot|||$ and space Y_r respectively by (3.6) and (3.7), of course with the Banach scale $\{X_\rho, \rho \in I_0\}$ defined by (3.17).

The map A we consider here is a nonlinear operator defined by

$$A[u] = a(\partial_x)u^2, \qquad (3.20)$$

where $a(\partial_x)$ is a linear pseudo differential operator. We assume that its symbol $a(\xi)$ enjoys the estimate

$$|a(\xi)| \leq a_0(1 + |\xi|)^\sigma, \quad \sigma \in [0, 1]. \qquad (3.21)$$

Thus, our $a(\partial_x)$ corresponds to a differential operator of order σ. The symbol of a first order partial differential operator (with constant coefficients) satisfies the condition (3.21) with $\sigma = 1$. It is easy to check that (3.20) fulfills the condition (A) including the case $\sigma = 1$, but we go differently.

Evidently, it is natural to consider (3.2) in the Fourier transform. Since the Fourier transform of (3.20) is given by

$$\mathcal{F}(A[u]) = a(\xi)\mathcal{F}(u^2) = a(\xi)(\hat{u} * \hat{u})(\xi), \qquad (3.22)$$

(3.2) becomes

$$\hat{u}(t) = \hat{N}[\hat{u}] \equiv \hat{u}_0 + \int_0^t a(\xi)\mathcal{F}(u(s)^2)ds.$$

Denote the last integral by $J(\hat{u})$. Recall that we defined the space Y_r as before by (3.7) but with the space X_ρ defined by (3.17). For any $u \in Y_r$, we get, by virtue of (3.18) with the choice $\rho = \rho(s)$ as in (3.9) and by (3.21),

$$|J(\hat{u})(t,\xi)| \leq \int_0^t |a(\xi)||\mathcal{F}(u(s)^2)(\xi)|ds$$

$$\leq a_0 c \int_0^t (1 + |\xi|)^\sigma w_{\rho(s)}(\xi)^{-1}||u(s)||_{\rho(s)}^2 ds$$

$$\leq a_1 \int_0^t (1 + |\xi|)^\sigma w_{\rho(s)}(\xi)^{-1}ds|||u|||^2$$

$$\leq a_2 w_{\rho(t)}(\xi)^{-1} \int_0^t (1 + |\xi|)^\sigma e^{-\gamma(t-s)(1+|\xi|)}ds$$

$$\leq c_1 \frac{a_2}{\gamma} w_{\rho(t)}(\xi)^{-1},$$

where c, a_0 are those in (3.19) and (3.21) respectively and $a_1 = a_0 c$, $a_2 = a_1 r^2$ and $c_1 > 0$ is a constant. In the above, we used the observation

$$w_{\rho(s)}(\xi)^{-1} = w_{\rho(t)}(\xi)^{-1}e^{-\gamma(t-s)(1+|\xi|)}.$$

As a consequence, we have

$$|||J(u)||| \leq c_1 \frac{a_2}{\gamma},$$

for all $u \in Y_r$. A similar computation gives

$$|||J(u) - J(v)||| \leq c_1 \frac{a_3}{\gamma}|||u - v|||,$$

with $a_3 = a_1 r$, for any $u, v \in Y_r$. Now, it is clear that the estimates (3.10) and (3.11) follow for our N with due modifications. Hence, the same procedure for the choice of r and γ as before leads to (3.12), and hence to Theorem 3.1 for (3.20) including the case $\sigma = 1$.

Remark 3.1. Theorem 3.2 is also true, but of course with the modification stated below that theorem.

Remark 3.2. Two observations which suggest the similarity with the classical Cauchy-Kovalevskaya theorem: The unbounded factor $(1 + |\xi|)^\sigma$ defines an unbounded operator, that is, a pseudo differential operator of order σ. In the above, this unboundedness is controlled by the smoothing effect of the integration in t. Since the triangle inequality is essential in estimating the nonlinear terms in (3.18), the above argument works only for $\sigma \in [0,1]$, and not for $\sigma > 1$ and thereby not for the (pseudo) differential operator of order > 1. This is the same as the original Cauchy Kovalevskaya theorem does not work for second or higher order partial differential equations like the heat equation.

Another observation is that the solution in the classical Cauchy - Kovalevskaya theorem is analytic both in x and t, and the convergence radius in x shrinks with time t. The same is true for our solution $u(t)$ since it belongs to the class (3.4) and $\rho(t)$, being the convergence radius, is a decreasing function of t. Indeed, the smoothing effect of the integration in t works only under the shrinking $\rho(t)$.

3.2 Example 2: Local Solutions of the Boltzmann Equation

Our technique provides also a simple method to construct local solutions to the Cauchy problem for the Boltzmann equation (2.21). In §4, the Boltzmann Grad limit will be established in the same spirit. The unbounded factor we should control now is the unbounded function

$$q(v, v_*, \omega) = |(v - v_*, \omega)|$$

appearing in the collision operator Q defined by (2.22), which is the *collision kernel* of the hard sphere gas. In this example, we deal with more general kernels so that our result can cover *Grad's cutoff potential* [13]. Thus, we assume that q is a general function satisfying

$$(q): \quad |q(v, v_*, \omega)| \leq q_0 (1 + |v| + |v_*|)^{\sigma}, \qquad (v, v_*, \omega) \in \mathbb{R}^n \times \mathbb{R}^n \times S^{n-1}, \quad \sigma \leq 2,$$

Evidently, the hard sphere gas satisfies this with $\sigma = 1$, and our existence theorem below is valid for $\sigma \leq 2$. Here, $q_0 \geq 0$ is a positive constant independent of (v, v_*, ω). Physically, q should be a nonnegative function but this is not necessary for our proof.

Recall that $x \in \mathbb{R}^n$ stands for the space position and $v \in \mathbb{R}^n$ for the velocity. Physically, $n = 3$ but mathematically, the Boltzmann equation (2.21) can be defined for arbitrary $n \in \underline{\nu}$. In the physical background, 1D and 2D Boltzmann equations means (2.21) with the space variable $x \in \mathbb{R}^d$ $(d = 1, 2)$ but with the velocity variable unchanged as $v \in \mathbb{R}^3$. We do not discuss this case, however, since the argument below also applies with minor modifications.

The Banach scale to be introduced here to control the unboundedness of the kernel q is, for $\rho \geq 0$,

$$X_\rho = \left\{ f(x, v) \in L_{loc}^\infty(\mathbb{R}^n \times \mathbb{R}^n) \mid \|f\|_\rho < \infty \right\} \tag{3.23}$$

$$\|f\|_\rho = \sup_{x, v \in \mathbb{R}^n} \left\{ w_\rho(v) |f(x.v)| \right\},$$

with the weight function

$$w_\rho(v) = e^{\rho(1+|v|^2)}. \tag{3.24}$$

Certainly, the condition (X) is fulfilled.

The semi-group $S(t)$ generated by the *transport operator* $-v \cdot \nabla_x$ in the whole space $\mathbb{R}^n$ without boundary conditions, which is appearing in (2.21), is formally given by

$$S(t)f_0 = f_0(x - tv, v), \tag{3.25}$$

for $f_0 = f_0(x, v)$. This is not a C_0-semi-group in our setting given below but has semi-group properties. Therefore, Duhamel's formula applies and the Cauchy problem for (2.21) with the initial condition (2.24) is rewritten in the form of the integral equation

$$f(t) = S(t)f_0 + a_0 \int_0^t S(t - s)Q[f(s)]ds, \tag{3.26}$$

The solution of (3.26) is the *mild* solution of the Cauchy problem for the Boltzmann equation (2.21).

Theorem 3.3. *Suppose that the condition* (q) *be fulfilled with* $\sigma \leq 2$ *and let* $\rho_0 > 0$. *Then, for any* $f_0 \in X_{\rho_0}$ *and for any constant* $a > 1$, *there exists a constant* $\gamma > 0$ *and the following holds. Define the positive constant* T *and a function* $\rho(t)$ *by*

$$T = \frac{\rho_0}{2\gamma}, \qquad \rho(t) = \rho_0 - \gamma t,$$

as in (3.3), *and the norm by*

$$|||f||| = \sup_{t \in [0,T]} ||f(t)||_{\rho(t)},$$

as in (3.6). *Then,* (3.26) *has a unique solution* f *satisfying*

$$f \in L^\infty([0, T]; X_0), \quad f(t) \in X_{\rho(t)} \quad (t \in [0, T]),$$
$$|||f||| \leq a||f_0||_{\rho_0}.$$

Proof. First, we note that

$$||S(t)f_0||_\rho = ||f_0||_\rho, \tag{3.27}$$

holds for any $\rho \in \mathbb{R}$.

Second, we shall derive an estimate of the collision operator Q in the space X_ρ. Let $Q(f, g)$ be the associated bilinear symmetric operator:

$$Q(f, g) = \frac{1}{2}\Big\{Q(f + g) - Q(f - g)\Big\} \tag{3.28}$$

$$= \frac{1}{2} \int_{\mathbb{R}^3 \times S^2} q(v, v_*, \omega)(f'g'_* + f'_*g' - fg_* - f_*g)dv_*d\omega,$$

where g, g', g_*, g'_* are the same as in (2.22). Clearly, $Q(f) = Q(f, f)$.

In order to derive an estimate of Q, we write

$$|f'g_*'| + |f_*'g'| \leq 2\Big\{ w_\rho(v') w_\rho(v_*') \Big\}^{-1} \|f\|_\rho \|g\|_\rho,$$

$$|fg_*| + |f_*g| \leq 2\Big\{ w_\rho(v) w_\rho(v_*) \Big\}^{-1} \|f\|_\rho \|g\|_\rho.$$

Notice the conservation law of energy $|v|^2 + |v'|^2 = |v'|^2 + |v_*'|$ (cf. (2.6)), to deduce

$$w_\rho(v') w_\rho(v_*') = w_\rho(v) w_\rho(v_*),$$

which, then, yields, under the assumptions (q),

$$|Q(f,g)(t,x,v)| \leq \frac{c_0}{w_\rho(v)} \int_{\mathbb{R}^n} (1 + |v| + |v_*|)^\sigma e^{-\rho|v_*|^2} dv_* \|f\|_\rho \|g\|_\rho \quad (3.29)$$

$$\leq \frac{C_\rho(1 + |v|)^\sigma}{w_\rho(v)} \|f\|_\rho \|g\|_\rho,$$

for all $\rho > 0$, with some positive constant C_ρ such that C_ρ is decreasing in $\rho > 0$ and $C_\rho \to \infty$ as $\rho \to 0$.

Denote the last integral of (3.26) by $J[f]$. If we choose $\rho = \rho(s)$, we have, thanks to the estimate (3.29),

$$|J[f](t,x,v)| \leq \int_0^t |Q(f,f)(s, x - (t-s)v, v)| ds \quad (3.30)$$

$$\leq \int_0^t C_{\rho(s)} (1 + |v|)^\sigma e^{-\rho(s)(1+|v|^2)} \|f(s)\|_{\rho(s)}^2 ds$$

$$\leq \frac{C_{c_1}}{w_{\rho(t)}(v)} \int_0^t (1 + |v|)^\sigma e^{-\gamma(t-s)(1+|v|^2)} ds \|\|f\|\|^2$$

$$\leq \frac{c_1}{w_{\rho(t)}(v)} \frac{(1 + |v|)^\sigma}{\gamma(1 + |v|^2)} \|\|f\|\|^2$$

$$\leq \frac{c_2}{\gamma w_{\rho(t)}(v)} \|\|f\|\|^2,$$

for $\sigma \leq 2$, where

$$c_1 = C_{\rho_0/2}, \qquad c_2 = cc_1,$$

with a constant $c \geq 0$. Here, we have used that

$$\rho_0/2 = \rho(T) \leq \rho(s) \leq \rho_0 = \rho(0), \qquad s \in [0, T],$$

with the choice of T as in the theorem and that C_ρ is decreasing with ρ.

The right hand side of (3.26) is written as

$$N[f] \equiv S(t)f_0 + J[f],$$

for which (3.30), together with (3.27), gives

$$||N[f](t)||_{\rho(t)} \leq ||f_0||_{\rho_0} + \frac{a_0 c_2}{\gamma}|||f|||^2$$

for $t \in [0,T]$. This is just like (3.10). If we note that $Q(f) - Q(g) = Q(f + g, f - g)$, we also have

$$|||N[f](t) - N[g](t)||_{\rho(t)} \leq \frac{a_0 c_2}{\gamma}|||f + g||||||f - g|||,$$

with the same constants c_2, γ. This is just like (3.11), and the rest of the argument is the same to conclude that N is a contraction map, and thus, Theorem 3.3 follows.

Remark 3.3. The smallness conditon on the initials f_0 is not necessary.

Remark 3.4. There is no differentiability of f because $S(t)$ is not a strongly continuous (C_0-) semi-group in the space X_ρ defined by (3.23). It becomes a C_0-semi-group in spaces of more smooth functions, and the regularity theorem of f then follows.

Remark 3.5. The unbounded factor to be controlled here is $(1 + |v|)^\sigma$ coming from the assumption (q), and as seen in (3.30), again the smoothing effect of integration in t combined with the shrinking in time weight function is a key ingredient. The restriction $\sigma \leq 2$ comes from the choice of the factor $1 + |v|^2$ in the exponent of the weight function (3.24), which is chosen to be compatible with the conservation laws (2.6) and hence cannot be replaced by other factors. Thus, it is not possible to relax the restriction $\sigma \leq 2$.

Remark 3.6. It is easy to see that the constant γ should be chosen to be proportional to $a_0 = 1/\epsilon$ (see (2.23)), and hence that the life span T of our mild solution goes to 0 as the mean free path ϵ tends to 0. Thus, Theorem 3.3 does not provide uniform estimates required for the multi scale analysis of the Boltzmann equation depicted in Fig.5.

Remark 3.7. We close this example with two remarks: The non-negativity of solutions and the backward solvability in time. First, the H-function (2.31) is defined only for nonnegative functions. It is not hard to show that solutions from Theorem 3.2 are nonnegative if the collision kernel $q(v, v_*, \omega)$ is nonnegative as well as the initials f_0, and therefore, the H-theorem (2.32) can be verified. On the other hand, as in the classical Cauchy-Kovalevskaya theorem, our thechnique is valid to solve the evolution equation (3.2) backward in time, as well. This does not contradict the H-theorem, however, because the non-negativity does not propagate backward in time.

4 The Boltzmann-Grad Limit

The idea proposed in [5], [2] to establish the Boltzmann-Grad limit (2.17) is to compare the three Cauchy problems for the BBGKY hierarchy (2.12), the Boltzmann hierarchy (2.18) and the Boltzmann equation (2.21). Lanford [8] carried out this for the mild solutions, and his proof has been achieved in [2]. In this section, we will present his theorem, and reproduce the part of his proof on the uniform estimates of solutions by using the Cauchy-Kovalevskaya technique introduced in the previous section. For other parts of the proof, including the most difficult part on the pointwise convergence, the reader is referred to [2].

4.1 Integral Equations

We have already established the local existence theorem of mild solutions to the Cauchy problem for the Boltzmann equation (2.21). Here, we shall establish the same theorems for the BBGKY and Boltzmann hierarchies, and furthermore, we shall derive estimates of their solutions uniform for k, N and σ which are needed to prove the existence of the Boltzmann-Grad limit.

First, we shall rewrite the relevant Cauchy problems in the form of the integral equations. Let

$$z^{(k)} = (z_1, \cdots, z_k), \quad z_i = (x_i, v_i)$$

be as in §2, and let

$$Z_\sigma^{(k)}(t, s, z^{(k)}) \in \Lambda_{k,\sigma}$$

denote the trajectory of (2.3) for $N = k$ which obeys the boundary condition of the elastic collision and passes the point $z^{(k)} \in \Lambda_{k,\sigma}$ at time $t = s$ (initial condition), and put

$$U_\sigma^{(k)}(t)g = g(Z_\sigma^{(k)}(0, t, z^{(k)})), \tag{4.1}$$

for $g = g(z^{(k)})$. Formally, this is the semi-group generated by the transport operator of (2.12),

$$-\sum_{i=1}^k v_i \cdot \nabla_{x_i}, \tag{4.2}$$

in the domain $\Lambda_{k,\sigma}$ associated with the boundary condition of elastic collision on the boundary $\partial\Lambda_{k,\sigma}$. Now, the Cauchy problem to (2.12) with the initial condition (2.16) can be formally rewritten as

$$P_{N,\sigma}^{(k)}(t) = U_\sigma^{(k)}(t)P_{N,\sigma,0}^{(k)} + (N-k)\sigma^2 \int_0^t U_\sigma^{(k)}(t-s)Q_\sigma^{(k+1)}(P_{N,\sigma}^{(k+1)}(s))ds, \tag{4.3}$$

for $k = 1, 2, \cdots, N$.

Similarly, with the free stream of k balls of diameter 0 (point particle)

$$z_i(t) = (x_i - tv_i, v_i), \qquad z^{(k)}(t) = (z_1(t), z_2(t), \cdots, z_k(t)), \qquad (4.4)$$

(a free trajectory of k balls), we define

$$S^{(k)}(t)g = g(z^{(k)}(t)). \qquad (4.5)$$

which is a formal semi-group generated by the same transport operator as in (4.2), but for (2.18) in the whole space $\mathbb{R}^{6k}$ without boundary conditions. Now, the Cauchy problem for (2.18) with the initial condition (2.19) becomes

$$f^{(k)}(t) = S^{(k)}(t)f_0^{(k)} + a_0 \int_0^t S^{(k)}(t-s)Q_0^{(k+1)}(f^{(k+1)}(s))ds, \quad k \in \mathbb{N}. \quad (4.6)$$

4.2 Local Solutions and Uniform Estimates

We begin with (4.3). We shall introduce a suitable Banach scale and solve (4.3) uniformly there.

The operator $Q_\sigma^{(k+1)}$ in (2.13) has two unbounded factors to controll:

- The unbounded number of terms coming from $\sum_{i=1}^k$ as $k \to \infty$.
- The unbounded function $|(v_i - v_*) \cdot \omega|$ as $|v_i| \to \infty$.

For their control, therefore, we need a Banach scale with double suffix. For $k \in \underline{\nu}$ and $\alpha, \rho \in > 0$, define the weight function in $v^{(k)} = (v_1, v_2, \cdots, v_k) \in \mathbb{R}^{3k}$ by

$$w_{\alpha,\rho}^{(k)}(v^{(k)}) = \exp\left(-\alpha k + \rho|v^{(k)}|^2\right), \quad |v^{(k)}|^2 = \sum_{i=1}^k |v_i|^2, \qquad (4.7)$$

and introduce the Banach spaces

$$X_{\alpha,\rho}^{(k)} = \left\{h \in L_{loc}^\infty(\Lambda_{k,\sigma}) \mid \|h\|_{k,\alpha,\rho} < \infty\right\}, \qquad (4.8)$$

with the norms

$$\|h\|_{k,\alpha,\rho} = \sup_{z^{(k)} \in \Lambda_{k,\sigma}} w_{\alpha,\rho}^{(k)}(v^{(k)})|h(z^{(k)})|. \qquad (4.9)$$

This definition involves σ but for simplifying the notation, we drop it from the suffix to $X_{\alpha,\rho}^{(k)}$ and $\|\cdot\|_{k,\alpha,\rho}$. The space $X_{\alpha,\rho}^{(k)}$ is increasing in α and decreasing in ρ.

Recall $U_\sigma^{(k)}$ of (4.1).

Lemma 4.1. *Let $k \in \underline{\nu}$, $\alpha, \rho \in \mathbb{R}$ and $\sigma > 0$. For any $h \in X_{\alpha,\rho}^{(k)}$ and $t \in \mathbb{R}$, we have*

$$\|U_\sigma^{(k)}(t)h\|_{k,\alpha,\rho} = \|h\|_{k,\alpha,\rho}.$$

Proof. Write the trajectory $Z_\sigma^{(k)}(t, s, z^{(k)})$ in (4.1) in components of positions and velocities as $(X^{(k)}, V^{(k)})$. Write simply $w^{(k)} = w_{\alpha,\rho}^{(k)}$ for (4.7). Since our dynamical system conserves the energy (*cf.* (2.6)), we get

$$|V^{(k)}| = |v^{(k)}|, \quad w^{(k)}(V^{(k)}) = w^{(k)}(v^{(k)}), \tag{4.10}$$

whence the lemma immediately follows.

Recall also the integral operator $Q_\sigma^{(k+1)}$ of (2.13).

Lemma 4.2. *There is a constant $C > 0$ such that for any $k \in \underline{\nu}$, $\alpha, \rho, \sigma > 0$ and $g \in X_{\alpha,\rho}^{(k+1)}$, we have in $\Lambda_{k,\sigma}$*

$$|Q_\sigma^{(k+1)}(g)(z^{(k)})| \le Ce^\alpha \rho^{-5/2}(1 + \rho)(k + |v^{(k)}|^2)\{w_{\alpha,\rho}^{(k)}(v^{(k)})\}^{-1}\|g\|_{k+1,\alpha,\rho}$$

Proof. Let $w^{(k)}$ be as before. The conservation law (2.6) applies to (v_i, v_*, v_i', v_*') of (2.15) so that, like (4.10),

$$w^{(k+1)}(v^{(k)}, v_*) = w^{(k+1)}(v_1, \cdots, v_i', \cdots, v_k, v_*'),$$

holds, while

$$w^{(k+1)}(v^{(k+1)}) = w^{(k)}(v^{(k)})w^{(1)}(v_{k+1}) = w^{(k)}(v^{(k)})e^{-\alpha + \rho|v_{k+1}|^2}.$$

whence, by the definition (2.14),

$$|g'|, |g_*| \le \{w^{(k)}(v^{(k)})\}^{-1}e^{\alpha - \rho|v_*|^2}\|g\|_{k+1,\alpha,\rho},$$

so that $Q_\sigma^{(k+1)}$ is majorized by

$$8\pi\left\{\sum_{i=1}^k \int_{\mathbb{R}^3}(|v_i| + |v_*|)e^{\alpha - \rho|v_*|^2}dv_*\right\}\{w^{(k)}(v^{(k)})\}^{-1}\|g\|_{k+1,\alpha,\rho,\sigma},$$

which proves the lemma.

Since the factor $k + |v^{(k)}|^2$ is unbounded as $k \to \infty$ and $|v^{(k)}| \to \infty$, this lemma implies that $Q_\sigma^{(k+1)}$ is unbounded as an operator from $X_{k+1,\alpha,\rho}$ to $X_{k,\alpha,\rho}$, and therefore, so is the integral

$$L^{(k+1)}(g)(t) = \int_0^t U_\sigma^{(k)}(t-s)Q_\sigma^{(k+1)}(g(s))ds \quad (g = g(t, z^{(k+1)})), \tag{4.11}$$

which is appearing in the right hand side of (4.3). Like in Example 2 in $\S3$, however, this integral becomes bounded if α, ρ are varied with time t, thanks to the smoothing effect of integration in t. To see this, let C be the constant in Lemma 4.2 and put

$$C_{\alpha,\rho} = Ce^\alpha \rho^{-5/2}(1 + \rho). \tag{4.12}$$

Lemma 4.3. *For any $\alpha_0, \rho_0, \gamma > 0$ and for any $k \in \underline{\nu}$,*

$$||L^{(k+1)}(g)(t)||_{k,\alpha(t),\rho(t)} \leq \frac{1}{\gamma} C_{2\alpha_0,\rho_0/2} \left\{ \sup_{0 \leq s \leq T} ||g(s)||_{k+1,\alpha(s),\rho(s)} \right\}$$

holds for $t \in [0,T]$, where

$$\alpha(t) = \alpha_0 + \gamma t, \quad \rho(t) = \rho_0 - \gamma t, \quad T = \min(\frac{\alpha_0}{\gamma}, \frac{\rho_0}{2\gamma}). \tag{4.13}$$

Proof. First, note that $U_\sigma^{(k)} g \leq U_\sigma^{(k)} h$ if $g \leq h$. Applying this to Lemma 4.2 and setting $\alpha = \alpha(s)$ and $\rho = \rho(s)$, we deduce, together with Lemma 4.1,

$$|U_\sigma^{(k)}(t-s)Q_\sigma^{(k+1)}(g(s))(z^{(k)})|$$
$$\leq C_{\alpha(s),\rho(s)}(k + |v^{(k)}|^2)\{w_{\alpha(s),\rho(s)}^{(k)}(v^{(k)})\}^{-1}||g(s)||_{k+1,\alpha(s),\rho(s)}.$$

Next, let T be as in (4.13). Then, by definition (4.12), $C_{\alpha(s),\rho(s)} \leq C_{2\alpha_0,\rho_0/2}$ on $[0.T]$. Further, by definition (4.7), we get for $t > s > 0$,

$$w_{\alpha(s),\rho(s)}^{(k)} = w_{\alpha(t)}^{(k)} w_{-\gamma(t-s),\gamma(t-s)}^{(k)} = w_{\alpha(t)}^{(k)} e^{\gamma(t-s)(k+|v^{(k)}|^2)}.$$

Finally, noting that

$$\int_0^t (k + |v^{(k)}|^2) e^{-\gamma(t-s)(k+|v^{(k)}|^2)} ds \leq \frac{1}{\gamma} \quad \text{for all } t \geq 0, \ k \in \underline{\nu},$$

we are done.

Remark 4.1. The final integral exhibits the smoothing effect of integration in t in the sense that it controls the unbounded factor $k + |v^{(k)}|^2$ in Lemma 4.2. In [2], an almost same estimate is derived and the method is also similar. The only difference is that the above lemma gives an explicit t-dependency of the estimate.

We are now ready to introduce the Banach scale suitable for (4.3). It is,

$$Y_{\alpha,\rho}^{(N)} = \prod_{k=1}^N X_{\alpha,\rho}^{(k)}, \quad ||g||_{\alpha,\rho} = \sup_{k=1,2,\cdots,N} ||g^{(k)}||_{k,\alpha,\rho}, \tag{4.14}$$

where

$$g = (g^{(1)}(z^{(1)}), g^{(2)}(z^{(2)}), \cdots, g^{(N)}(z^{(N)})).$$

This scale is increasing in α and decreasing in ρ.

Now, go back to (4.3) and regard it as an equation on the space $Y_{\alpha,\rho}^{(N)}$ in the form,

$$g = Ug_0 + Lg, \tag{4.15}$$

where

$$g_0 = (P_{N,\sigma,0}^{(1)}(z^{(1)}), P_{N,\sigma,0}^{(2)}(z^{(2)}), \cdots, P_{N,\sigma,0}^{(N)}(z^{(N)})),$$
$$g = (P_{N,\sigma}^{(1)}(t, z^{(1)}), P_{N,\sigma}^{(2)}(t, z^{(2)}), \cdots, P_{N,\sigma}^{(N)}(t, z^{(N)})),$$

and

$$U g_0 = (\ U_\sigma^{(k)}(t) P_{N,\sigma,0}^{(k)}\ )_{k=1}^N,$$
$$L g = (\ (N-k)\sigma^2 L^{(k+1)}((P_{N,\sigma}^{(k+1)}))\ )_{k=1}^N, \qquad (L^{(N+1)} = 0).$$

Note that the equation (4.15) is linear.

With α_0, ρ_0, γ, given in Lemma 4.3, define the norms

$$\|g_0\| = \|g_0\|_{\alpha_0,\rho_0}, \quad \||g|\| = \sup_{0 \le t \le T} \|g(s)\|_{\alpha(t),\rho(t)},$$

with the norm in (4.14). We have

$$\||U g_0|\| \le \|g_0\|,$$

by Lemma 4.1, while, assuming, for simplicity, $(N-k)\sigma^2 \le a_0$. in view of (2.17), we have

$$\||Lg|\| \le \frac{1}{\gamma} a_0 C_{2\alpha_0,\rho_0/2} \||g|\|,$$

by Lemma 4.3. This shows that if γ is chosen so large that

$$\mu = \frac{1}{\gamma} a_0 C_{2\alpha_0,\rho_0/2} < 1, \tag{4.16}$$

(4.15), being linear, can be solved by the contraction argument, or equivalently, by the Neumann series, as $g = (I - L)^{-1} U g_0$ with the estimate

$$\||g|\| \le (1-\mu)^{-1}\|g_0\|, \tag{4.17}$$

which is uniform, in particular, in N. Summarizing, we have proved

Theorem 4.1. *For any positive numbers α_0, ρ_0, there exists a positive number γ and the following holds with $\alpha(t), \rho(t), T$ defined in (4.13). For any numbers $N \in \underline{\nu}$, $\sigma \in (0,1)$, and for any initial $g_0 \in Y_{\alpha_0,\rho_0}^{(N)}$, the equation (4.15) has a unique solution g belonging to the function class*

$$g(t) \in Y_{\alpha(t),\rho(t)}^{(N)}, \qquad t \in [0,T], \tag{4.18}$$

and satisfying the estimate (4.17).

In this way, we have established the uniform existence and estimates, in N and σ, of solutions to the BBGKY hierarchy (4.3).

Theorem 4.1 is valid also for (4.6). To see this, first, redefine the space $X^{(k)}_{\alpha,\rho}$ given in (4.8) and its norm (4.9) by replacing $\Lambda_{k,\sigma}$ by $\mathbb{R}^{6k}$, as

$$X^{(k)}_{\alpha,\rho} = \left\{ h(z^{(k)}) \in L^\infty_{loc}(\mathbb{R}^{6k}) \mid ||h||_{k,\alpha,\rho} < \infty \right\}, \qquad (4.19)$$

$$||h||_{k,\alpha,\rho} = \sup_{z^{(k)} \in \mathbb{R}^{6k}} w^{(k)}_{\alpha,\rho}(v^{(k)})|h(z^{(k)})|,$$

where the weight function $w^{(k)}_{\alpha,\rho}$ is the same as in (4.7).

Clearly, $S^{(k)}$ of (4.5) enjoys Lemma 4.1 with the above norm. Similarly, if we define the integral

$$M^{(k+1)}(h)(t) = \int_0^t S^{(k)}(t-s)Q^{(k+1)}_0(h(s))ds \quad (h = h(t, z^{(k+1)})), \quad (4.20)$$

then, we have again Lemma 4.3 for $M^{(k+1)}$ with the norm (4.19), and what is important, with the same constants.

Next, with $X^{(k)}_{\alpha,\rho}$ and $||g||_{\alpha,\rho}$ defined in (4.19), we introduce the Banach scale

$$Z_{\alpha,\rho} = \prod_{k=1}^\infty X^{(k)}_{\alpha,\rho}, \qquad ||g||_{\alpha,\rho} = \sup_{k \in \nu} ||g^{(k)}||_{k,\alpha,\rho}, \qquad (4.21)$$

$$g = (g^{(1)}(z^{(1)}), g^{(2)}(z^{(2)}), \cdots).$$

and regard (4.6) as an equation on the space $Z_{\alpha,\rho}$ in the form

$$h = Sh_0 + Mh, \qquad (4.22)$$

where

$$h_0 = (f_0^{(1)}(z^{(1)}), f_0^{(2)}(z^{(2)}), \cdots), \qquad h = (f^{(1)}(t, z^{(1)}), f^{(2)}(t, z^{(2)}), \cdots),$$

and

$$Sh_0 = (\, S^{(k)}f_0^{(k)} \,)_{k=1}^\infty, \qquad Mh = (\, a_0 M^{(k+1)}(h^{(k+1)}) \,)_{k=1}^\infty.$$

Finally, with the constants α_0, ρ_0, γ, given in Lemma 4.3, define the norms

$$||h_0|| = ||h_0||_{\alpha_0,\rho_0}, \qquad |||h||| = \sup_{0 \le s \le T} ||g(s)||_{\alpha(s),\rho(s)},$$

with the norm in (4.21), and deduce, as before,

$$|||Sh_0||| \le ||h_0||, \qquad |||Mh||| \le \frac{1}{\gamma}a_0 C_{2\alpha_0,\rho_0/2}|||h|||.$$

Consequently, with exactly the same choice of γ in (4.16), (4.22) has a unique solution $h = (I - M)^{-1}Sh_0$ with the estimate

$$|||h||| \le (1 - \mu)^{-1}||h_0||, \qquad (4.23)$$

whence follows

Theorem 4.2. *For any positive numbers* α_0, ρ_0, *let* γ, $\alpha(t)$, $\rho(t)$ *and* T *be those of Theorem 4.1. Then, for any initial* $h_0 \in Z_{\alpha_0,\rho_0}$, *the equation* (4.22) *has a unique solution* h *belonging to the function class*

$$h(t) \in Z_{\alpha(t),\rho(t)}, \qquad t \in [0,T], \tag{4.24}$$

and satisfying the estimate (4.23).

Remark 4.2. Our argument is essentially in the spirit of the abstract Cauchy-Kovalevskaya theorem. Indeed, since $se^{-\delta s} \leq 1/(e\delta)$ holds for $s \geq 0$ and $\delta > 0$, Lemma 4.2 gives

$$\|Q_\sigma^{(k+1)}(g)\|_{k,\alpha+\delta,\rho-\delta} \leq \frac{1}{e\delta} C_{\alpha,\rho} \|g\|_{k+1,\alpha,\rho}.$$

This is just the case $\sigma = 1$ in the assumption (A2) of §3, which is the key assumption for the abstract evolution equation discussed in [31], [27], [28]. Thus, we could follow their argument to establish the same conclusion. However, a much simpler computation is enough for our problem, as presented, in particular, in the proof of Lemma 4.3.

4.3 Lanford's Theorem

The life spans T of Theorems 4.1 and 4.2 are the same if the parameters α_0 and ρ_0 the same. Thus, we conclude the uniform estimates for both the BBGKY and Boltzmann hierarchies, which are what Lanford [8] used to establish the Boltzmann-Grad limit:

Theorem 4.3. (Lanford [8]) *Suppose that with some positive constants* α_0, ρ_0 *and* r_0, *the initials to* (4.3) *and* (4.6) *satisfy*

$$g_0 = (P_{N,\sigma,0}^{(k)})_{k=1}^N \in Y_{\alpha_0\varphi_0}^{(N)}, \quad \|g_0\|_{N,\alpha_0,\rho_0} \leq r_0, \qquad \text{for each } N \in \underline{\nu},$$

$$h_0 = (f_0^{(k)})_{k=1}^\infty \in Z_{\alpha_0,\rho_0}, \quad \|h_0\|_{\alpha_0,\rho_0} \leq r_0.$$

Suppose further that they be continuous in $\Lambda_{k,\sigma}$ *and in* $\mathbb{R}^{6k}$ *respectively and that the convergence*

$$P_{N,\sigma,0}^{(k)} \to f_0^{(k)} \quad (N \to \infty, \sigma \to 0) \qquad \text{locally uniformly in } \Lambda_{k,\sigma_0}, \tag{4.25}$$

holds for each $k \in \underline{\nu}$ *and* $\sigma_0 > 0$. *Then, in the Boltzmann-Grad limit* (2.17), *the solution* $g = (P_{N,\sigma}^{(k)})_{k=1}^N$ *of Theorem 4.1 converges to the solution* $h = (f^{(k)})_{k=1}^\infty$ *of Theorem 4.2 in the sense*

$$P_{N,\sigma}^k \to f^k \quad \text{a.e. in } \mathbb{R}^{6k} \times [0,T] \quad \text{for each } k \in \underline{\nu}, \tag{4.26}$$

where T *is the common life span of the solutions of Theorems 4.1 and 4.2.*

Remark 4.3. The convergence (4.26) of the solutions is much weaker than (4.25) of initial data. This cannot be strength to a (semi-) norm convergence. As is shown In [2], this loss in the convergence quality is significant to explains the emergence of time irreversibility of the Boltzmann stated in Remark 2.2. This will be discussed after establishing Theorem 4.6 on the validity of the ansatz of the molecular chaos (Remark 4.6)

Sketch of the proof. Let U, L, S, M be as in (4.15) and (4.22). The proof of the theorem is easy once we admit the

Theorem 4.4. ([8], [2]). *Under the assumption of Theorem 4.3, $L^p U g_0$ converges to $M^p S h_0$ pointwise for each $p \in \underline{\nu}$.*

Remembering that we constructed the solutions of (4.15) and (4.22) by the Neumann series. we write for every $q \in \underline{\nu}$.

$$g = \sum_{p=0}^{q} L^p U g_0 + V_q, \qquad h = \sum_{p=0}^{q} M^p S h_0 + W_q, \qquad (4.27)$$

where V_q, W_q are the remainders given by

$$V_q = \sum_{p=q+1}^{\infty} L^p U g_0, \qquad W_q = \sum_{p=q+1}^{\infty} M^p S h_0.$$

which are bounded as

$$|||V_p||| \le \frac{\mu^p}{1-\mu} ||g_0||, \qquad |||W_p||| \le \frac{\mu^p}{1-\mu} ||h_0||.$$

μ being the constant < 1 given by (4.16). In (4.27), take the Boltzmann-Grad limit first, and then let p tend to ∞, whence the theorem follows.

Remark 4.4. The proof of Theorem 4.4 is the most difficult part. The theorem for $p = 0$ means that the elastic collision boundary condition can be ignored in the Boltzmann-Grad limit. However, the pointwise convergence cannot be strengthened to norm convergences. This is because the limit of the boundary condition for $U_{N,\sigma}^{(k)}$ is not empty. Also, the proof for the case $p \ge 1$ is not quite trivial because, although L and M are bounded linear operators, the pointwise convergence is not the norm convergence. The structures of the operators should be examined carefully, using the 'collision trees'. All these details are found in [2].

The validity of the Boltzmann's ansatz (2.20) can be established with the aid of Theorems 3.3 and 4.2. Let $f_0 \in X_{\rho_0}$. Theorem 3.3 assures the existence of a unique solution f to (3.26) for the initial f_0. With them, define $h_0 = (f_0^{(k)})_{k \in \underline{\nu}}$ by

$$f_0^{(k)}(z^{(k)}) = \prod_{i=1}^{k} f_0(z_i), \qquad k \in \underline{\nu}, \tag{4.28}$$

and $h = (f^{(k)})_{k \in \underline{\nu}}$ by

$$f^{(k)}(t, z^{(k)}) = \prod_{i=1}^{k} f(t, z_i), \qquad k \in \underline{\nu}. \tag{4.29}$$

It is easy to check that this h satisfies (4.22) with h_0 as the initial. On the other hand, if we put,

$$\alpha_0 = \log \|f_0\|_{\rho_0},$$

then, clearly,

$$h_0 \in Z_{\alpha_0, \rho_0},$$

and hence Theorem 4.2 assures the existence of a solution h to (4.22) in the class (4.18). On the other hand Theorem 3.3 assures, with a change of the constant γ if necessary, that the factorized h in (4.29) is also in the class (4.18). By virtue of the uniqueness established in Theorem 4.2, therefore, these two h are identical to each other. Thus, we proved,

Theorem 4.5. *Let f_0 and f be those of Theorem 3.3. Then, the solution h of (4.22) is factorized by f as in (2.20) or (4.29) on the time interval $[0, T]$ if and only if the initial h_0 is factorized by f_0 as in (4.28).*

Finally, combining these two theorems confirms the validity of the diagram in Figure 4.

Theorem 4.6. *The solution to the BBGKY hierarchy (4.15) which is equivalent to the Newton equation (2.3) converges to the solution to the Boltzmann equation (3.26) in the sense that the convergence (4.25) with the limit h_0 given by (4.28) assures the convergence (4.26) with the limit h given by (4.29).*

Remark 4.5. The ansaz of the molecular chaos asserts that particles are random, that is, stochastically independent, which is possible for a rarefied gas because collisions are rare. However, this is not a consequence of the deterministic mechnical law. Theorem 4.5 reveals the nature of this randomness: It is satisfied for the time interval $[0, T]$ if it is at the initial time $t = 0$.

Remark 4.6. As proved in [2], this theorem also provides a mathematical reasoning of the loss of time reversibility in the Boltzmann equation stated in Remark 2.2. It is due to the weaker convergence in (4.26) compared with (4.25). More precisely, a stronger convergence for (4.26) leads to a contradiction to the H-theorem. To show this by contradiction, admit for a moment that Theorem 4.3 be strengthened enough that the condition (4.26) results in the convergence of the same strength for (4.26). Consider the situation in Theorem 4.5, and at time $t = t_0$, reverse the sign of velocities v_i in the solution $P_{N,\sigma}^{(k)}(x^k, v^k, t)$, namely, make $P_{N,\sigma}^{(k)}(x^k, -v^k, t_0)$. With this as the initial data, solve again the BBGKY hierachy. By our assumtion, Theorem 4.3 again applies and a unique solution, say $R_{N,\sigma}^{(k)}(x^k, v^k, t)$, exists and enjoys the convergence property (4.26). Denote its limit by $g^{(k)}$. Obviously, it is also in a factorized form with a unique solution g of the Boltzmann equation for the initial data $f(x, -v, t_0)$ The point is now clear: If the convergence is weaker, the existence of such a factorized g is not assured. Since the classical mechanics is time reversible, it must hold that $R_{N,\sigma}^{(k)}(x^k, v^k, t_0) = P_{N,\sigma}^{(k)}(x^k, -v^k, 0)$. From the uniqueness of the limit, it follows that $g(x, v, t_0) = f(x, -v, 0)$. Now we have the contradiction stated in Remark 2.2.

5 Fluid Dynamical Limits

Roughly speaking, the existing methods for establishing macroscopic limits of the Boltzmann equation depicted in Figure 5 may be classified to three different methods. Each method has its own advantage and disadvantage.

1. Method of spectral analysis [20], [13], [29], [32], [33]: This method starts with the study of the the semi-group generated by the linearized Boltzmann equation obtained from (2.39). The spectral analysis gives estimates both in t and ϵ sharp enough to give a strong control on the convergence of nonlinear problems. Moreover, it is not necessary to restrict the initial data to well prepared, that is, macroscopic data. For the incompressible limit, ill-prepared data may generate acoustic waves, but this method enables us to trace the pathological waves in detail and eventually, to reveal a mathematical mechanism of the development of the initial layer. A serious disadvantage of this method is that it requires the smallness condition in the nonlinear regime as well as strong regularity conditions both on initial data and solutions. In particular, this method has been successful for the compressible Euler limit only in the framework of analytic functions in x.

2. Method of the Hilbert-Chapman-Enskog expansion [21], [23]: This method starts with expanding the solution of (2.39) in a series of ϵ to some finite power. Then, for well-prepared data, the residual can be estimated uniformly. Regular solutions exist for (2.39) for sufficiently small ϵ as long as the corresponding limit macroscopic fluid equations have regular solutions, that is, before they devlop singularities including shock. Solutions and initial data must be regular enough but are not required to be analytic functions. The well-preparedness is necessary, so that this method deals with the case where initial layers do not develop.

3. Method of L^1 compactness. [25],[37] This method is developing recently. It relies on the relative entropy control (H-theorem) and the 'velocity averaging lemma'. The main result is the weak L^1 convergence of the Diperna-Lions renormalized solutions of the Boltzmann equation ([12]) to Leray's weak solutions of the Navier-Stokes equation. This is indeed a big

progress on this subject. Both the smallness and regularity conditions are not required. The convergence is not strong enough to make visible initial layers, however. When the data are well-prepared data, the convergence becomes the strong L^1 convergence.

In the below, we discuss only Method 1. For the other methods, the reader is referred to the references mentioned above and references therein.

5.1 Preliminary

We thus verify the diagram of Figure 5 by Method 1. To start, recall the scaled Boltzmann equation (2.39) and definitions (2.40), (2.41) of the operator L and Γ. We shall rewrite (2.39) in the form of the evolution equation

$$\begin{cases} \dfrac{dg^\epsilon}{dt} = B^\epsilon_\alpha g^\epsilon + \dfrac{1}{\epsilon^{1+\alpha-\beta}} \Gamma(g^\epsilon), \quad t > 0, \\ g^\epsilon(+0) = g_0, \end{cases} \tag{5.1}$$

where

$$B^\epsilon_\alpha = \frac{1}{\epsilon^{\alpha+1}}(-\epsilon v \cdot \nabla_x + L), \tag{5.2}$$

is the *linearlized Boltzmann operator*.

We shall consider the mild solution of (5.1), which is defined by transforming (5.1), by the Duhamel formula, into the integral form:

$$g^\epsilon(t) = U^\epsilon_\alpha(t)g_0 + \frac{1}{\epsilon^{\alpha-\beta}} J^\epsilon_\alpha[g^\epsilon](t), \tag{5.3}$$

where

$$U^\epsilon_\alpha(t) = e^{tB^\epsilon_\alpha}, \tag{5.4}$$

is the semi-group generated by the linearized Boltzmann operator (5.2), whereas

$$J^\epsilon_\alpha[g^\epsilon](t) = \frac{1}{\epsilon} \int_0^t U^\epsilon_\alpha(t-s)\Gamma[g^\epsilon(s)]ds, \tag{5.5}$$

is a nonlinear term. The unbounded factor ϵ^{-1} is included in J because it can be controlled by the smoothing effect of integration in t. Thus, the only unbounded factor is $\epsilon^{\beta-\alpha}$ for $\beta < \alpha$ in front of J.

To verify Figure 5, we need uniform estimates. According to Theorem 3.3, for each $\epsilon > 0$ fixed, (5.3) has a local solutions on the time interval $[0, T]$ for any initial, but the life span $T \to 0$ as $\epsilon \to 0$ (Remark 3.6). It is also known, [16], [17], that a global solution exists for a small initial (if limited to the strong solution), but the size of the initial specified in [16],[17] should be smaller with smaller ϵ. Thus, the existing results do not provide us with any uniform estimates.

The integral equation (5.3) tells us that g^ϵ is the sum of $U^\epsilon_\alpha(t)g_0$ and $\epsilon^{\beta-\alpha}J$. Thus, to establish the convergence and find the limit of g^ϵ, we should verify at least two following claims:

(1) The semi-group $U_\alpha^\epsilon(t)$ converges to a limit $U_\alpha^0(t)$ as $\epsilon \to 0$.

(2) The integrals $J_\alpha^\epsilon[g^\epsilon]$ are uniformly bounded for small $\epsilon > 0$.

Indeed, once these claims are justified, we can immediately deduce the following.

(a) The case $0 \le \alpha < \beta$: Since $\epsilon^{\beta-\alpha} \to 0$ as $\epsilon \to 0$, the second term of the right hand side of (5.3) vanishes, and hence, we have

$$g^\epsilon(t) \to g^0(t) \equiv U_\alpha^0(t)g_0 \quad (\epsilon \to 0). \tag{5.6}$$

by virtue of the claim (1). As will be discussed below, the reduced (limit) equations for the limits g^0 are linear equations.

(b) The case $0 \le \beta < \alpha$: Since the factor $\epsilon^{\beta-\alpha}$ tends to infinity, the second term of the right hand side of (5.3) diverges unless $J^\epsilon \to 0$, as $\epsilon \to 0$. In this case, g^ϵ has no limits, and hence no reduced equations.

(c) The case $0 \le \alpha = \beta$: This is the only possible case where the reduced equation becomes a nonlinear equation, if any.

Before going further, we mention a special case of the case (a) :

(a') The case $1 < \alpha \le \beta$. In this case, $U_\alpha^0(t) = 0$ for $t \ne 0$. Hence, so does $g^0(t)$, and the reduced equations are trivial. This will been seen in Theorem 5.1 below.

Now, we shall look at the limit g^0 of (5.6) more closely. To this end, we recall the null space $\mathcal{N}$ of L and the associated projectons P^0 defined by (2.46) and (2.57). Normarize the basis (2.45) as

$$\begin{cases} \psi_0(v) = M_0^{1/2}(v), \\ \psi_i(v) = v_i M_0^{1/2}(v) \quad (i = 1, 2, 3), \\ \psi_4(v) = \frac{1}{\sqrt{6}}(|v|^2 - 3)M_0^{1/2}(v), \end{cases} \tag{5.7}$$

to write P^0 explicitly in the form

$$P^0 g = \sum_{i=0}^{4} <g, \psi_i> \psi_i(v), \tag{5.8}$$

where $<,>$ is the inner product in $L^2(\mathbb{R}_v^3)$. For any fixed (t, x) and for any function $g(t, x, v)$, the orthogonal decomposition

$$g = P^0 g + (I - P^0)g$$

gives the "hydrodynamic (macroscopic)" part $P^0 g$ and the "kinetic (microscopic)" part $(I - P^0)g$. The hydrodynamic part plays a key role in diagrams of Figure 5. Indeed, it already appears in the convergence statement (5.6): It asserts that the limit of g^ϵ, denoted again by g^0, is in $\mathcal{N}$ with the form

$$g^0 = \eta\psi_0 + \sum_{i=1}^{3} u_i\psi_i + \theta\psi_4, \tag{5.9}$$

where the coefficients $(\eta, \ u = (u_1, u_2, u_3), \theta)$ are to solve the fluid equations in Figure 5. In view of the definition (5.8) of P^0,

$$\eta =< g^0, \psi_0 >, \quad u_i =< g^0, \psi_i > \quad (i = 1, 2, 3), \quad \theta =< g^0, \psi_4 > . \tag{5.10}$$

Now, a remark concerning the limit equation and its initial condition in the case (a). The limit $U_\alpha^0(t)$ given in (5.6) happens to be a semi-group generated by some linear operator, say, B_α^0. Then, the reduced equation for the limit g^0 is a linear evolution equation

$$\frac{d}{dt}g^0(t) = B_\alpha^0 g^0(t), \quad t > 0. \tag{5.11}$$

Since the convergence (5.6) suggests the convergence

$$P^0 g^\epsilon(t) \to P^0 g^0(t) = g^0(t) \quad (\epsilon \to 0),$$

(5.11) is an equation for the quantities (η, u, θ) in (5.10), and putting $t = 0$ in the above and since $g^\epsilon(0) = g_0$, it follows that the initial condition to (5.11) should be of the form

$$g^0(+0) = P^0 g_0. \tag{5.12}$$

This can be rewritten, using (5.10), as the initial condition for (η, u, θ);

$$(\eta, u, \theta)(+0) = (\eta_0, u_0, \theta_0), \tag{5.13}$$

where the initial data (η_0, u_0, θ_0) is determined by g_0 through (5.10),

$$\eta_0 =< g_0, \psi_0 >, \quad u_{0,i} =< g_0, \psi_i > \quad (i = 1, 2, 3), \quad \theta_0 =< g_0, \psi_4 > .$$

This causes the initial layer. Indeed, since $U_\alpha^\epsilon(0) = id$, so is the limit $U_\alpha^0(0)$, On the other hand, $U_\alpha^0(t)$ is continuous for $t > 0$ up to $t = 0$ and (5.12) implies $U_\alpha^0(+0) = P^0$. In other words, $U_\alpha^0(t)$ is discontinuous at $t = 0$ with

$$U_\alpha^0(+0) \equiv P^0 \neq id = U_\alpha^\epsilon(0), \tag{5.14}$$

which implies that the convergence (5.6) is not a uniform convergence near $t = 0$. This phenomenon is called the initial layer and depicted in Figure 6. Clearly, this happens also in the case (c).

5.2 Main Theorems

Here, we state the claims (1) and (2) in further detail. The claim (1) is substantiated in the space $L^2 = L^2(\mathbb{R}^6)$. Put

$$g^\epsilon(t) = U_\alpha^\epsilon(t)g_0.$$

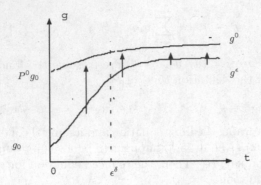

Fig. 6. Initial Layer

Theorem 5.1. (i) *For any initial $g_0 \in L^2$, the limit $g^0(t) = U_\alpha^0(t)g_0$ exists as $\epsilon \to 0$ for each $t \geq 0$ and $\alpha \geq 0$ in the strong topology of L^2. The convergence is locally uniform in $t > 0$, and so, the limit $g^0(t)$ is continuous in $t > 0$. Further, $g^0(t)$ is continuous up to $t = 0$ with a strong limit $g^0(+0)$. However, in general, the convergence is not uniform near $t = 0$ and the limit $U_\alpha^0(t)$ is discontinuous at $t = 0$: We have*

$$U_\alpha^\epsilon(0) \equiv id \ \neq \ U_\alpha^0(+0) = \begin{cases} 0, & \alpha > 1, \\ P^0, & 0 \leq \alpha \leq 1, \end{cases} \tag{5.15}$$

where P^0 is the projection defined by (5.8).

A necessary and sufficient condition that the convergence is uniform up to $t = 0$ and $g^0(t)$ is continuous at $t = 0$ is that $g_0 \in \mathcal{N}$.

(ii) *The limit $g^0(t)$ is in the space $\mathcal{N}$ with the form in (5.9) whose coefficients $(\eta, \ u = (u_1, u_2, u_3), \theta)$ solve the equations*
  - *(2.48) (LCE) if $\alpha = 0$,*
  - *(2.52) (LICE) if $0 < \alpha < 1$,*
  - *(2.54) (LINS) if $\alpha = 1$,*
for $t > 0$ and satisfy the initial condition (5.12). However, when $\alpha > 1$, $g^0(t)$ is identically 0 for $t > 0$.

Remark 5.1. The last part of (i) implies that when $\alpha \in [0, 1]$, the initial layer develops if and only if the initial $g_0 \notin \mathcal{N}$.

Remark 5.2. Thus, the initial layer comes from the linear semi-group $U_\alpha^\epsilon(t)$, and in this sense, we can say that the initial layer is a linear phenomenon. On the other hand, the boundary layer is a nonlinear phenomenon.

The claim (2) should be stated in different function spaces according to the value of α. We begin with the case $\alpha \in [0, 1)$. Let $X_\rho = X_\rho(\mathbb{R}^3)$ be the space (3.17) of functions $u(x)$ given in Example 1 of $3 with $n = 3$, and define

$$Y_\rho = L_v^\infty(\mathbb{R}^3, (1+|v|)^b dv; X_\rho(\mathbb{R}_x^3)) \quad (b > 3/2).$$

$$\|g\|_\rho = \sup_{v\in\mathbb{R}^3}\left\{(1+|v|)^b\|g(\cdot,v)\|_\rho\right\}$$

Also, with any constants $\rho_0, \gamma > 0$, put

$$T = \frac{\rho_0}{2\gamma}, \qquad \rho(t) = \rho_0 - \gamma t, \tag{5.16}$$

and define the space and norm,

$$Y = \{g(x,v,t) \mid g \in C^0([0,T];Y_0), \ g(t) \in Y_{\rho(t)}, \ t \in [0,T]\}, \tag{5.17}$$

$$\||g\|| = \sup_{t\in[0,T]} \|g(t)\|_{\rho(t)}.$$

Theorem 5.2. *Suppose $\alpha \in [0,1)$. For any $\rho_0, \gamma > 0$, the following holds with T defined by (5.16) and Y by (5.17).*

(i) *If $\{g^\epsilon\}_{\epsilon>0}$ is a bounded set in Y, so is $\{J_\alpha^\epsilon[g^\epsilon]\}_{\epsilon>0}$.*

(ii) *If g^ϵ has a nonzero limit in Y as $\epsilon \to 0$, so does $J_\alpha^\epsilon[g^\epsilon]$.*

This local estimate cannot be extended globally in t. Moreover, note that functions involved here must be analytic in x.

For the case $\alpha \geq 1$, on the other hand, a global in t estimate is possible for less smooth functions. Let $H^s = H^s(\mathbb{R}^3)$ be the usual L^2 Sobolev space and introduce

$$Z_T = L_{v,t}^\infty([0,T] \times \mathbb{R}^n, (1+|v|)^b dv dt; H^s(\mathbb{R}_x^3)), \qquad s, b \in \mathbb{R}.$$

Theorem 5.3. *Suppose $\alpha \in [1,\infty)$. For any $s > 3/2$, $b > 5/2$ and $T > 0$, the following holds.*

(i) *If $\{g^\epsilon\}_{\epsilon>0}$ is a bounded set in Z_T, so is $\{J_\alpha^\epsilon[g^\epsilon]\}_{\epsilon>0}$.*

(ii) *If $\alpha = 1$ and if g^ϵ has a nonzero limit in Z_T as $\epsilon \to 0$, so does $J_\alpha^\epsilon[g^\epsilon]$.*

(iii) *If $\alpha > 1$ and if $\{g^\epsilon\}_{\epsilon>0}$ is a bounded set in Z_T, $J_\alpha^\epsilon[g^\epsilon]$ converges to 0 as $\epsilon \to 0$ in Z_T.*

Here $T > 0$ is arbitrary, and the statements are valid also with $T = \infty$.

As noted above, these three theorems lead almost to the following theorem verifying the diagram of Figure 5.

Theorem 5.4. (i) *The case $0 \leq \alpha \leq \beta < 1$: Let $\rho_0 > 0$. Then, there exist positive numbers a_0 and γ such that for any initial*

$$g_0 \in Y_{\rho_0}, \qquad \|g_0\|_{\rho_0} \leq a_0,$$

the following holds with T defined by (5.16) and Y by (5.17).

(a) *For each $\epsilon > 0$, (5.3) has a unique solution $g^\epsilon \in Y$ on the time interval $[0, T]$ independent of ϵ and the convergence*

$$g^\epsilon(t) \to g^0(t) \quad (\epsilon \to 0) \quad \text{in } Y_{\rho(t)} \quad \text{for each } t \in [0, T] \qquad (5.18)$$

holds with a limit $g^0(t) \in Y_{\rho(t)}$.

(b) *The limit g^0 has the the form (5.9) and its coefficients $(\eta, \ u = (u_1, u_2, u_3), \theta)$ solve the linear or nonlinear fluid equations indicated in Figure 5 according to the choice of parameters (α, β).*

(c) *Unless $g_0 \subset \mathcal{N}$, the convergence (5.18) is locally uniform on $(0, T]$ but not uniform near $t = 0$, and so the initial layer develops as $\epsilon \to 0$.*

(ii) *The case $1 \le \alpha \le \beta$: If g_0 is sufficently small, the statement (i) including the initial layer is true in the space Y replaced by Z_T for any time interval $[0, T]$. Moreover, $g^0 = 0$ for $t > 0$ if $\alpha > 1$,*

(iii) *The case $\alpha > \beta$: g^ϵ diverges and not limit.*

Remark 5.3. In both (i) and (ii), the smallness condition is required on the initials. Of course, in case $\beta > \alpha$ strictly, then a_0 depends on ϵ in such a way that $a_0 \to \infty$ as $\epsilon \to 0$. This is natural because the limit equation is linear when $\alpha < \beta$.

The proofs of the above theorems are found in [29], [33], [20].... Here, we present their sketches.

5.3 Proof of Theorem 5.1

Our proof is based on the spectral analysis of the linearized Boltzmann operator B_α^ϵ given in [24], [16], [17], [3]. It gives a semi-explicit formula of the Fourier transform (in x) of the semigroup $U_\alpha^\epsilon(t)$. For any function $u = u(x, v)$, let

$$\mathcal{F}(u)(x, \xi) = \frac{1}{(2\pi)^{3/2}} \int_{\mathbb{R}^3} e^{i\xi \cdot x} u(x, v) dx, \quad \xi \in \mathbb{R}^3 \quad (i = \sqrt{-1})$$

be its Fourier transform with respect to x. Then,

$$\mathcal{F}(B_\alpha^\epsilon u) = \frac{1}{\epsilon^{\alpha+1}} (-i\epsilon v \cdot \xi + L)\mathcal{F}(u).$$

This implies that if

$$\hat{B}_\alpha^\epsilon(\xi) = \frac{1}{\epsilon^{\alpha+1}} (-i\epsilon v \cdot \xi + L)$$

is taken to be an operator in the space $L^2 = L^2(\mathbb{R}_v^3)$ with $\xi \in \mathbb{R}^3$ as a parameter, its semigroup gives a parametric representation of the Fourier transform of the semigroup $U_\alpha^\epsilon(t)$. Denote it by $\Phi(\xi, t)$;

$$\Phi(\xi, t) = e^{t\hat{B}_\alpha^\epsilon(\xi)} = \mathcal{F}(U_\alpha^\epsilon(t)).$$

For simplicity of notation, α, ϵ are dropped from the symbol Φ.

Theorem 5.5. *The following formula holds.*

$$\Phi(\xi,t) = \sum_{j=0}^{4} e^{\lambda_j(\xi)t} P_j(\xi)\chi(\xi) + \phi(\xi,t). \tag{5.19}$$

The meanings of notations are as follows. $(\lambda_j(\xi), P_j(\xi))$ *are the eigenvalues and eigenprojections of* $\hat{B}_\alpha^\epsilon(\xi)$:

$$\hat{B}_\alpha^\epsilon(\xi)P_j(\xi) = \lambda_j(\xi)P_j(\xi), \tag{5.20}$$

$P_j(\xi)$ *being a projection (not necessarily orthogonal) on the space* L^2. *Put*

$$\xi = \kappa\tilde{\xi}, \quad \kappa = |\xi|, \qquad \tilde{\xi} = \frac{1}{|\xi|}\xi \in S^{n-1}.$$

There exists a positive number κ_0, *and* $(\lambda_j(\xi), P_j(\xi))$ *has an asymptotic expansion*

$$\lambda_j(\xi) = \frac{1}{\epsilon^\alpha}\{i\lambda_{j,1}\kappa - \epsilon\lambda_{j,2}\kappa^2 + O(\epsilon^2|\kappa|^3)\}, \tag{5.21}$$

$$P_j(\xi) = P_{j,0}(\tilde{\xi}) + \epsilon\kappa O(1),$$

for $|\kappa| \leq \kappa_0/\epsilon$, *with some constants*

$$\lambda_{j,1} \in \mathbb{R}, \quad \lambda_{j,2} > 0,$$

and

$$\sum_{j=0}^{4} P_{j,0}(\tilde{\xi}) = P^0$$

holds for all $\tilde{\xi} \in S^{n-1}$. *Under a suitable numbering for* j, *we have*

$$\lambda_{j,1} \neq 0 \quad \text{for } j = 0, 4, \qquad \lambda_{j,1} = 0 \quad \text{for } j = 1, 2, 3. \tag{5.22}$$

Further, $\chi(\xi)$ *is the characteristic function for* $|\xi| \leq \kappa_0/\epsilon$.

Finally, $\phi(\xi,t)$ *is a linear bounded operator on* $L^2(\mathbb{R}_v^3)$ *with the operator norm*

$$\|\phi(\xi,t)\| \leq c\exp\left(-\frac{\sigma_0 t}{\epsilon^{\alpha+1}}\right), \quad \xi \in \mathbb{R}^3, \ t \geq 0, \tag{5.23}$$

with some constants $c, \sigma_0 > 0$ *independent of* ϵ, ξ *and* t.

Sketch of proof. Here, we follow [24]. First, the matrix representation of the operator $A(\tilde{\xi}) = P^0 v \cdot \tilde{\xi} P^0$ in the basis (5.7) is a 5×5 matrix given by

$$\sqrt{\begin{pmatrix} 0 & \tilde{\xi} & 0 \\ \tilde{\xi}^t & 0 & \gamma\tilde{\xi}^t \\ 0 & \gamma\tilde{\xi} & 0 \end{pmatrix}}$$

where $\gamma = \sqrt{2/3}, \tilde{\xi} = (\tilde{\xi}_1, \tilde{\xi}_2, \tilde{\xi}_3)$ (row vector) and ^{t} means the transposition. It is easy to compute its eigenvalues and normalized eigenvectors as

$$
\begin{cases}
\lambda_{0,1} = c, & \varphi_{0,1}(\tilde{\xi}) = \dfrac{1}{\sqrt{2}c}(\psi_0 + \tilde{\xi} \cdot \psi' + \gamma\psi_4), \\[2mm]
\lambda_{1,1} = 0, & \varphi_{1,1}(\tilde{\xi}) = \dfrac{1}{c}(\gamma\psi_0 - \psi_0), \\[2mm]
\lambda_{j,1} = 0, & \varphi_{j,1}(\tilde{\xi}) = C^j(\tilde{\xi}) \cdot \psi' \quad (i = 2,3), \\[2mm]
\lambda_{0,4} = -c, & \varphi_{4,1}(\tilde{\xi}) = \dfrac{1}{\sqrt{2}c}(\psi_0 - \tilde{\xi} \cdot \psi' + \gamma\psi_4),
\end{cases}
$$

where ψ' and C^j are 3-dimensional normalized vectors such that

$$
\psi' = (\psi_1, \psi_2, \psi_3) = v\psi_0(v), \quad C^2(\tilde{\xi}) \cdot C^3(\tilde{\xi}) = \tilde{\xi} \cdot C^j(\tilde{\xi}) = 0 \quad (j = 2,3).
$$

Now we consider the eigenvalue problem (5.20) for $\epsilon = 1$ in the form

$$
B_\alpha^1(\xi)\varphi = -\kappa\lambda\varphi. \tag{5.24}
$$

Decomposing $\varphi = P^0\varphi + (I - P^0)\varphi = \varphi^0 + \varphi^1$ and applying P^0 and $P^1 = I - P^0$ to (5.24), we get

$$
iP^0 v \cdot \tilde{\xi}\varphi^0 + iP^0 v \cdot \tilde{\xi}\varphi^1 = \lambda\varphi^0,
$$
$$
-i\kappa P^1 v \cdot \tilde{\xi}\varphi^0 - i\kappa P^1 v \cdot \tilde{\xi}\varphi^1 + L\varphi^1 = -\kappa\lambda\varphi^1,
$$

where we have used properties of L in Proposition 2.1. Since L^{-1} exists on $\mathcal{N}^\perp$, and since κ may be assumed small in our situation, the second equation can be solved as

$$
\varphi^1 = i\kappa(L + \kappa\lambda P^1 - i\kappa P^1 v \cdot \tilde{\xi}P^1)^{-1} P^1 v \cdot \tilde{\xi}\varphi^0.
$$

Insert this to the first equation to get

$$
\lambda\varphi^0 = iA(\tilde{\xi})\varphi^0 - i\kappa P^0 v \cdot \tilde{\xi}P^1(L + \kappa\lambda P^1 - i\kappa P^1 v \cdot \tilde{\xi}P^1)^{-1} P^1 v \cdot \tilde{\xi}\varphi^0. \tag{5.25}
$$

If we plug the forms

$$
\lambda = i\lambda_1 + \kappa\lambda_2, \qquad \varphi^0 = \varphi_0^0 + \kappa\varphi_1^0
$$

and put $\kappa = 0$, we get the eigenvalue problem for $A(\tilde{\xi}); \lambda_1\varphi_0^0 = A(\tilde{\xi})\varphi_0^0$. Thus we have already obtained $\lambda_1 = \lambda_{j,1}$ and $\psi_0 = \psi_{j,1}, j = 0, \cdots, 4$. Expand (5.25) in the power series of κ. The κ^1 term gives

$$
\lambda_2\psi_{j,1} + i\lambda_{j,1}\varphi_1^0 + = iA(\tilde{\xi})\varphi_1^0 - P^0 v \cdot \tilde{\xi}P^1 L^{-1} P^1 v \cdot \tilde{\xi}P^0\psi_{j,1}.
$$

Take the inner product in $L^2(\mathbb{R}_v^3)$ of this and $\varphi_{j,1}$, we arrives at

$$
\lambda_2 = \lambda_{j,2} = - < L^{-1} P^1 v \cdot \tilde{\xi}\psi_{j,1}, P^1 v \cdot \tilde{\xi}\psi_{j,1} > \quad (j = 0, \cdots, 4)
$$

It is now easy to see that $\lambda_{j,2}$ is nothing but viscosity coefficient ν or thermal diffusivity μ given in (2.62). Actually, we have

$$\lambda_{0,2} = \lambda_{4,2} = \mu, \qquad \lambda_{i,2} = \nu \quad (i = 1, 2, 3),$$

and

$$P_{j,0} h = < h, \psi_{j,1} > \psi_{j,1}$$

for the expansion (5.21). The proof of (5.23) is lengthy and omitted. See [3], [16], [17].

We now look at the first term of (5.19).

Lemma 5.1. *For each fixed $t > 0$ and $\kappa \neq 0$, the following limits exist in the sense of distribution.*
(i) *For $\lambda_{j,1} \neq 0$,*

$$\lim_{\epsilon \to 0} e^{\lambda_j(\xi)t} \chi(\xi) = \begin{cases} e^{i\lambda_{j,1}\ t}, & \alpha = 0, \\ 0 & \alpha > 0. \end{cases}$$

(ii) *For $\lambda_{j,1} = 0$,*

$$\lim_{\epsilon \to 0} e^{\lambda_j(\xi)t} \chi(\xi) = \begin{cases} 1, & 0 \leq \alpha < 1, \\ e^{-\lambda_{j,2}|\xi|^2 t}, & \alpha = 1, \\ 0, & \alpha > 1. \end{cases}$$

Proof. (5.21) yields

$$e^{\lambda_j(\xi)t} = \exp\left\{ i\frac{\lambda_{j,1}\kappa t}{\epsilon^\alpha} \right\} \exp\left\{ -\epsilon^{1-\alpha}\kappa^2(\lambda_{j,2} + \epsilon O(|\kappa|))t \right\}.$$

Suppose $\lambda_{j,1} \neq 0$. If $\alpha = 0$, the last exponential function tends to 1 as $\epsilon \to 0$ for each fixed κ, whence the limit is just the first exponential function on the right hand side. On the other hand, if $\alpha > 0$, the same first exponential function goes to 0 by virtue of the Riemann-Lebesque theorem, in the distribution sense, if $\kappa t \neq 0$, whereas the second exponential function on the right hand side tends to 1 or is bounded or tends to 0 according to $\alpha \in (0, 1)$, $\alpha = 1$ or $\alpha > 1$, for fixed κ. Here, the fact $\lambda_{j,2} > 0$ is crucial. Thus, (i) follows.
    Similarly, if $\lambda_{j,1} = 0$, we have

$$e^{\lambda_j(\xi)t} = \exp\left\{ -\epsilon^{1-\alpha}\kappa^2(\lambda_{j,2} + \epsilon O(|\kappa|))t \right\},$$

whence, since $\lambda_{j,2} > 0$ again, (ii) is straight forward.

Remark 5.4. We can see from where are coming the differential operators ∇_x and $\triangle_x$ in the linear fluid equation (2.48),(2.52) and (2.54) in §2.5. First, $e^{-\nu\ ^2 t}(\nu > 0)$ is the symbol of a heat kernel, i.e. a symbol of the semi-group generated by $-\nu\triangle$. On the other hand, $e^{-ic\ t}$ gives rise to a symbol of a semi-group generated by the operator $c\nabla$. This is because the linearized collision operator L is invariant with respect to the rotation in $\mathbb{R}^3$ and hence so is the eigenprojection in (5.21). This observation, combined with Theorem 5.5 and Lemma 5.1, now implies almost Theorem 5.1 as well as the forms of the linear fluid equations in Figure 5.

Stronger convergences are known for (i) of Lemma 5.1. This is seen from the following lemma based on the stationary phase method. For any function $w_0(x), x \in \mathbb{R}^3$, define

$$w(x,t) = \mathcal{F}^{-1}(e^{i|\xi|t}\hat{w}_0).$$

Lemma 5.2. *For any $\ell > 3$, there are positive numbers C, δ such that for any w_0,*

$$|w(x,t)| \leq C(1+t)^{-\delta}(\|w_0\|_{H^\ell(\mathbb{R}^3)} + \|w_0\|_{L^1(\mathbb{R}^3)}),$$

holds for all $t \geq 0$.

Proof. Let $R > 0$ be a constant to be defined later, and decompose w as

$$w(x,t) = c_0 \int_{\mathbb{R}^3} e^{ix\cdot\xi} e^{ic|\xi|t} \hat{w}_0(\xi)$$

$$= c_0 \int_{|\xi|\leq R} + c_0 \int_{|\xi|>R} = w_1 + w_2,$$

where $c_0 = (2\pi)^{-3/2}$. It is immediate that

$$|w_2(x,t)| \leq C(1+R)^{\ell-3}\|w_0\|_{H^\ell(\mathbb{R}^3)}.$$

Write

$$w_2(x,t) = c_1 \int_{\mathbb{R}^3} H(x-y,t)w_0(y)dy \quad (c_1 = c_0^2),$$

where

$$H(x,t) = \int_{|\xi|\leq R} e^{i(x\cdot\xi+|\xi|t)}d\xi.$$

In the polar coordinate, with $r = |\xi|$,

$$H(x,t) = 2\pi \int_0^R e^{irt} \frac{\sin|x|r}{|x|r} r^2 dr,$$

and by integration by parts,

$$= \frac{2\pi}{it}\left\{\left[e^{irt}\frac{\sin|x|r}{|x|r}r^2\right]_0^R - \int_0^R e^{irt}\frac{\sin|x|r - r|x|\cos|x|r}{|x|}dr\right\}.$$

Hence,

$$|H(x,t)| \leq C(1+t)^{-1}R^3,$$

which, in turn, gives

$$|w_2(x,t)| \leq C(1+t)^{-1}R^3\|w_0\|_{L^1(\mathbb{R}^3)}.$$

Choosing $R = (1+t)^{3/(\ell-3)}$ concludes the lemma.

Remark 5.5. This lemma strengthens the convergence in Lemma 5.1 for the case (i) with $\alpha > 0$. In fact, we have,

$$|w(x,\frac{t}{\epsilon^\alpha})| \leq C\frac{\epsilon^\alpha}{t}(\|w_0\|_{H^\ell(\mathbb{R}^3)} + \|w_0\|_{L^1(\mathbb{R}^3)}).$$

Since $w(x,0) = w_0(x)$, this proves that the convergence is not uniform near $t = 0$.

5.4 Proof of Theorems 5.2 and 5.3

Recall the definition (5.5) of the integral J_α^ϵ. It has two unbounded factors: $1/\epsilon$ in front of the integral and $|v|$ included in the nonlinear operator Γ which is coming from the collision kernel $q = |(v - v*) \cdot \omega|$ as has been seen in Example 2 of §3. We will discuss the hard sphere case only, but can do the same thing for Grad's cutoff hard potential. If the soft potential is concerned, the unbounded factor in v does not appear.

Unfortunately, it is not possible to control both unbounded factors simultaneously. We first show how to control the factor $|v|$. To this end, we shall recall the classical decomposition of the linearized collision operator L for Grad's cutoff potential [5],

$$L = -\nu(v) \times +K, \tag{5.26}$$

where $\nu(v)$ is a function such that

$$0 < \nu_0 \leq \nu(v) \leq \nu_1(1 + |v|),$$

with some positive constants ν_0, ν_1, while K is an integral (in v) operator having the smoothing effects

$$\|Kg\|_{b+1} \leq C\|g\|_b, \quad \|Kg\|_0 \leq C\|g\|_{L^2}, \tag{5.27}$$

for any $b \geq 0$, where the norm is

$$\|g\|_b = \sup_{v\in\mathbb{R}^3} (1 + |v|)^b|g(v)|.$$

The space equipped by this norm will be denoted by L_b^∞. (5.27) is a motivation for the bootstrapping argument we shall use to control the above mentioned unboundedness.

Γ is an unbounded nonlinear operator, but is controlled by the decomposition (5.26) in such a way that

$$||\nu(v)^{-1}\Gamma(g,h)||_b + ||K\Gamma(g,h)||_b \leq C||g||_b||h||_b. \tag{5.28}$$

We use this smoothing effect as follows. Define the operator

$$A_\alpha^\epsilon = -\frac{1}{\epsilon^{\alpha+1}}(\epsilon v \cdot \nabla + \nu(v)\times) \tag{5.29}$$

which is the free streaming term of B_α^ϵ of (5.4), and denote its semi-group by

$$S_\alpha^\epsilon(t) = e^{tA_\alpha^\epsilon}.$$

Then, Duhamel's formula holds;

$$U_\alpha^\epsilon(t) = S_\alpha^\epsilon(t) + \frac{1}{\epsilon^{\alpha+1}}\int_0^t S_\alpha^\epsilon(t-s)KU_\alpha^\epsilon(s)ds.$$

Substitute this into (5.5) to deduce

$$J_\alpha^\epsilon[g^\epsilon](t) = \frac{1}{\epsilon}\int_0^t S_\alpha^\epsilon(t-s)\Gamma[g^\epsilon(s)]ds + \frac{1}{\epsilon^{\alpha+1}}\int_0^t S_\alpha^\epsilon(t-s)KJ_\alpha^\epsilon[g_\alpha^\epsilon](s)ds \tag{5.30}$$

$$\equiv J_1(t) + J_2(t).$$

Here and hereafter, the symbols ϵ, α in the suffix are dropped occasionally, to simplify the notation. The last term J_2 contains K and hence has a gain of regularity due to (5.27) if J_α^ϵ has less regularity. J_1 has also a smoothing effect. To see this, define the operator

$$M(t)h = \int_0^t S(t-s)h(s)ds.$$

Lemma 5.3. *For any $b \geq 0$, there is a constant $C \geq 0$ such that for all $\epsilon > 0$ and $\alpha \geq 0$,*

$$[[M(t)h]]_b \leq C\epsilon^{\alpha+1} \sup_{s\in[0,t]} [[\nu(v)^{-1}h(s)]]_b,$$

holds for all $t \geq 0$, where the norm is that of the space $L^\infty(\mathbb{R}_x^3; L_b^\infty)$,

$$[[h]]_b = \sup_{x\in\mathbb{R}^3} ||h(x,\cdot)||_b.$$

Proof. We can easily have the explicit formula

$$S(t)h(x,v) = \exp\left(-\frac{\nu(v)t}{\epsilon^{\alpha+1}}\right)h(x-vt,v),$$

whence

$$|S(t)h(x,v)| \leq \exp\left(-\frac{\nu(v)t}{\epsilon^{\alpha+1}}\right)(1+|v|)^{-b}\nu(v)\sup_{x\in\mathbb{R}^3}||\nu(v)^{-1}h(x,\cdot)||_b.$$

Substituting this into $M(t)$ yields

$$(1+|v|)^b|M(t)g(x,v,t)| \leq \nu(v)\int_0^t \exp\left(-\frac{\nu(v)(t-s)}{\epsilon^{\alpha+1}}\right)ds$$
$$\times \sup_{s\in[0,t],x\in\mathbb{R}^3}||\nu(v)^{-1}h(x,\cdot,s)||_b.$$

This completes the proof of the lemma.

Thus, M can control both $\epsilon^{-(\alpha+1)}$ and $|v|$. Indeed, applying this lemma to

$$J_1(t) = \epsilon^{-1}M(t)\Gamma(g), \tag{5.31}$$

yields

$$[[J_1(t)]]_b \leq C\epsilon^\alpha \sup_{s\in[0,t]}[[\nu(v)^{-1}\Gamma(g)]]_b \leq C\epsilon^\alpha \sup_{s\in[0,t]}[[g(s)]]_b^2. \tag{5.32}$$

The estimate of J_2 is more delicate. Apply again Lemma 5.3 to J_2,

$$J_2 = \frac{1}{\epsilon^{\alpha+1}}M(t)KJ_\alpha^\epsilon[g], \tag{5.33}$$

to deduce

$$[[J_2(t)]]_b \leq C \sup_{s\in[0,t]}[[\nu(v)^{-1}KJ_\alpha^\epsilon[g](s)]]_b \leq C \sup_{s\in[0,t]}[[J_\alpha^\epsilon[g](s)]]_{b-2}. \tag{5.34}$$

This controls the factor $|v|$ but not ϵ^{-1} in J_α^ϵ.

To overcome this difficulty, we shall go back to Theorem 5.1. Define

$$V_\alpha^\epsilon(t) = U_\alpha^\epsilon(t)(I - P_0),$$

and write its Fourier transform as

$$\Psi(\xi,t) = \mathcal{F}(V_\alpha^\epsilon(t))\mathcal{F}^{-1}. \tag{5.35}$$

Using (5.19), we decompose this as

$$\Psi(\xi,t) = \sum_{j=0}^4 e^{\lambda_j(\xi)t}E_j(\xi)\chi(\xi) + \varphi(\xi,t)(I - P^0) \tag{5.36}$$
$$\equiv \Psi_1(\xi,t) + \Psi_2(\xi,t),$$

with

$$E_j(\xi) = P_j(\xi)(I - P^0) = \epsilon|\xi|O(1),$$

the last term coming from the expansion (5.21).

Recall (5.5) and notice that thanks to Proposition 2.1 (ii),

$$J_\alpha^\epsilon[g^\epsilon](t) = \frac{1}{\epsilon} \int_0^t V_\alpha^\epsilon(t-s)\Gamma[g^\epsilon(s)]ds, \qquad (5.37)$$

holds. The Fourier transform of this is, with the decomposition (5.36),

$$\mathcal{F}(J_\alpha^\epsilon(t))(\xi, v) = \frac{1}{\epsilon} \int_0^t \Big(\Psi_1(\xi, t-s)) + \Psi_2(\xi, t-s)\Big) G(\xi, s)ds \qquad (5.38)$$
$$= \hat{J}_3 + \hat{J}_4,$$

where

$$G(\xi, t) = \mathcal{F}(\Gamma[g^\epsilon(t)])(\xi). \qquad (5.39)$$

The point is to control the unbounded factor ϵ^{-1}. First, we see

$$\epsilon^{-1}\Psi_1(\xi, t) = \sum_{j=0}^4 |\xi|e^{\lambda_j(\xi)t}\chi(\xi)O(1).$$

Thus, the unbounded factor ϵ^{-1} is replaced by the unbounded symbol $a(\xi) = |\xi|$. Since this $a(\xi)$ satisfies the condition (3.21) on the symbol in Example 1 of §3, J_3 can be controlled, similarly, in the space of functions which are real analytic in x and belongs to L_{b-2}^∞ in v. Finally, we recall (5.33) and have

$$\epsilon^{-(\alpha+1)}|||M(t)KJ_3(t)||| \le \frac{1}{\gamma}|||g|||^2. \qquad (5.40)$$

Here, the norm is that of (5.17) and the constant $\gamma > 0$ will be chosen later. The proof is lengthy and omitted here.

It remains to estimate J_4. (5.19) of Theorem 5.1 gives

$$\frac{1}{\epsilon}|| \int_0^t \Psi_2(\xi, t-s)h(s)ds|| \le C\frac{1}{\epsilon} \int_0^t \exp(-\frac{\sigma_0(t-s)}{\epsilon^{1+\alpha}})||h(s)||ds$$

$$\le C\epsilon^\alpha \sup_{s \in [0,t]} ||h(s)||$$

uniformly in $\epsilon > 0, \alpha \ge 0$, with constants $c > 0$ and $\delta \in [0, 1)$. The norm here is that of $L^2(|R^3)$. By virtue of another smoothing effect (5.27) of K, we then have

$$\epsilon^{-1}|||M(t)KJ_4||| \le C\epsilon^\alpha|||g|||^2, \qquad (5.41)$$

with a choice $b = b_0 > 3/2 + 1$ so that $L_b^\infty \subset L^2$. The proof of this final step is again too lengthy to present here.

Finally, the estimate (5.32) can be also put in the same norm as in (5.40) and (5.41) (the proof is also lengthy), so that we get

Lemma 5.4. *For any ρ_0, there is positive numbers γ, C such that*

$$|||J_\alpha^\epsilon[g]||| \le C(1 + \epsilon^\alpha + \frac{1}{\gamma})|||g|||^2,$$

holds for all $\epsilon > 0$ and for all $\alpha \ge 0$.

This lemma gives short time estimates, but for $\alpha \ge 1$ global estimates are possible as stated in Theorem 5.4.

To see this, we notice that if the limit of $e^{\lambda_j(\xi)t}\chi(\xi)$ is like $e^{-a|\xi|^2}, (a > 0)$ as is seen in Lemma 5.10, we get

$$|\xi|e^{-a|\xi|^2 t} \le ct^{-1/2}e^{-a'|\xi|^2 t}$$

with some $0 < a' < a$ and $c > 0$, and thereby J_1 has a uniform estimate in $\epsilon > 0$ in the space Z_T. J_2 can be estimated as before, whence Lemma 5.4 also holds in Z_T with a minor modification, and Theorem 5.3 follows. The detail is omitted.

As for Theorem 5.4 (i), we sketch the proof of the existence assertion. First, consider the case $0 \le \alpha < 1, \alpha \le \beta$.

We note that the semi-group $U_\alpha^\epsilon(t)$ has the uniform estimate both in $\epsilon \ge 0$ and $t \ge 0$,

$$\|U_\alpha^\epsilon(t)g_0\|_{\rho_0} \le C_0\|g_0\|_{\rho_0},$$

where the norm is as in (5.17). Next, We define the nonlinear operator $N[g^\epsilon]$ by the right hand side of (5.3):

$$N[g^\epsilon](t) = U_\alpha^\epsilon(t)g_0 + \frac{1}{\epsilon^{\alpha-\beta}}J_\alpha^\epsilon[g^\epsilon](t), \tag{5.42}$$

and apply the contraction mapping principle. If $\alpha \in [0,1)$, Lemma 5.4 gives the estimate

$$|||N[g]||| \le C_0\|g_0\|_{\rho_0} + C_1\epsilon^{\beta-\alpha}(1 + 1/\gamma)|||g|||^2.$$

and similarly,

$$|||N[g] - N[h]||| \le C_1\epsilon^{\beta-\alpha}(1 + 1/\gamma)|||g + h||| \, |||g - h|||.$$

Here, $C_1 > 0$ is independent of $\epsilon > 0$ and $g \in Y$. Fix $\gamma > 0$. It is standard that these two estimates imply that N is a contraction map on Y if $\|g_0\|_{\rho_0}$ is small. It is seen that this smallness condition cannot be relaxed with larger γ.

For $\alpha = 1 \le \beta$, similar estimates hold in the space Z_T, and we conclude existence of global solutions for small initials.

Clearly, in both cases, if $\beta > \alpha$ strictly, the size a_0 of the smallness condition in Theorem 5.4 can be chosen to go to ∞ as $\epsilon \to 0$.

If $1 < \alpha \le \beta$, everything goes to 0.

Acknowlegement The author would like to express his sincere thanks to the referee for valuable comments and suggestions.

References

General

1. L. Boltzmann, Lectures on Gas Theory (Transl. by S. G. Brush), Dover Publ. Inc. New York, 1964.
2. C.Cercignani, R.Illner and M.Pulvirenti, The Mathematical Theory of Dilute Gases, Appl. Math. Sci. Vol. 109, Springer-Verlag, New-York-Berlin, 1994.
3. S.Ukai, Solutions of the Boltzmann equation, Pattern and Waves – Qualitave Analysis of Nonlinear Differential Equations (eds. M.Mimura and T.Nishida), Studies of Mathematics and Its Applications **18**, pp37-96, Kinokuniya-North-Holland, Tokyo, 1986.
4. Y.Sone, Kinetic Theory and Fluid Dynamics, Publisher:Birkhauser Boston, 2002.

Boltzmann-Grad Limit

5. H.Grad, On the kinetic theory of rarefied gases, Comm. Pure Appl. Math, **2**(1949), 331-407.
6. R.Illner and M.Pulvirenti, Global validity of the Boltzmann equation for a two-dimensional rare gas in vacuum, Commun. Math. Phys., **105**(1986), 189-203.
7. R.Illner and M.Pulvirenti, Global validity of the Boltzmann equation for two- and three-dimensional rare gases in vacuum: Erratum and improved result, Commun. Math. Phys., **121**(1989), 143-146.
8. O.Lanford III, Time evolution of large classical system, Lec. Notes in Phys. **38** (ed. E.J.Moser), pp1-111, Springer-Verlag, New York, 1975.
9. K.Uchiyama, Derivation of the Boltzmann equation from particle dynamics, Hiroshima Math. J.,**18**(1988), 245-297.
10. S.Ukai, The Boltzmann-Grad limit and Cauchy-Kovalevskaya theorem, Japan J. Indust. Appl. Math., **18**(2001), 383-392.

Existence Theorems for the Boltzmann Equation

11. K.Asano, Local solutions to the initial and intial-boundary value problem for the Boltzmann equation with an external force, I, J.Math.Kyoto Univ., **24**(1984), 225-238.
12. R.Diperna and P.L.Lions, On the Cauchy problem for the Boltzmann equation: Global existence and weak stability, Ann. Maths., **130**(1989), 321-366.
13. H.Grad, Asymptotic equivalence of the Navier-Stokes and nonlinear Boltzmann equations, Proc.Symp. Appl.Math. **17** (ed. R.Finn), pp154-183, AMS, Providence.
14. K. Hamdache, Existence in the large and asymptotic behavior for the Boltzmann equation, Japan J. Appl. Math., **2**(1985), 1-15.
15. T.Nishida and K.Imai, Global solutions to the initial value problem for the nonlinear Boltzmann equation, Publ.RIMS, Kyoto Univ., **12**(1976), 229-239.
16. S.Ukai, On the existence of global solutions of a mixed problem for the Boltzmann equation, Proc.Japan Acad., **50**(1974),179-184.
17. S.Ukai, Les solutions globales de l'équation de Boltzmann dans l'espace tout entier et dans le demi-espace, C.R.Acad.Sci.Paris, **282A**(1976), 317-320.

Fluid DI namical Limits of the Boltzmann Equation

18. C.Bardos, F.Golse and D.Levermore, Fluid dynamical limits of kinetic equations: Formal derivation, J.Stat.Phys., **63**(1991), 323-344.
19. C.Bardos, F.Golse and D.Levermore, The acoustic limit of the Boltzmann equation, Arch. Raional Mech. Anal., **153**(2000), 177-204.
20. C.Bardos and S.Ukai, The classical incompressible Navier-Stokes limit of the Boltzmann equation, Math. Models and Methods in Appl. Sci., **1**(1991), 235-257.
21. R.Caflisch, The fluid dynamic limit of the nonlinear Boltzmann equation, Commun. Pure Appl. Math., **33**(1980), 651-666.
22. S.Chapman and T.G.Cowling, The Mathematical Theory of Non-uniform Gases, Cambridge University Press, Cambridge, 3rd ed., 1970.
23. A.DeMassi, R.Esposito and J.L.Lebowitz, Incompressible Navier-Stokes and Euler limits of the Boltzmann equation, Comm.Pure Appl.Math., **42**(1989), 1189-1214.
24. R. S. Ellis and M. A. Pinsky, The first and seconf fluid approximations to the linearized Boltzmann equation, J. Math. Pures Appl., **54**(1972), 1825-1856.
25. F. Golse and Saint-Raymond, Laure, The Navier-Stokes limit for the Boltzmann equation., C. R. Acad. Sci. Paris Ser. I Math., **333**(2001), 897–902.
26. M.Lachowicz, On the hydrodynamical limit of the Enskog equation, Publ. RIMS Kyoto Univ., **34**(1998), 191-210.
27. L. Nirenberg, An abstract form of the nonlinear Cauchy-Kowalewski theorem. Collection of articles dedicated to S. S. Chern and D. C. Spencer on their sixtieth birthdays. J. Differential Geometry, **6** (1972), 561–576.
28. T.Nishida, A note on the theorem of Nierenberg, J.Diff.Geometry, **112**(1977), 629-633.
29. T.Nishida, Fluid dynamical limit of the nonlinear Boltzmann equation to the level of the compressible Euler equation, Comm. Math. Phys., **61**(1978), 119-148.
30. S. Olla, S.R.S. Varadhan, and H.-T. Yau, Hydrodynamical limit for a Hamiltonian system with weak noise. Comm. Math. Phys. **155** (1993), no. 3, 523–560.
31. L. V. Ovsjannikov, Abstract form of the Cauchy-Kowalewski theorem and its application. (Russian) Partial differential equations (Proc. Conf., Novosibirsk, 1978) (Russian), pp. 88–94, 250, "Nauka" Sibirsk. Otdel., Novosibirsk, 1980.
32. S.Ukai, The incompressible limit and the initial layer of the compressible Euler equation, J. Math. Kyoto Univ., **24** (1986), 323-331.
33. S.Ukai and K.Asano, The Euler limit and initial layer of the Boltzmann equation, Hokkaido Math.J., **12**(1983), 311-332.

Boundarl Lal ers of the Boltzmann Equation

34. C. Bardos t R. E. Caflish t B. Nicolaenko, *The Milne and Kramers problems for the Boltzmann equation of a hard sphere gas*, Comm. Pure Appl. Math. **49.** (1986), 323-352.
35. F. Coron, F. Golse, t C. Sulem , *A classification of well-posed kinetic layer problems,* Commun. Pure Appl. Math. **41**, (1988), 409–435.
36. F. Golse t F. Poupaud, *Stationary solutions of the linearized Boltzmann equation in a half-space*, Math. Methods Appl. Sci. **11**, (1989), 483-502,

37. N. Masmoudi and L. St.-Raymond, From the Boltzmann equation to the Stokes-Fourier System in a bounded domain, Commun. Pure Appl. Math., **LVI**(2003), 1263-1293.
38. S. Ukai, T. Yang and S.-H. Yu, Nonlinear boundary layers of the Boltzmann equation: I, Existence, Commun. Math. Phys., **236**(2003), 373-393.
39. S. Ukai, T. Yang and S.-H. Yu, Nonlinear stability of boundary layers of the Boltzmann equation : I, The case $\mathcal{M}^{\infty} < -1$, Commun. Math. Phys., **244**(2004), 99-109.

List of Participants

1. Annalisa Ambroso
   CEA, Saclay, France
   ambroso@cea.fr
2. Fabio Ancona
   Università di Bologna, Italy
   ancona@ciram.unibo.it
3. Arnaud Basson
   ENS Paris, France
   arnaud.basson@ens.fr
4. Jessica Benthuysen
   University of Washington, USA
   jbenthuy@u.washington.edu
5. Stefano Bianchini
   IAC-CMR Roma, Italy
   bianchini@iac.rm.cmr.it
6. Lorenzo Brandolese
   Université Lyon 1, France
   brandolese@igd.univ-lyon1.fr
7. Marco Cannone
   Université de Marne-la-Vallée
   France
   cannone@math.univ-mlv.fr
   (editor)
8. Tomasz Cieslak
   Warsaw University, Poland
   t.cieslak@students.mimuw.edu.pl
9. Peter Constantin
   University of Chicago, USA
   const@math.uchicago.edu
   (lecturer)
10. Aleksander Cwiszewski
    Nicholas Copernicus University
    Poland
    aleks@mat.uni.torun.pl
11. Benedetta Ferrario
    Università di Pavia, Italy
    ferrario@dimat.unipv.it
12. Yoshiko Fujigaki
    Kobe University, Japan
    fujigaki@math.kobe-u.ac.jp
13. Giulia Furioli
    Università di Bergamo, Italy
    giulia.furioli@unibg.it
14. Giovanni Gallavotti
    Università di Roma 1, Italy
    giovanni.gallavotti@roma1.infn.it
    (lecturer)
15. Leonid Gershuni
    Chalmers University, Sweden
    leonid@math.chalmers.se
16. Leo Miguel Gonzàlez
    Universidad Pontificia de Camillas in Madrid, Spain
    lmgonzalez@dfc.icai.upco.es
17. Giovanna Guidoboni
    Università di Ferrara, Italy
    gio@dm.unife.it
18. Naoyuki Ishimura
    Hitotsubashi University, Japan
    ishimura@math.hit-u.ac.jp

19. Petr Kaplicky
    MFF UK, Prague, Czech
    Republic
    kaplicky@karlin.mff.cuni.cz

20. Alexandre Kazhikhov
    Insitute of Hydrodynamics
    Siberia Branch of RAS, Russia
    kazhikhov@hydro.nsc.ru
    (lecturer)

21. Pawel Konieczny
    Warsaw University, Poland
    p.konieczny@students.mimuw.
    edu.pl

22. Zhanna Korotkaya
    Lappeenranta University of
    Technology, Finland
    korotkay@lut.fi

23. Galina Kouteeva
    Saint-Petersburg State
    University, Russia
    star@gk1662.spb.edu

24. Maria Carmela Lombardo
    Università di Palermo, Italy
    lombardo@math.unipa.it

25. Yves Meyer
    CMLA, ENS-CACHAN, France
    ymeyer@cmla.ens-cachan.fr
    (lecturer)

26. Tetsuro Miyakawa
    Kanazawa University, Japan
    miyakawa@kenroku.kanazawa-
    u.ac.jp (editor)

27. Masaaki Nakamura
    Nihon University, Tokyo, Japan
    nakamura@math.cst.nihon-
    u.ac.jp

28. Michal Olech
    University of Wroclaw, Poland
    molech@wp.pl

29. Alejandro Otero
    Universidad de Buenos
    Aires-Faeultad, Argentina
    aotero@fi.uba.ar

30. Mikhail Perepelitsa
    Northwestern University, USA
    mikhailp@math.nwu.edu

31. Ermanna Piras
    Università di Ferrara, Italy
    piras@dm.unife.it

32. Piotr Przytycki
    Warsaw University, Poland
    p.przytycki@students.mimuw.
    edu.pl

33. Marco Romito
    Università di Firenze, Italy
    romito@math.unifi.it

34. Walter Rusin
    Warsaw University, Poland
    w.rusin@students.mimuw.edu.pl

35. Okihiro Sawada
    Technology University
    Darmstadt, Germany
    sawada@mathematik.tu-
    darmstadt.de

36. Vincenzo Sciacca
    Università di Palermo, Italy
    sciacca@math.unipa.it

37. Robert Skiba
    University UMK in Torun Poland
    robo@math.unitorun.pl

38. Elide Terraneo
    Università di Milano, Italy
    terraneo@mat.unimi.it

39. Seiji Ukai
    Yokohoma National University
    Japan
    ukai@mathlab.sci.ynu.ac.jp
    (lecturer)

40. Dmitry Vorotnikov
    Voronezh State University, Russia
    mitvorot@mail.ru

41. Irmela Zentner
    ONERA, ENPC, France
    zentner@onera.fr

42. Ali Zhioua
    Université d'Evry, France
    zhiouaa@yahoo.fr

LIST OF C.I.M.E. SEMINARS

Lecture Notes in Mathematics

For information about earlier volumes
please contact your bookseller or Springer
LNM Online archive: springerlink.com

Recent Reprints and New Editions